Copper Plate Photogravure

Copper Plate Photogravure: Demystifying the Process

David Morrish

and

Marlene MacCallum

Routledge
Taylor & Francis Group
LONDON AND NEW YORK

First published 2003 by Focal Press

Published 2022 by Routlege
2 Park Square, Milton Park, Abingdon, Oxon OX14 4RN
605 Third Avenue, New York, NY 10017

Routledge is an imprint of the Taylor & Francis Group, an informa business

Library of Congress Cataloging-in-Publication Data

A catalog record for this book is available from the Library of Congress.

ISBN 13: 978-0-240-80527-6 (pbk)

British Library Cataloguing-in-Publication Data

A catalogue record for this book is available from the British Library.

Contents

Preface

The desire to write this book came out of our own experiences and struggles while learning photogravure. In the course of researching this wonderful process, we discovered that current comprehensive English language literature on the technique was scarce. It is our aim in producing this book to assist you in learning the complex process of photogravure, making it a little less mysterious and a little easier to master. We recommend that you read the entire text first. Then, before attempting each step, reread the appropriate chapter and have the outline of steps beside you in the studio.

As visual artists who also teach photography and printmaking in the visual arts program at Sir Wilfred Grenfell College, Memorial University's Corner Brook campus, we have always been interested in the printed image, especially one that uses the visual language of photography. We decided to learn the photogravure process for our own art practice. We are mainly interested in photogravure's unlimited potential for combining photographic fidelity with the surface quality and visual language of printmaking.

One of the things that we feel has helped us in our research is the fact that we have separate technical backgrounds—printmaking and photography. As artists we both have had some experience in each other's medium, but it is our collaboration that has made it possible for us to learn this complex process.

Living and working on the west coast of the island of Newfoundland is an experience that enriches our lifestyle and our art practice. In spite of our seemingly isolated location, we have made many connections with other artists from across Canada, the northeastern United States, and even Ireland and England. We really have no sense of isolation in that context, but we do feel alone in our practice as photogravure printers. In Canada you can count those who use this medium on your fingers and still have a few to spare. Luckily for us, there are many more practitioners in the United States and abroad, and the Canadian numbers are beginning to grow.

We decided to work with the traditional copper plate process because we found it malleable, responsive, and durable in nature. Therefore, in this book, we will limit the discussion to the traditional hand-pulled flat plate photogravure on copper rather than the new photopolymergravure, which does not meet our personal requirements.

We started our technical research ten years ago. An early photogravure print distributed by *Aperture* (printed by Richard Benson) hangs on our wall: *Iris*, 1928, by Paul Strand. It has acted as our benchmark while we labored to match its subtle tonal variations and fine detail. We began to experiment but were soon frustrated by mysteriously insurmountable failures. At this point we realized that more historical research was needed. We sought out hard-to-find resources, the challenge being to weed out the useful from the misleading or vague, the obsolete from the classic. We compiled a bibliography of as many of the English and American manuals, treatises, and books on photogravure as we could still find. (See the Reference Bibliography.) After reading and comparing these sources, we were able to confirm our methodology to a point but were still stymied by inexplicable inconsistencies. The next step was to seek advice from current practitioners. But where were they? We contacted Jon Goodman, an expert photogravure printer who, amongst other things, prints portfolios for *Aperture* and *21st*. His telephone advice to two desperate novices was what was needed to make another leap forward.

A serendipitous discovery was that a symposium on photogravure was to be held at GraphicStudio at the University of South Florida in Tampa, Florida, in March of 1995. We felt that this opportunity should not be missed. David attended the symposium and witnessed first hand the working methods and actual prints of such accomplished senior gravure printers as Deli Sacilotto, Jon Goodman, Johan de Zoete, and Paul Taylor. There were many practitioners from all over the United States, with one or two each from Sweden, England, and Canada. This symposium provided first-hand information and helped us solve many of our problems. It was good to see that we were not alone. Above all else, it encouraged us by illustrating the dedication of those attending and the growth of interest in the medium. We resumed our testing and soon achieved even more successful prints. Encouraged, we arranged funding for an advanced, private workshop with Jon Goodman. He came to Newfoundland in the summer of 1995 and led us through the finer points of the process. This solidified our understanding of the etching process, the most crucial part of photogravure. When we resumed our research using this new understanding along with our own obsessive working methods, we were soon making successful prints with consistency and predictability. We had finally reached the point where we could call ourselves *photograveurs*. This book reflects the knowledge we have gathered from historical texts, contemporary practitioners in gravure and related fields, and our own testing and working methods. We wrote it in order to provide a clear and detailed methodology for the dedicated practitioner who wants to rediscover this wonderful image-making process.

David Morrish and Marlene MacCallum, 2003

Acknowledgments

We would like to thank everyone who has helped us along the long and tortuous path to learning the beautiful photogravure process and the equally tortuous path of putting together this book. Those who helped us bring this book to a higher level of usefulness deserve our utmost thanks. We thank Jon Goodman for his patience and advice and for sharing his in-depth knowledge of the process. Sandy King deserves our gratitude for the enormous amount of work he did adjusting his carbon printing tissue for use as gravure tissue. His chapter on making one's own tissue is a valuable inclusion in this book. In a very short time Richard Benson, Dean of the Yale School of Art, enabled us to re-evaluate our technique and showed us how to see a photogravure print in a broader tonal scale, allowing us to see beyond the usually dark tonal scale we work with. To those artists who shared their wonderful images and allowed them to be included in this book, we are greatly appreciative: Jon Goodman, Steve Dixon, and Lothar Osterburg. We wish to thank Suzy Taraba, University Archivist and Head of Special Collections, for her generous assistance in providing access to the Special Collections at Wesleyan University. Closer to home, we have always appreciated the amazing library assistance we have been given by Elizabeth Behrens, Associate University Librarian at Sir Wilfred Grenfell College. The chemistry staff and faculty at Sir Wilfred Grenfell College have been generous in so many ways. In particular, Dr. Geoff Rayner-Canham, Professor, Environmental Science (Chemistry), helped us decipher the chemistry of etching. We thank Memorial University's Office of Research for a Subvention Grant for additional color images within the text. Dr. Holly Pike, Associate Professor, English, gave us important assistance with writing, structure, and clarity in earlier drafts. Thanks for advice and information over the years go to Jon Goodman; Kent Jones, Professor in Visual Arts; and Dr. Geoff Rayner-Canham. Mark Katzman, a true believer in the beauty of photogravure, was most generous with his time and collection. For additional image assistance and access to their collections, we thank Steven Albahari from *21st*, Vincent FitzGerald, and Grant Ball. Ted McLachlan, Associate Professor, Landscape Architecture, University of Manitoba, our proofreader and process guinea pig, deserves many thanks. We also greatly appreciate the support and encouragement given by Diane Wurzel, Associate Editor at Focal Press.

Introduction

What is a photogravure print? If you can imagine a photo-real mezzotint or aquatint print with the finest grain possible, you can get a fair picture of what a photogravure print might look like. Photogravure is the only intaglio process, other than the Woodburytype, that gives an apparently continuous tone image. It should be made clear that photogravure is not photo-etching. This latter process is made up of dots of ink on paper—a halftone pattern. The photo-etching resist is made up of tiny areas that are totally protected from the etch and areas that are totally exposed to the etch. There is no half-way, no migration of ferric chloride through gelatin, no smooth tones. Although it gives the semblance of tone, it is not a continuous tone process. There is often confusion with the terminology, especially when discussing photogravure with a Francophone printmaker. In French, *gravure* is virtually any intaglio process, and *photogravure* is used to describe what we call photo-etching in English. When describing photogravures to a European audience, it is necessary to use the local terms. In France or Québec, for example, it is *héliogravure*, and in Germany, *fotogravüre* or simply *gravüre*.

Photogravure is a photo-imaging technique that remains one of the most satisfyingly beautiful image-making processes. The rich depths and detailed tonalities of a photogravure print are unparalleled. It gives a combination of the best traits of both intaglio and photography on one archival support. The range of possibilities of ink color and paper qualities is endless. Unlike most other photo-imaging processes, copper plate photogravure allows the print-artist the opportunity to rework, adjust, alter, and present the image in unique ways. The potential for the physical alteration of the copper plate provides yet another realm of expressive variation. Besides the advantage of unlimited possibilities for interpretation of the image, the photogravure plate can be absolutely faithful to the information on the original negative. The subject matter and its treatment are as variable and broad as photography itself. When manipulated it can be transformed by the artist's hand into something unattainable by any other means. The hand-pulled, flat plate photogravure process was virtually abandoned by the latter half of the twentieth century due to its difficulty, impracticality, and expense.

Photogravures are now being seen more often as artists recognize their unique visual and tactile qualities in this digital age. The revitalization of the traditional photogravure print has been slow and arduous and requires the dedication of modern practitioners who appreciate its unique qualities.

Artists who have used copper plate photogravure as the final presentation of their work look upon the gravure print as a singular work of art, not simply a reproduction. Images are made with the gravure print in mind as their final form. It is a translation of the information in the original negative in the same way that, most often, the silver gelatin or platinum print is considered the final translation of the negative.

Photogravures (most often rotogravures) used in publication are different things. They arose from a need for the mass production of photographic images that were archival, or at least longer lasting than nineteenth century salt and albumen prints. They made possible the inclusion of images and text on a single page and have allowed high quality images to appear in print beginning at a time in history when halftones were inferior or nonexistent. Two other main advantages of rotogravures are the speed of production and the durability of the rotogravure cylinder. Since the nineteenth century, many images have been published using rotogravure or sheet-fed mechanized gravure (also called *mezzogravure*). The quality of these publications varies from poor to excellent, but more often than not, they exceed what one would normally expect of black and white half tone reproductions.

Is the hand-pulled copper plate photogravure process as difficult as many make it out to be? This book is an exploration of the technical processes involved in making a photogravure using currently available materials. Its purpose is to demystify and clarify what is ultimately a complex but altogether *do-able* photomechanical process. Anyone with dedication and some basic knowledge of photography and printmaking can hope to achieve respectable results. The information that follows builds on many of the English language texts printed since the 1890s. Not a lot has changed over time; even many of the specialized materials tend to be the same. Through experimentation, research, and practice we have sought to find the most practical and effective procedures needed to achieve the finest results. The text that follows is by no means definitive, but we hope it provides a solid grounding and a clear explanation of a process that deserves to live on.

A BRIEF DESCRIPTION OF THE PROCESS

Photogravure is a positive working photomechanical intaglio process. Making a photogravure can be described as a series of discrete stages: making the positive, sensitizing the gelatin tissue, making the gelatin resist, etching the copper plate, and, finally, printing the plate. Briefly, the steps are: A transparent continuous tone film positive—usually enlarged—is made from a camera negative, or, in the case of direct gravure, a drawing is made onto a translucent surface. To create an extremely fine pattern or texture, a hard-dot screen is exposed to a presensitized sheet of gelatin-coated paper (tissue), or an aquatint is applied to a polished copper plate. The positive is then exposed onto this gelatin tissue, which is then adhered onto the copper plate and developed with warm water. After development, the gelatin resist is dried and the plate is etched in a series of ferric chloride baths, each bath being of a different density or Baumé. Once the etch is completed, the gelatin resist is removed from the plate. The plate is cleaned, inked, wiped, and printed on an intaglio press, transferring the image in ink onto paper.

From the original negative to the final print, there are five generations: negative, positive, gelatin tissue/resist, etched plate, and print. Each layer has its own subtle but distinct language by virtue of the materials and their handling. These layers can be used to transform the information on the negative (closer to a printmaking aesthetic) or used to create a true facsimile of the negative (closer to a photographic aesthetic or even photomechanical reproduction). The process is bracketed by stages that give the artist a range of aesthetic options: producing the negative and the positive, working the copper plate, and interpreting the image with ink on paper.

SAFETY ISSUES

The photogravure process involves various steps that can be harmful to those unaware of the potential hazard. The process as a whole, however, is relatively safe. The most dangerous toxic materials are: the sensitizing agent, potassium dichromate (which is required); powdered asphaltum; and powdered ferric chloride (both of which we suggest not using). Careful attention to safety is advised when working with any chemicals or solvents. Other substances used in photogravure and the printing of the plates are sometimes the cause of sensitivities and allergic reactions. Solvents and inks are commonly used by printmakers and have been discussed in manuals on printmaking safety. Ferric chloride is one of the least problematic mordants available when in solution. Nevertheless, all materials and equipment have their risks, especially when not used with caution and common sense.

See Appendix A for a list of safety considerations and advice. Please be aware that this advice does not pretend to be definitive and should always be researched further if you have concerns. The authors make no claims as to the safety or risk of using the materials and methods described in this book. We urge the reader to research these materials and practice appropriate caution when using any potentially harmful materials or equipment.

1 A Brief History

ORIGINS

The medium of photography is evolving toward ever more immediate and ethereal images that often barely exist as digital data. The ease of reproducibility has increased and the scope of dissemination has become instantly global. It was not long ago, however, when photography was mainly a chemical process—an image formed on paper or, more recently, a plastic support. When photography was invented in the 1820s, the image-making process was even more physical. A unique image was etched onto a metal plate through an acid resistant layer. Joseph Nicéphore Niépce (French: 1765–1833) is credited with the first permanent photographic images using sensitized bitumen of Judea on a pewter plate—images that could ultimately be etched and reproduced as intaglio plates. He saw the potential of this process for quick, accurate reproduction of existing engravings (Figure 1-1). Niépce called these first successful photomechanical reproductions *heliogravures*. These prints, however, did not reproduce any of the smooth continuous tones we now associate with a photograph.

A partner of Niépce, Louis Jacques Mandé Daguerre (French: 1787–1851), developed his own version of the photographic process after Niépce's death. After the announcement of Daguerre's invention of the daguerreotype in 1839, the process was immediately tested in order to make the one-of-a-kind daguerreotype plate printable as photomechanical intaglio plates. Hippolyte Fizeau (French: 1819–1896) devised a method using aquatint, etching, and even electroplating to create a printable daguerreotype plate. Dr. Alfred Donné (French: 1801–1878) published details of his process in June of 1840 after patenting his method of etching daguerreotype plates. He displayed his pale prints from etched daguerreotypes to the French Academy of Science in the same year. His process utilized the natural grain and acid-resisting properties of the mercury amalgam that forms the highlights and light tones of the image to etch the silver plating from the open shadow areas on the surface of the plate. Dr. Joseph Berres of Vienna made darker and richer images from daguerreotypes. He attained a deeper etch by using solid silver plates and building up the highlights with varnish.

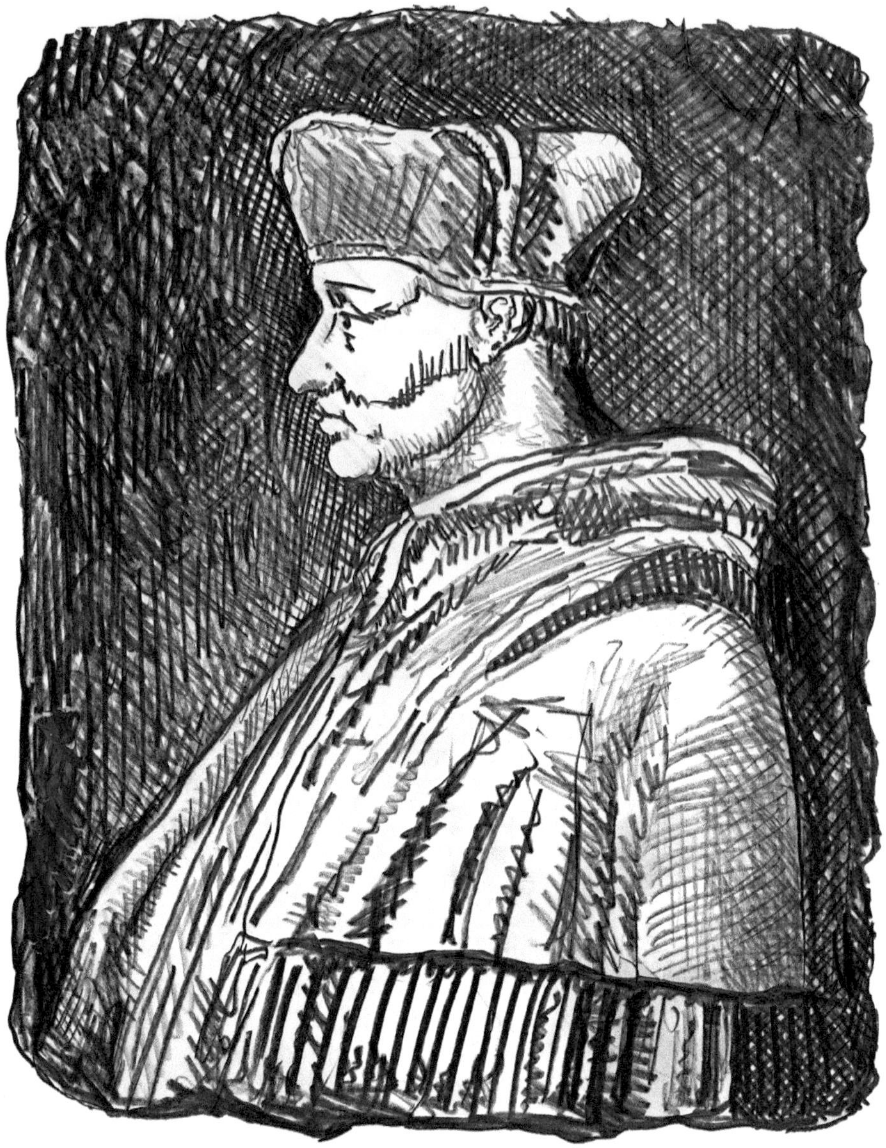

Figure 1-1 Niépce used a waxed line-engraving as a positive and exposed it onto a pewter plate so it could be etched and printed. (Drawn after an image of the original of 1826.)

William Henry Fox Talbot (English: 1800–1877) is credited with the development of negative/positive photography. He made multiple positive prints from the paper negatives he produced in his "mouse-trap" cameras. In 1844, Talbot was the first to publish a book illustrated with photographs. *The Pencil of Nature* contained actual tipped-in salt prints, which, much to Talbot's dismay, proved to be impermanent. He sought another way of making a more stable photographic image. The well-established fact that ink on paper was permanent led him to explore the idea of photographically producing etched plates that could be printed. In 1852, Talbot found that normally soluble colloids such as gum arabic, albumen, and gelatin become insoluble when mixed with potassium dichromate and exposed to light. Utilizing this hardening or tanning effect, Talbot developed an etching resist over which he used a screen of black crepe to help

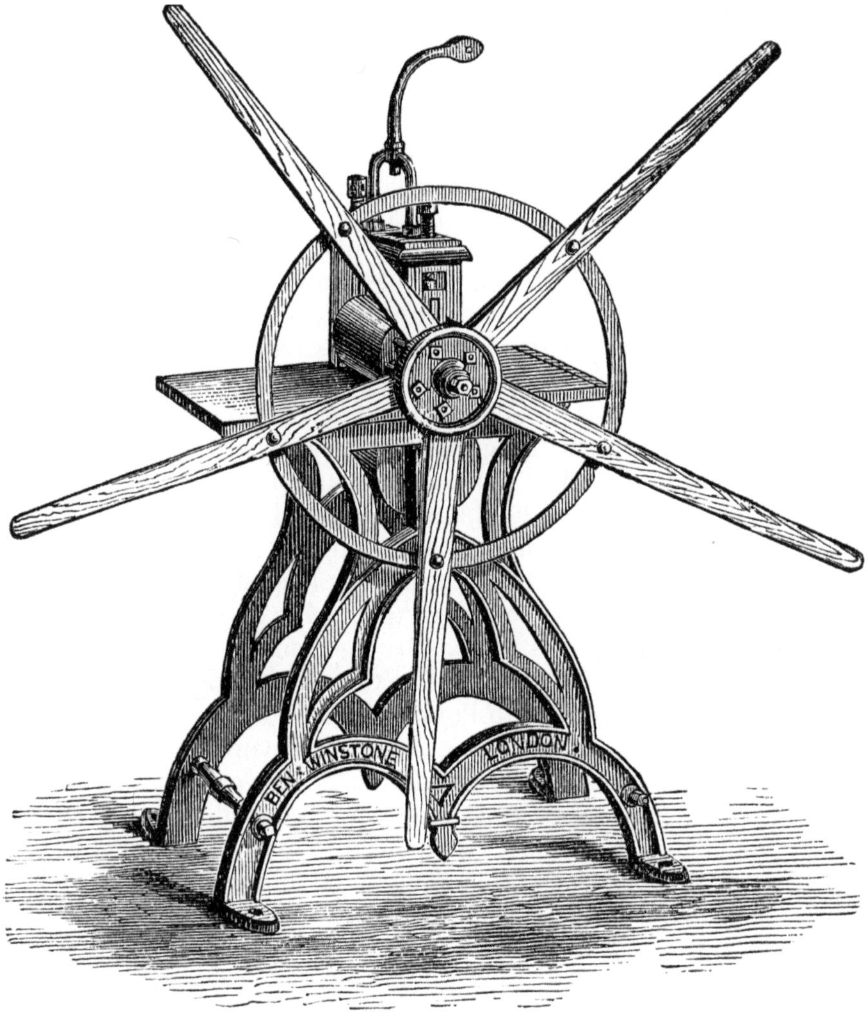

Figure 1-2 Henry Fox Talbot used an intaglio press very much like this late 19th century press.
Illustration by W. L. Colls, © Iliffe & Son, London, 1890.

with the translation of tonal values in the etched plate. This negative screen (a network of crossed lines) is the forebear to the positive screen used in modern rotogravure. Talbot's *photoglyphic engravings* are sharp and detailed but lack the smooth gradation of tone associated with other photographic representations, including his own calotypes (Figure 1-2). He continued to improve on his technique as he moved from iron plates to copper and etched with ferric chloride instead of platinum chloride. He also etched with three baths and greatly improved the tonal scale on later tests. He felt the resulting prints were well suited to book illustration and had good commercial value (Buckland 1980, p. 114).

In 1855, Alphonse Louis Poitevin (French: 1819–1882) patented the first carbon process, in which he added carbon black to the colloid (gelatin) and dichromate mixture and coated it on paper. Again, the image left on the paper was formed by the insolubility of the exposed gelatin (Crawford 1979, p. 70). By 1856, eight different carbon processes were announced, but none were capable of capturing a full and gradual tonal scale. After numerous failed attempts, Joseph W. Swan (English: 1828–1914) solved the tone reproduction problem in 1864 when he and John W. Sawyer patented their version of carbon-gelatin tissue and the carbon transfer process. Adolphe Braun of Dornach (Alsace) bought the rights to this patent and began publishing carbon transfer reproductions of paintings and Old Master drawings (Crawford 1979, p. 72). In 1858, Talbot changed his process by adding a rosin aquatint to the surface of the exposed gelatin prior to etching the plate. This was then etched with ferric chloride through the underlying bichromated gelatin layer. He was able to obtain richer plates with more tonal graduation due to the delicate grain (Mertle and Monsen 1957, p. 325).

Early printers were driven to search for a more permanent image-making process due to the impermanence of silver-based images such as salt and albumen prints. There was also a need to reproduce photographic images, either as independent entities or as integral parts of publications. The fact that printing ink on rag paper was archival was key to those who tried to perfect the photogravure process.

Charles Nègre (French: 1820–1880) produced the first published reproduction of a small "proto-photogravure" within a page of text in 1854 in *La Lumière*. From 1856 to 1867, Nègre was competing in a drawn-out competition sponsored by Honoré d'Albert, duc de Luynes for the best way to reproduce an image in a totally mechanical way. Nègre had to submit a sampler showing a range of textures and tonalities. His gravure submission came in second to Poitevin's winning planographic entry because the jury suspected Nègre used hand-work on his plate and because his process was slow and complicated. Nevertheless, the Duc de Luynes appreciated Nègre's work and commissioned him to produce a large work called *La Mer Morte* (started in 1865 and completed in 1868). Nègre also produced large-scale architectural studies of the restoration of Chartres cathedral using his gravure-based process. Although very detailed and richly printed, these images were not like modern photogravures in their technique nor their tonal range. They lacked the smooth transitions from one tone to another. Some examples seem almost posterized. They were also heavily retouched. Façades were lightened, shadows were opened up with hand work, and skies were added in a painterly fashion. A key difference in the technique was that the images were printed from steel plates rather than copper. Steel plates have a self-graining effect when etched, eliminating the need for an aquatint. In contrast, the copper plates used for classic photogravure require the application of a dust-grain aquatint in order to maintain tone.

Although these developments formed the foundation of modern photogravure printing, the process as we know it today was actually devised in 1879 by Karl Wenzel Klič (Karel Václav Klitsch) (Czech Republic then Arnau: 1841–1926). Utilizing an asphaltum aquatint under the sensitized and developed gelatin-coated pigment paper resist, he combined Talbot's etching procedure with these new materials to produce a true photogravure print. Klič's procedure differed from Talbot's in that the resinous powder was applied directly to a copper plate and then covered with the sensitized carbon tissue. The Talbot-Klič process of photogravure was born. After this point, the production of hand-pulled, flat plate photogravures continued to improve slightly using a technology that has not appreciably changed since its invention.

The main commercial development was the advent of rotogravure, a mechanized commercial process invented by Klič and mastered as early as 1890. Rotogravure is still used today by the printing industry. For the purposes of this text, and the discussion of photogravure as an artist's medium, we will not address the particulars of rotogravure.

ARTIST-PRACTITIONERS

Historically, the photogravure process as used by printers and publishers was determined by the balance between image quality and production economy. Meanwhile, artists wanted to reproduce their work by capitalizing on photogravure's inherent aesthetic qualities. Most had previously worked with platinum, albumen, or silver-gelatin. The photogravure print more closely resembled a mezzotint than a halftone and therefore had more cachet as a fine print when included in a publication. It was clear that no other reproductive medium could come as close to the artists' aesthetic vision. The appreciation of the malleability of the medium superceded its amazing verisimilitude and soon artists were using photogravure to express themselves in ways that traditional photographic means could not. Many photographers became printmakers when they realized this potential.

Peter Henry Emerson (American working in Britain: 1856–1936), the preeminent figure of the *naturalistic school* of nineteenth-century photography, created many publications that utilized photogravure to echo his atmospheric platinum prints. His pale, low-contrast, but fully toned images were reproduced with photogravure more and more successfully from one publication to the next. *On English Lagoons* (1893) was printed by Emerson from plates he etched himself. *Marsh Leaves* (1895) was his last self-produced album of gravure prints. In his book *Naturalistic Photography* (1889), Emerson states his preference for photogravure over other photographic media, including platinum prints. He states that it is the ideal medium with which to present pure photography because of the flexibility of choice in ink and the range of available papers (Coe and Haworth-Booth 1983, p. 100). *The Compleat Angler, or the Contemplative Man's Recreation. Being a discourse of Rivers, Fish Ponds, Fish and Fishing written by Izaac Walton, and Instructions How to Angle for a Trout or Grayling in a Clear Stream by Charles Cotton,* was reprinted in its 100th edition in 1888 by the publisher/editor R. B. Marston. This two-volume publication contained 54 photogravure illustrations on special India paper, of which 27 were by Emerson. Emerson made the original negatives along the Lea River in the spring of 1887. The resulting publication was an outstanding example of Emerson's aesthetic and skill (Figure 1-3 and Color Plate 1).

In 1968 and again in 1877, before *Pictorialism* became the dominant aesthetic posture of artist-photographers, Thomas Annan (Scottish: 1829–1887) was commissioned by the Glasgow Improvement Trust to photograph the closes, wynds and buildings slated for demolition in the city center. *The Old Closes and Streets of Glasgow* was first published as carbon prints but was republished in 1900 as *Old Closes and Streets, a Series of Photogravures, 1868–1899*, in two editions of 100, each containing 50 photogravures, some of which were heavily manipulated (Figure 1-4). Annan's work stands out for its rich clarity and skillful printing by James Craig Annan (1864–1946), Thomas Annan's son. James Craig was the printing firm's expert on photogravure, having been tutored in the process by Klič himself (Crawford 1979, p. 250). J. C. Annan's own work appeared in a portfolio entitled *Venice and Lombardy: A Series of Original Photogravures*, published in 1898 in an edition of 75 copies (Figure 1-5).

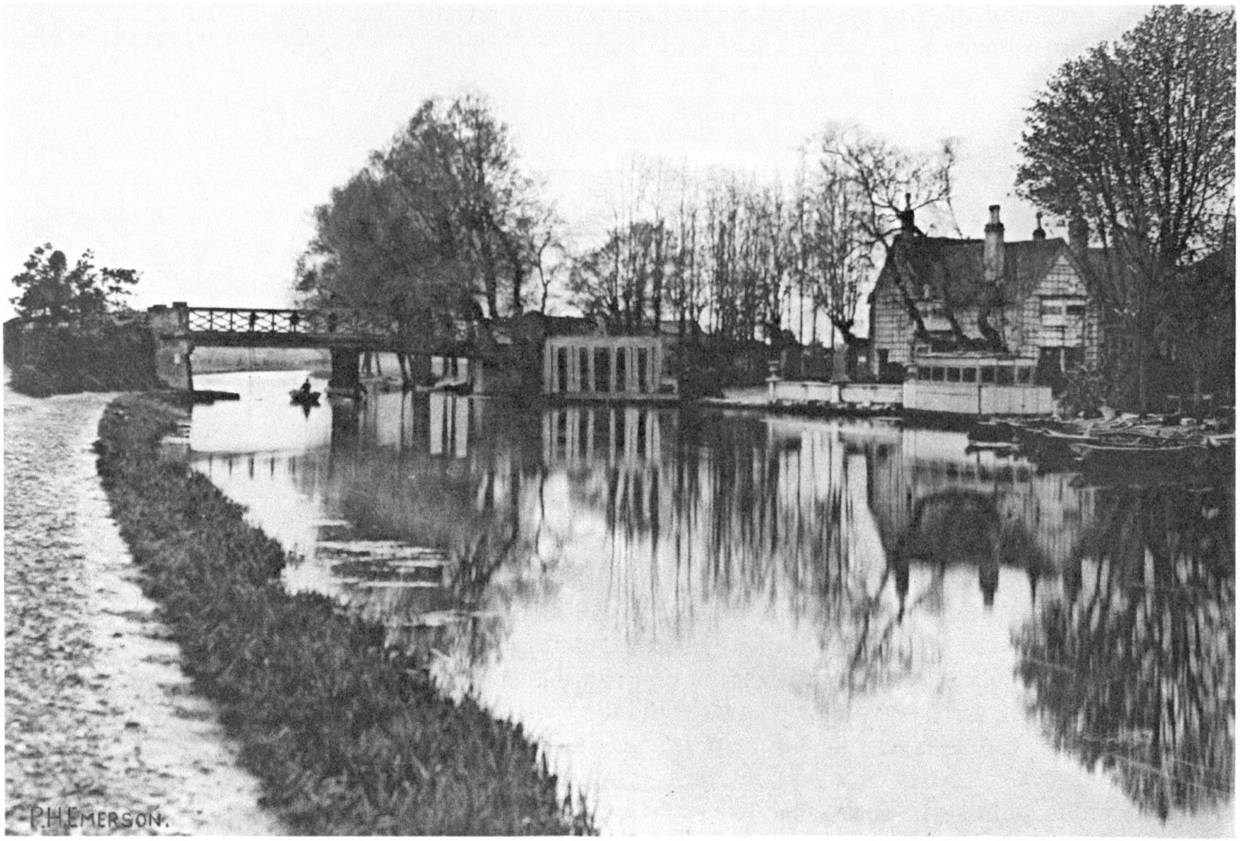

Figure 1-3 Peter Henry Emerson. *The Old Rye House Inn*, Plate XV (12.7 × 19.5 cm) from Volume One of the 100th edition of Izaak Walton's *The Compleat Angler*, 1888.

Photo by Mark Katzman, Ferguson and Katzman. From private collection.

He is also credited with the reappearance of many of David O. Hill and Robert Adamson's calotypes in both gravure and carbon. J. C. Annan was a member of the Linked Ring in Britain and a friend of Alfred Stieglitz (American: 1864–1946). Annan's work was featured in Stieglitz's *Camera Work*, some of the photogravures being of Annan's own images.

The first notable 20th century publication to use photogravure at the highest artistic standard was Alfred Stieglitz's periodical, *Camera Work*, which was introduced in 1903 and was published until 1917 (Figure 1-6). Of the 544 illustrations published in the complete run of *Camera Work*, 416 were printed using copper plate photogravure. Many of the images were atmospheric and, in early issues, quite pictorial in the treatment and technical application of the medium. This seemed in contradiction to the photo-secessionist mandate. Later issues were increasingly *straight*, with a sharper formal emphasis. From the start, each issue was complex and multi-layered. The gravure images were often printed on thin Japan paper and backed with colored papers that showed through (see Color Plate 2). Few of the featured photographers were skilled practitioners of the photogravure process, with some exceptions. Alvin Langdon Coburn's (American, Naturalized British: 1882–1966) own photogravure work was featured in Stieglitz's *Camera Work*. Coburn also presented his work in all forms of gravure, from the 83 plates he personally etched and steel-faced from 1909–1914 to a rotogravure supplement in *Pall Mall* (Weaver 1986,

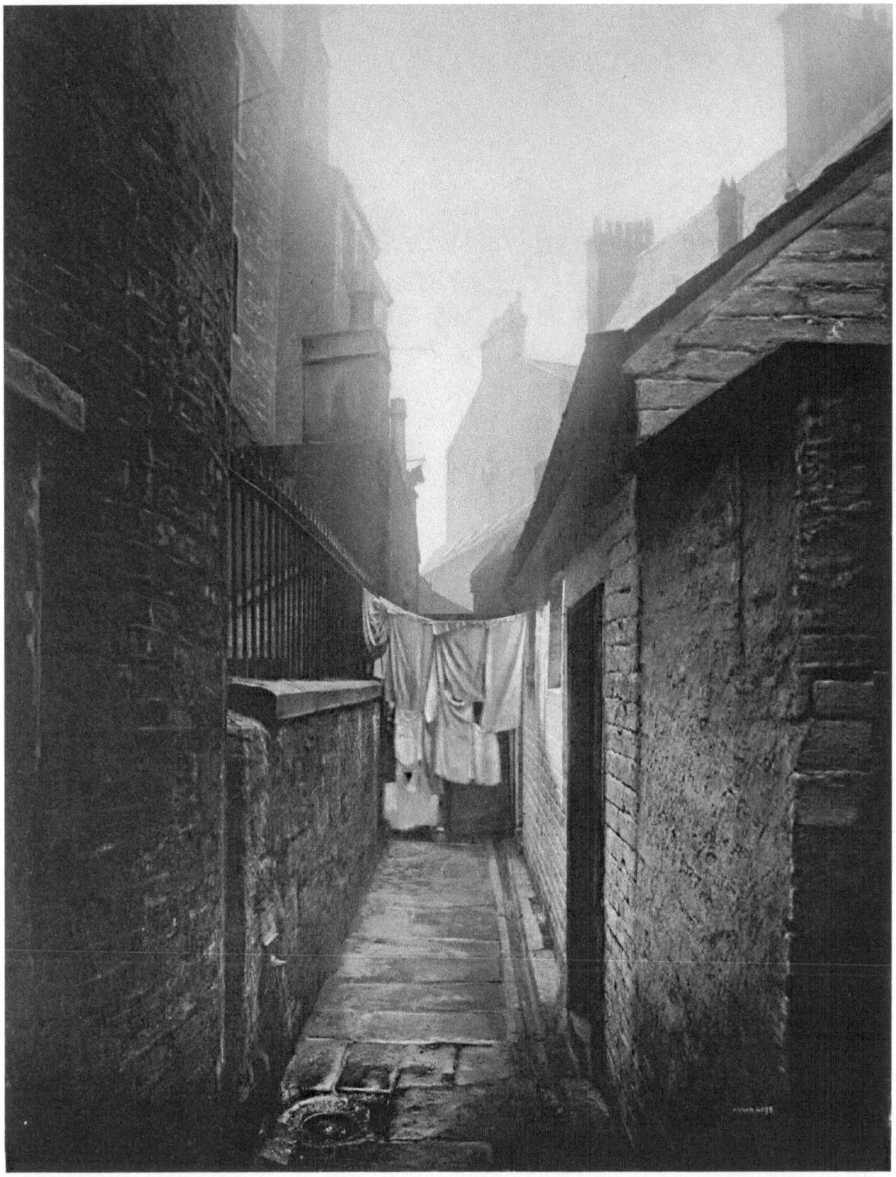

Figure 1-4 *Close No. 11 Bridgegate, 1897* (21.6 × 17.1 cm). Plate from *Old Closes and Streets, a Series of Photogravures, 1868–1899 by Thomas Annan.* Note the evidence of retouching.

Photo by Mark Katzman, Ferguson and Katzman. From private collection.

pp. 48–49). Publications such as *London* (1909), *New York* (1910), *The Door in the Wall and Other Stories* by H. G. Wells with gravure images by Alvin Langdon Coburn (1911), and *Men of Mark* (1913) are fine examples of his talent and skill (Figures 1-7 and 1-8). Both the contrast and somewhat coarse grain of his images are evident in his photographs and in his gravures and are typical of his evolving aesthetic at the time. The photogravures within these publications and in *Camera Work* are often surrounded by a dark gray aquatint band about 5 mm wide (much wider on the bottom). They are bleed-trimmed to this edge and tipped onto heavy paper of various shades and hues of marbled gray (see Color Plate 3).

A magnificent example of twentieth century photogravure and bookbinding, especially in its elegance and scope, was produced by Edward

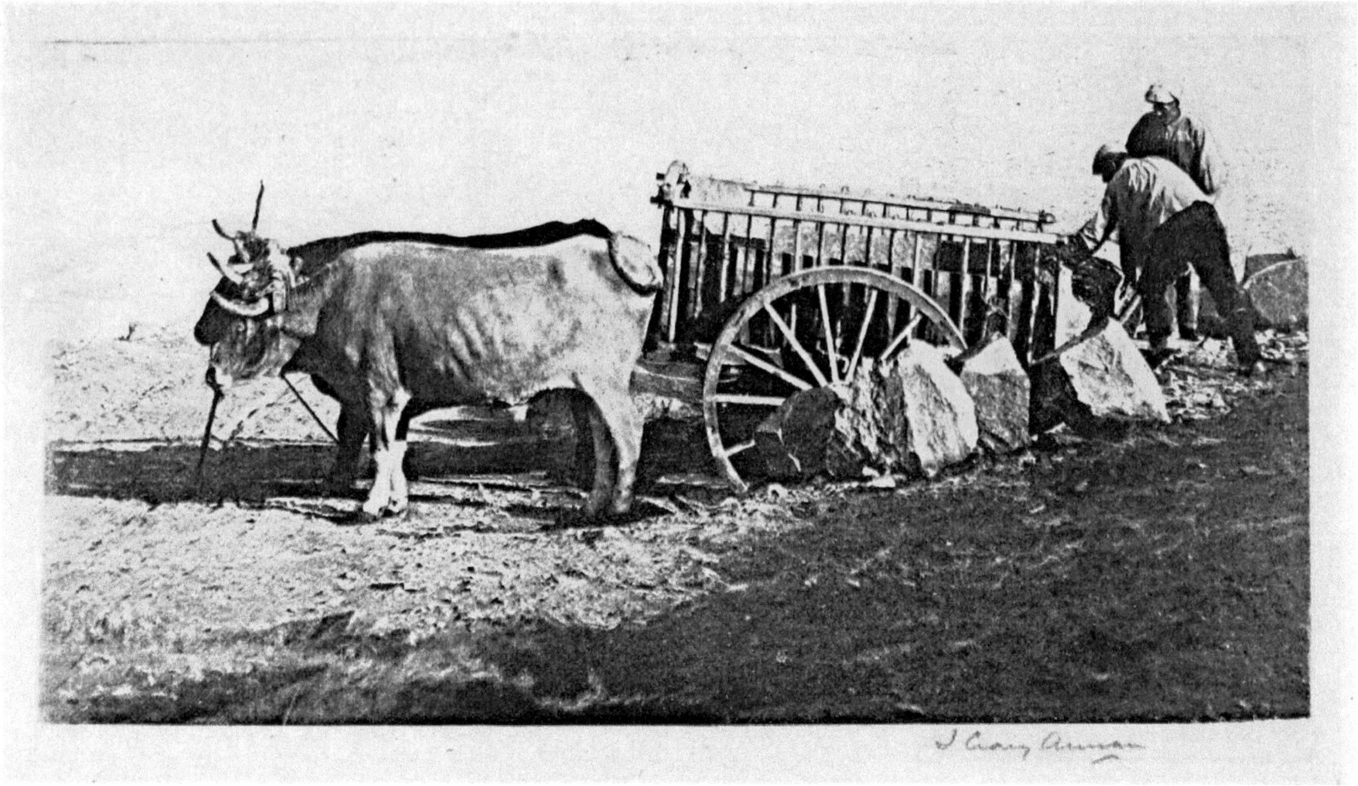

Figure 1-5 James Craig Annan. *Bullock Cart, Toledo*, n.d. (11.5 × 20.4 cm). Photo by Mark Katzman, Ferguson and Katzman. From private collection.

Sheriff Curtis (American: 1868–1952). His treatise entitled *The North American Indian* was printed in an edition of 500 leather-bound, 20-volume sets with 1500 photogravure plates bound in (see Color Plates 4 and 5). Each volume is made up of an equal number of pages of text and full page photogravures. Each gravure is just under 20 cm × 25 cm (8″ × 10″) and is printed in sepia or a rich chocolate brown. Volume I was introduced in 1907, Volume X was published in 1915, and Volume XX finally appeared in 1930. Along with these bound volumes were companion portfolios of larger, loose gravures of 722 supplemental images in the same volume divisions. These larger photogravures were up to 46 cm × 56 cm (18″ × 22″). The subject matter is often treated in a pictorial style in spite of the sociological or anthropological tone of the text. This style appears to evolve from one volume to another as pictorialism gained and then lost favor between 1907 and 1930.

Examples of the finest publications of photography are found in Andrew Roth's *The Book of 101 Books: Photographic Books of the Twentieth Century* (PPP Editions with Roth Horowitz, LIC: New York, 2001). It notes how some of the featured publications from Coburn, Curtis, and Doris Ullman were printed in hand-pulled grain gravure, whereas others by Brassaï, Man Ray, Eugène Atget, Karl Blossfeldt, Henri Cartier-Bresson, William Klein, Helen Levit, Eikoh Hosoe, Robert Frank, Bill Brandt, and many more were printed in rotogravure. This text catalogues the most important photography books of this century and it is

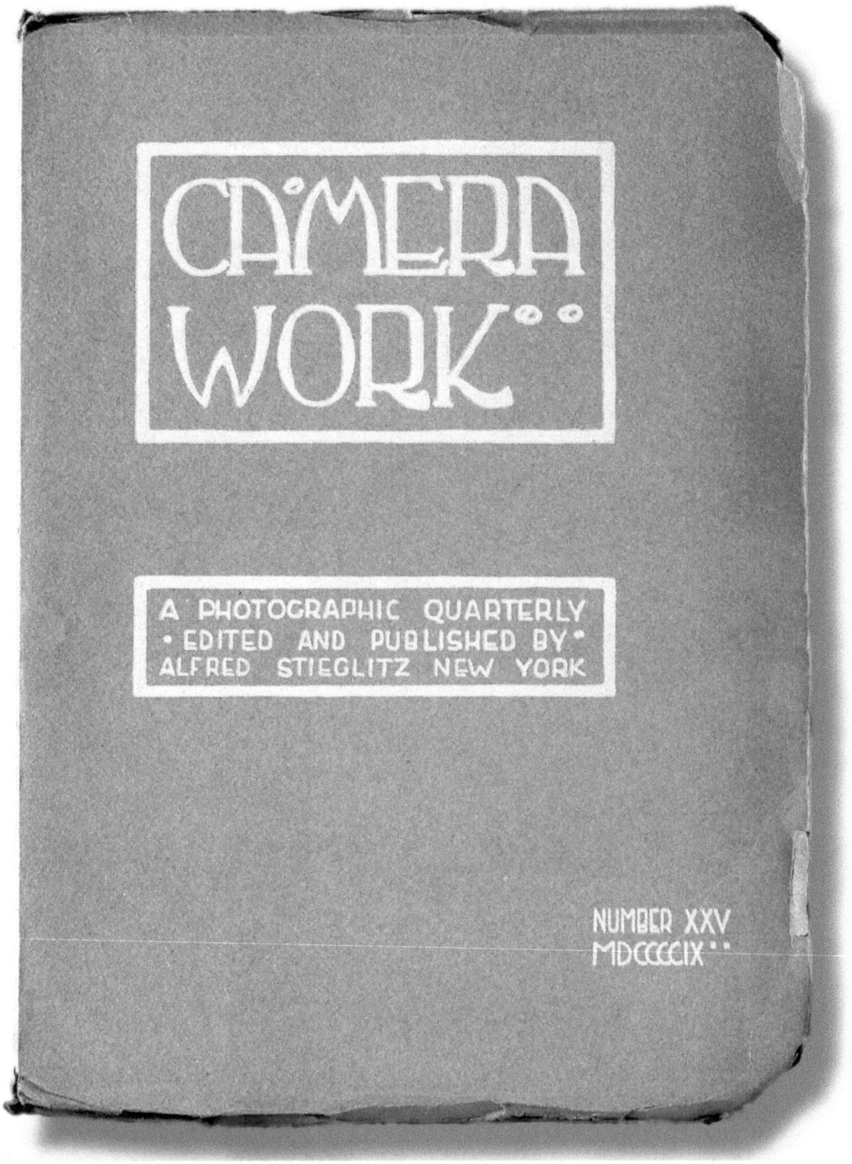

Figure 1-6 Cover of *Camera Work* (30.5 × 21.6 cm).
Photo by Mark Katzman, Ferguson and Katzman. From private collection.

no surprise that photogravure played an important role in the production of so many high quality publications.

Paul Strand's (American, Naturalized French: 1890–1976) *The Mexican Portfolio* is one of the most powerful mid-century collections of fine photogravure. It was first published by Virginia Stevens in an edition of 250 as *Photographs of Mexico* in 1940. It was printed by Charles Furth of the Photogravure and Color Company. The second edition was reissued as *The Mexican Portfolio* by DeCapo Press in 1967 in an edition of 1000. It was hand printed with great skill by Albert DeLong of the Anderson Lamb Company of Brooklyn, New York. Strand thought that this second edition was superior to the first even though both were printed by the most skilled gravure printers alive at the time (Crawford 1979, p. 251). The flawless images are rich and extremely detailed (see Color Plate 6).

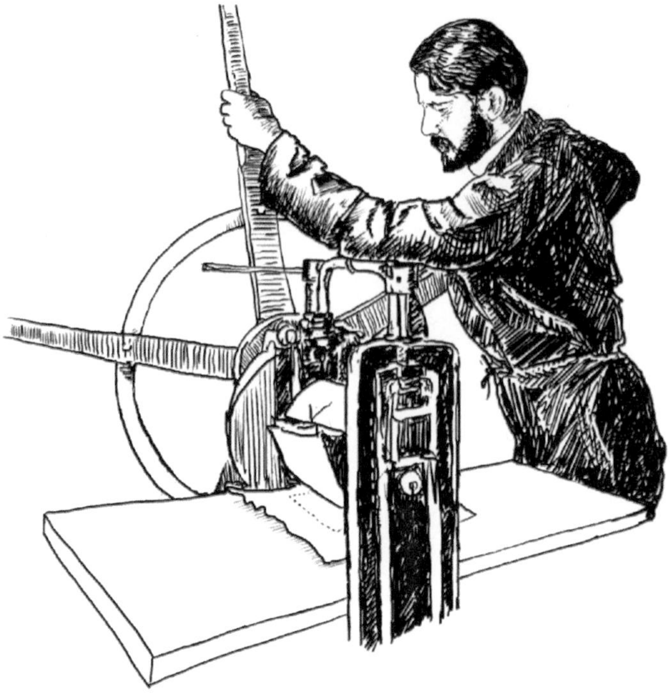

Figure 1-7 Coburn working at his press. (Drawing based on a photograph.) Copyright © David Morrish.

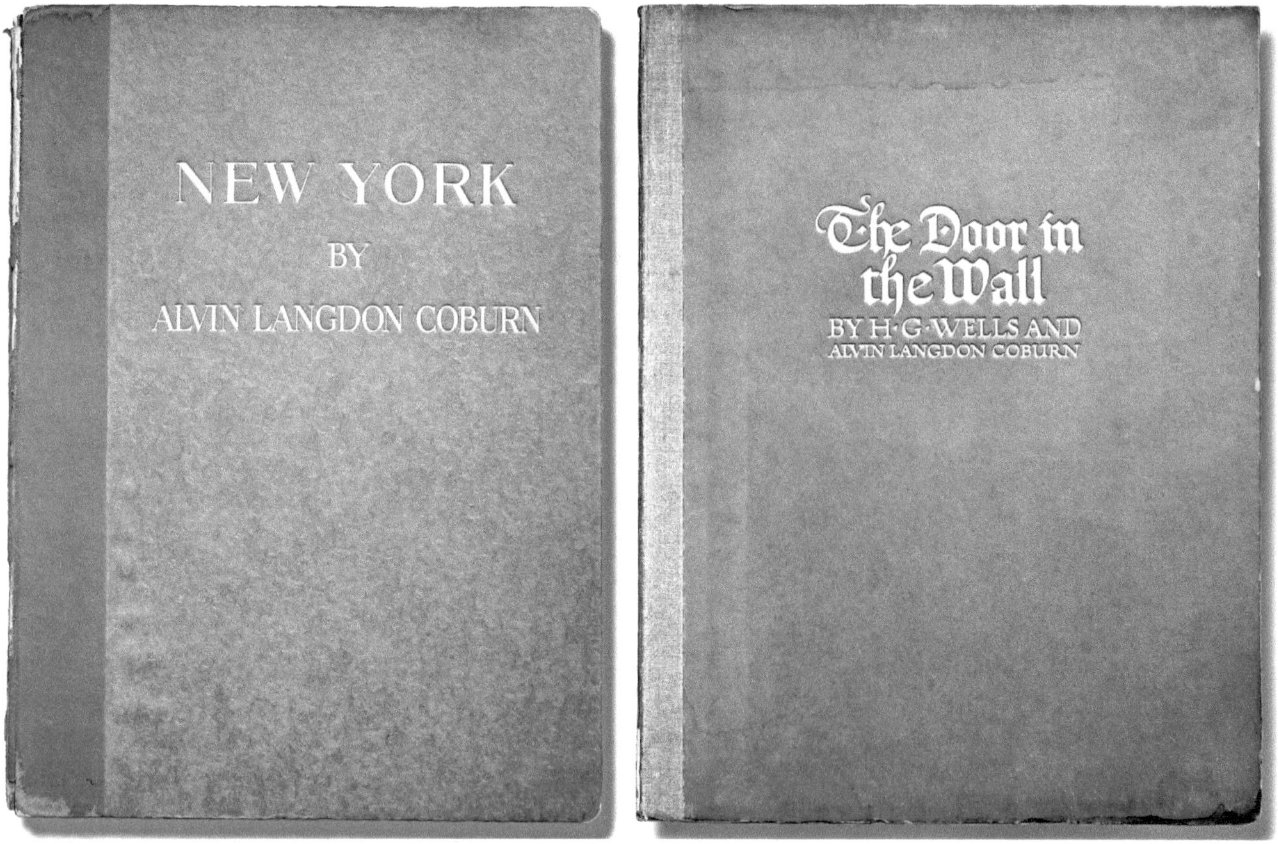

Figure 1-8 Covers of Alvin Langdon Coburn's *New York* (1910) and *The Door in the Wall* (1911). Photo permission of Special Collections and Archives, Wesleyan University, Middletown, CT.

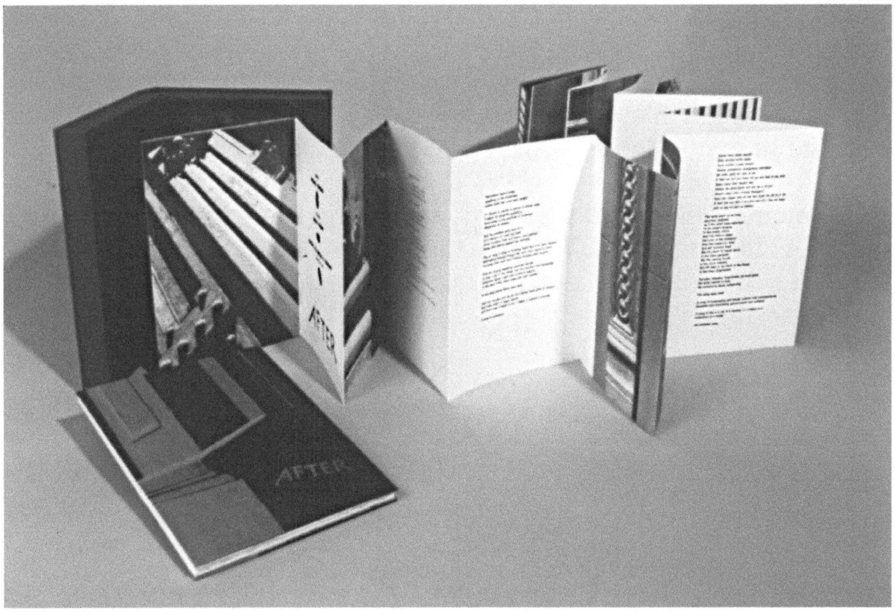

Figure 1-9 *After*, by Michael Feingold and Judith Turner. Published by Vincent FitzGerald and Company: New York, 1993.

All of Strand's photogravures, even to this day, are spray lacquered to enhance the richness of the blacks. The images are small (slightly less than 20 cm × 25 cm or 8″ × 10″) and presented as loose sheets without any text.

In recent years, photogravures have reappeared in portfolios and fine press books. Paul Taylor's Renaissance Press and Vincent FitzGerald and Company are only two of many companies using photogravure. In 1993, FitzGerald produced the book *After*, a letterpress accordion-fold book with poems by Michael Feingold and images by Judith Turner. The images were printed in gravure using plates made by Jon Goodman (Figure 1-9).

The year 1998 marked the appearance of a periodical inspired by *Camera Work* and dedicated to the same aesthetic goals and production quality; it was entitled *21st: The Journal of Contemporary Photography*. Edited by John Wood and published by Steven Albahari, its deluxe editions include bound-in photogravures (printed by Jon Goodman). The museum editions include complete sets of signed, unbound photogravure prints (see Color Plate 7). In 2002, Volume V was released, a continuation of the goal to publish a collection of the finest contemporary photographic imagery and aesthetic discourse.

Modern practitioners are as varied in their use of photogravure as were historical practitioners. Many photographers have their work reproduced in gravure as an alternative to their standard prints, whereas others produce negatives that are intended to be photogravures from the start. Those print-artists who appreciate the inherently fine quality and character of a photogravure image have been re-establishing photogravure as a form of expression that will survive the digital age, as does stone lithography for many printmakers (see Color Plates 27 to 36). Since the 1970s many photographers have been attracted to the older, more difficult technologies, as Lyle Rexer demonstrates in his book *Photography's*

Antiquarian Avant-Garde: The New Wave in Old Processes (Abrams, 2002). Deli Sacilotto, one of North America's pre-eminent photogravure printers, has been instrumental in maintaining the viability of photogravure through his collaborations with artists. He has brought the medium to new heights in his use of large scale and multi-color printings. Growing numbers of artists are making photogravures for themselves or are working with professional photogravure printers in ateliers that provide this service. We hope that this book will be helpful to those who wish to explore the potential of an exquisite medium and apply it to their aesthetic practice.

2 Making the Film Positive

Normally, a photographic image starts as a film negative. Most photographic processes work directly from this original matrix. Photogravure, however, is a positive working photomechanical intaglio process. It requires a continuous tone film positive from which an etching resist is made for the copper plate. Although you do not have to use a continuous tone image as the source, it is significant that photogravure can accommodate and reproduce true continuous tone as no other photomechanical process can. Originally, a carbon print on glass served as this positive. Today, a continuous tone photographic film positive can be produced by traditional darkroom methods (discussed here) or by high resolution digital output.

THE PROCESS

The most basic method of obtaining a black and white continuous tone film positive is to enlarge or contact print an existing negative onto another sheet of film. The quality of the original negative largely determines the contrast range, full tonal scale, detail, and grain characteristics. These have a direct effect on the positive and, ultimately, the photogravure print. The film's grain itself may present a problem when making the resist if it interferes with the screen texture. Some photographers or printmakers may prefer to accept the grain of the film as a part of the image and can make fine gravures from 35 mm negatives. If smooth tones and a high degree of resolution are desired, large, fine-grained negatives are best. Another consideration is that the contrast of the negative should not be excessive because of the difficulty in producing a positive that is supposed to maintain detail in both the shadows and the highlights.

The traditional film for making positives from existing negatives has been a continuous tone orthochromatic film. Professional-grade sheet films of this type are usually expensive special order products, many of which are now discontinued. Bergger Products, Inc. still makes various continuous tone films that work well. We have found that a very practical method for making enlarged continuous tone positives with the least expensive materials is by using low-contrast processing lith film. A less expensive graphic arts film such as Freestyle's Arista APH film can be processed to a long, smooth tonal scale using a variety of film and paper developers at higher than normal dilution. One can control and predict the contrast by being accurate and systematic with exposure, development time, temperature, and dilution.

Use a basic black and white photographic enlarger to enlarge negatives onto the lith film. A condenser enlarger will produce slightly more contrast than a cold light or diffusion enlarger. A graphic arts copy camera is well suited to making enlarged positives because the lens sharpness and the flat film planes normally found in graphic arts equipment are far superior to those in the basic enlarger. Whereas enlargers can make drastic enlargements with a minimum of intermediate steps, the copy cameras are ideal for making sharp positives from large or medium format negatives. Another method of making positives from negatives is to contact print the negative directly to film. The only really important requirements here would be dust-free tight contact between the negative and the positive film and even illumination from a timed light source.

EQUIPMENT AND SUPPLIES

The usual equipment found in a black and white darkroom is needed in the production of a film positive. Access to temperature-controlled water is very helpful. Other equipment such as red (1A) safelights, a darkroom timer, photographic trays, graduated cylinders, tongs, and chemicals normally used to process black and white paper and film are required to process film positives using orthochromatic and blue-sensitive films.

The standard black and white darkroom safelights must be filtered with graphic arts 1A light red, not the OC yellow/orange-green safelights used for black and white photographic paper. The latter will fog graphic arts films. Lith films are quite slow and do not easily fog under 1A safelight illumination. The continuous tone blue-sensitive films, however, are much more sensitive and will develop a serious base fog if held under 1A safelights for more than a minute or so. When using these films, we use one safelight, facing the opposite direction from the tray, and turn it off during the open tray development to minimize the effect. If the base density plus fog reads more than 0.10, we suggest that you take steps to all but eliminate the use of safelights. A base plus fog density of up to about 0.09 is acceptable. Test your situation with a piece of film placed on the counter with a coin on it. Allow it to sit for five minutes under safelight conditions. Develop in the dark.

Highly dilute solutions of paper developers such as Clayton P20 Print Developer, Kodak's Polymax, Dektol, or Selectol-Soft can be used to achieve a continuous tone positive from normally high contrast lith films. You can also mix a developer from dry ingredients that is tailored to low contrast processing of lith films. See Dave Soemarko's "Lith Film in Continuous Tone" in *Post-Factory Photography*, Issue #2, Oct. 1998. See also *The Film Developing Cookbook* by Steve Anchell and Bill Troop, Focal Press, 1998. The continuous tone Bergger films, such as BPFB-18, can be developed in several film developers including Kodak HC-110, Dilution C.

A Kodak No. 2 Step Scale or a numbered Stouffer 21-Step Scale No. T2115 can be used to compare the densities achieved on the positives to a known density and will be needed for judging the etch as well (Figure 2-1).

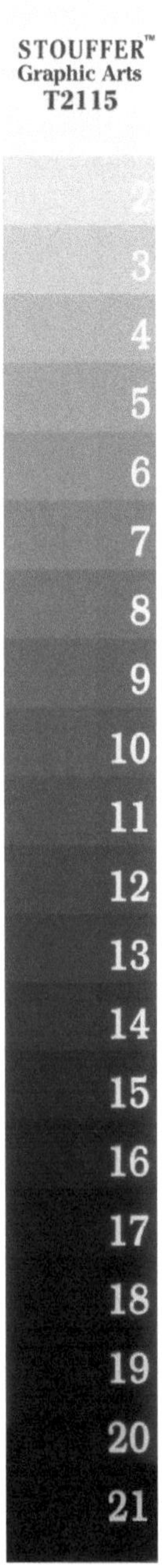

Figure 2-1 Stouffer 21-Step Scale No. T2115.

The uncalibrated version is far less expensive and is fine for our purposes. A transmission densitometer is an invaluable aid and eliminates the guesswork when it comes to determining exact densities and the contrast range.

PROCEDURE

Exposure

Place the negative in the enlarger's negative carrier with the emulsion side set facing up rather than down toward the easel. Focus the negative with an enlarger onto the easel bed in the same way as you would for a black and white enlargement. The image should appear on the easel as mirror-imaged or laterally reversed from the original scene. In this way, the correct right–left orientation of the original scene in the final gravure print is maintained, if that is your intent. All other stages of making the photogravure maintain the emulsion-facing-emulsion rule. It is important to establish the image orientation at this first stage (Figure 2-2).

Use a standard photographic printing easel to center the image on a sheet of lith film. Be sure to place the film with emulsion side up in the easel. The emulsion side of most thin graphic arts material is lighter in tone than the plastic or glossy side. During exposure, a piece of black, red, or goldenrod paper is required on the easel under the film to prevent halation if the easel is white. Because you are making a positive image, expose for the highlights (thin areas) and develop for the shadows (dense areas). Make test strips to check for proper exposure and density range before using a whole sheet of film. During the final exposure, you can dodge and burn various areas of the image to control or adjust the bright areas and shadow details in the same way you would when making a photographic print.

Alternatively, a copy camera can be used if it is equipped with a diffuse backlight under the negative on the lower exposure table. The negative is placed on this lower, backlit copy plane, emulsion side down. Be sure to block off all the surrounding white light that backlights the original negative so as to prevent flare, lowered contrast, and highlight fogging in the positive.

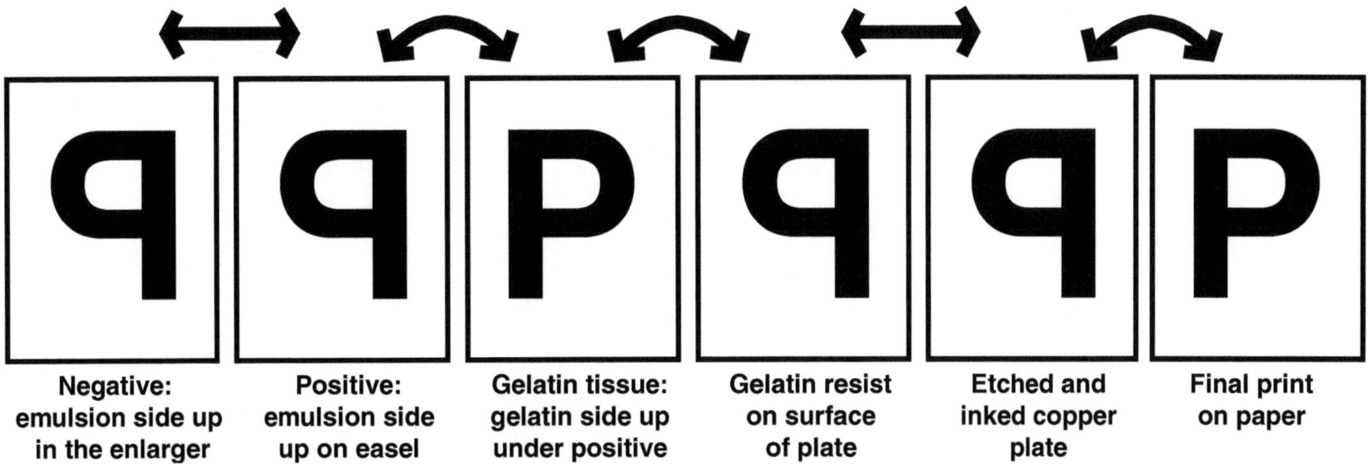

Figure 2-2 Orientation of negative to final print.

Cut the final film at least 2.5 cm (1 inch) larger than the image on all sides to act as a safe-edge for handling. Cleanliness is very important in all stages to prevent dust spots and fingerprints from interfering with the image's integrity. Use compressed air at every stage and avoid handling the film and tissue except by the extreme edges, off the image area.

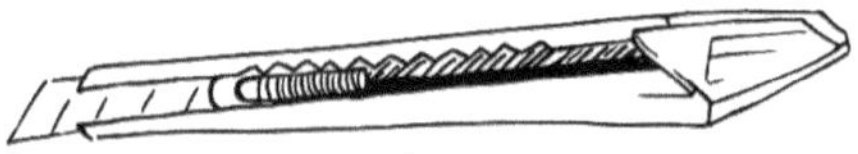

Development

You must force the normally high contrast lith film to produce continuous tone—a gradation of tones from clear to full density. Many paper developers can achieve this when highly diluted. Some have problems with low maximum density, mottling, or unpredictable contrast. The ideal developer is one that predictably achieves the continuous tone needed, with the densities and contrast in a workable range, and keeps the developing times reasonable to avoid mottling or premature exhaustion.

Developer *dilution* is used to adjust contrast or affect shadow density by way of developer activity. The more dilute the developer, the lower the contrast (and the shadow details remain less dense). The less dilute the developer, the higher the contrast (and the shadow details become more dense). Keep in mind that the more a developer is diluted, the greater the risk of its premature exhaustion, even during the processing of one sheet of film. Larger quantities are needed and each new sheet of film requires fresh developer. Overly dilute, exhausted, or cold developer can result in brownish or greenish densities, mottled instead of smooth tones, and totally unpredictable and unrepeatable results. It is important to maintain a neutral color in the film positive because color plays a role in filtering the exposure to the tissue and affects contrast (Smeil 1975, p. 70).

Freshness is vitally important with the highly dilute developers needed for this process. The activity of the developer in a large volume of water is quickly compromised over time and use. A mixed volume of working solution will not keep beyond the average session nor will it be useful for more than one piece of film in a given volume. Do not mix developer a day ahead of time. Mix one or two liters of developer in a graduated cylinder as you need it. Pour off measured amounts for each test strip and full sheet of film and remember to discard the used developer after each test strip or sheet. It is advisable to use trays of different sizes for each size sheet or test strip of film so that the developer-to-film ratio remains somewhat constant.

Development *time* is used to control shadow density and thereby overall contrast. Increasing the development time increases the shadow density of the positive without having as much effect on the highlights.

The developer *temperature* plays a major role, and if it varies the results will be inconsistent and unpredictable. Stabilize the temperature, normally at 20°C (68°F). Process all test strips and full size film in the same temperature and dilution developer.

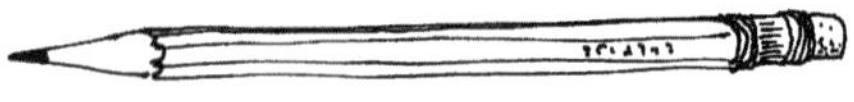

The *agitation* of the film in the developer is also a factor in evenness of development and the final contrast. Be sure, above all else, to be consistent from test to test and to final sheet. Slow, steady rocking of the tray is best. Alternate the rocking from side to side and end to end in regular intervals.

The ratio of *volume* of developer to area of film should be kept constant from the test strip to the full sheet. Use a 5″ × 7″ tray for a 2″ × 5″ test strip, then a larger tray for the full sheet of film with appropriate amounts of fresh developer in each. Use a tray at lease one size larger than the sheet of film to ensure proper agitation (i.e., use an 11″ × 14″ tray for an 8″ × 10″ sheet of film).

Follow the development with a 30-second stop-bath and then 3 to 4 minutes in rapid fix diluted 1:3 for film, not paper. For test strips, fix briefly (2–3 minutes) and rinse in running water for a couple of minutes. Squeegee and rapidly dry with a hair dryer before evaluation. For final positives, wash for 5–10 minutes in running water at 18°–24°C. (65°–75°F) and use a wetting agent in the final rinse (optional). It is possible to gently squeegee the film positive before hanging. Be sure the squeegee is soft and without nicks or embedded grit in order to avoid scratching the emulsion. Hang from one corner and dry in a dust-free environment.

Various developers differ in activity and development times. For lith films we generally use Kodak Polymax T Paper Developer, diluted 1:19 and even 1:24. Development times are usually 1 to 2 minutes. Clayton Paper developer is also useful, but diluted at 1:30. Dektol is useful at 1:9, but gives more contrast. Develop for 2 minutes or more in this developer. Selectol-Soft is useful for making soft, low contrast positives but must be developed for at least 3 minutes at a higher temperature (24°C or 75°F). You can even combine a soft positive processed in Selectol-Soft with a thin *shadow mask* positive, made in Dektol. The two layers of film are combined to make a sandwich with the correct densities. Be sure to use a loupe when taping them together so that alignment is perfect. Check the combined densities on a densitometer.

Table 2-1 21-Step Scale Ideal Density Chart

*Step Number**	*Ideal Density•*
1	0.05
2	0.20
3	0.35
4	0.50
5	0.65
6	0.80
7	0.95
8	1.10
9	1.25
10	1.40
11	1.55
12	1.70
13	1.85
14	2.00
15	2.15
16	2.30
17	2.45
18	2.60
19	2.75
20	2.90
21	3.05

* Step Numbers correspond to Stouffer Scale Numbers; the Kodak scale is un-numbered.
• As read using a transmission densitometer. These values are ideal and may vary slightly from actual readings.

CONTRAST RANGE

The most accurate method of determining the contrast range of a film negative or positive is with a transmission densitometer. If one is not available, side-by-side visual comparison with a step tablet of known densities will give a rough approximation, but is not entirely accurate. You can construct a viewing device out of two stiff black cards, each with a 4 mm (1/8″) hole punched in the center. Place a numbered step tablet of known densities, such as a Stouffer 21-Step Scale No. T2115 (see Figure 2-1), under one hole and a portion of the positive being read under the other. Place it on an evenly lit light table to evaluate. Move one hole up and down over the step scale in order to find the closest visual match to the hole over a spot in the image (Figure 2-3). A visual comparison is far more accurate when these tones are isolated in this way from everything else. All readings or comparisons should be done when the film is dry, both for accuracy and to prevent damage to the wet film surface. For test strips, use a hair dryer to speed things up.

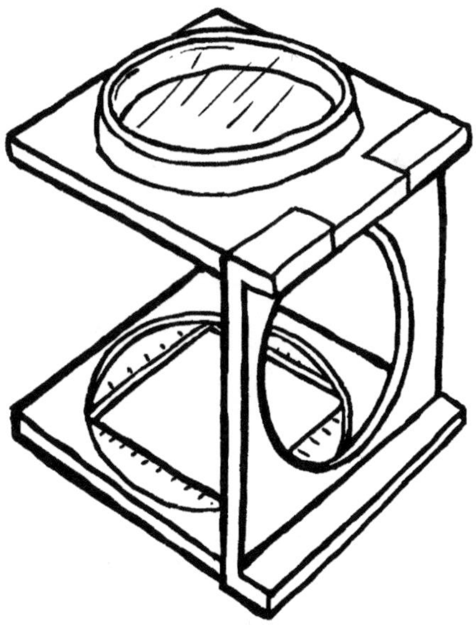

Full detailed highlights are paramount so that the gelatin highlight resist will not be too dense relative to the shadow details. The detailed highlight density should be within a range from 0.40 to 0.50 (steps #3+ to #4 on the Stouffer Scale). Any less indicates a problem with the exposure of the positive. Make sure that the highlight area chosen to read is in an area of bright texture detail, not simply clear film (spectral highlights).

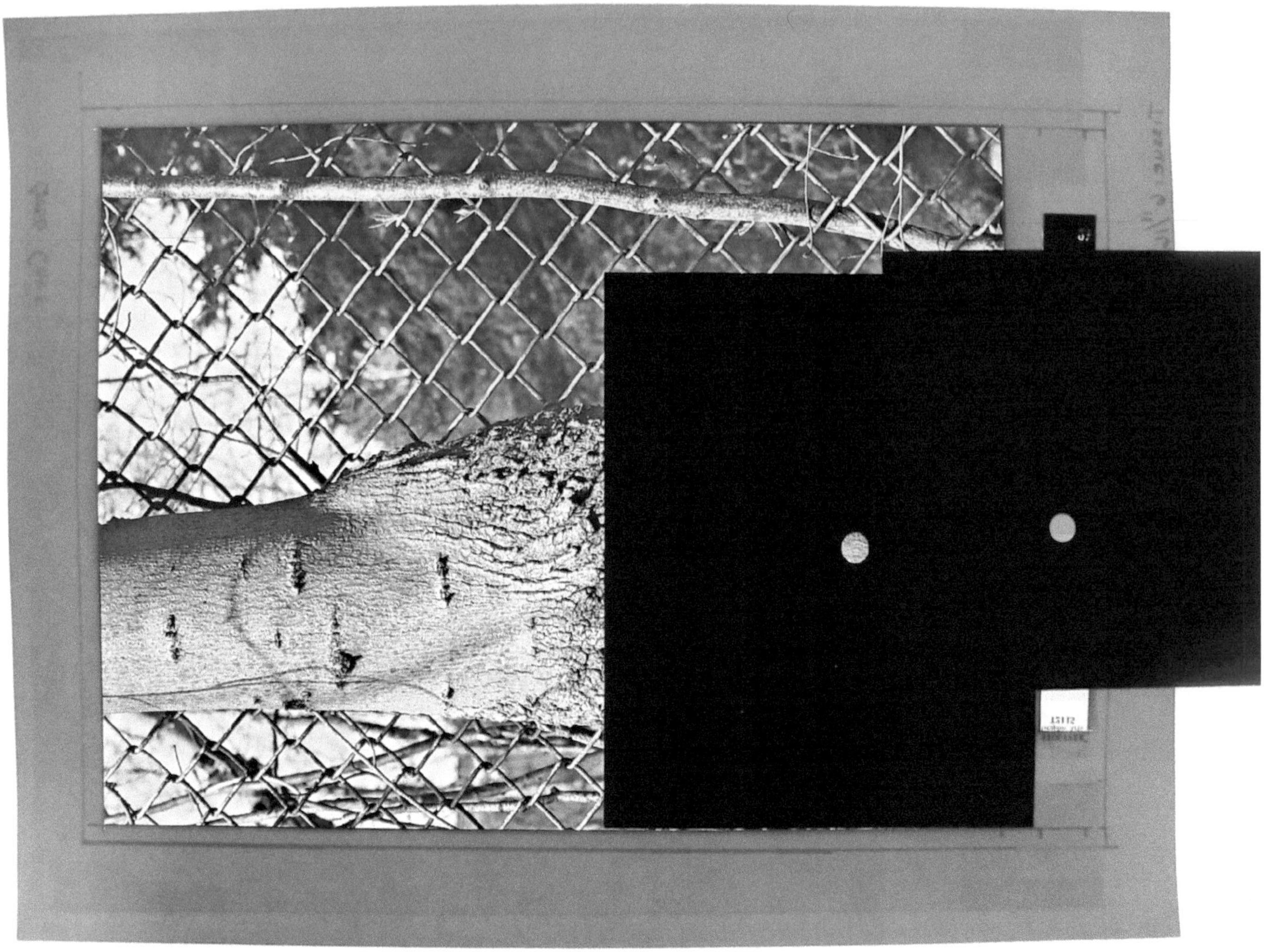

Figure 2-3 Comparison by eye between film positive and 21-Step Scale of known densities.

Highlight areas will appear darker in the positive than one would expect in a silver print.

Open shadow details should approximate the density of steps #12 to #13 on the Stouffer scale (1.70–1.85). Only the most dense, featureless blacks will be darker than step #13 or #14. The detailed shadow density should remain between 1.65 and 1.85. Anything more or less than that could indicate a problem with film positive development or exposure.

When a positive is read by a transmission densitometer (Figure 2-4), the readings of shadow details minus highlight details should be within the acceptable range from 1.20 to 1.45. Based on a normal image with a complete range of tones, anything less is too flat; anything more has too much contrast and will produce a resist that may be difficult to etch properly. Contrast will also increase by virtue of the etching process; therefore, it is preferable that the positives should appear to be slightly flat.

A good way to visually evaluate a film positive is to view it by reflected light by holding it up to a brightly lit white wall. A light table, although

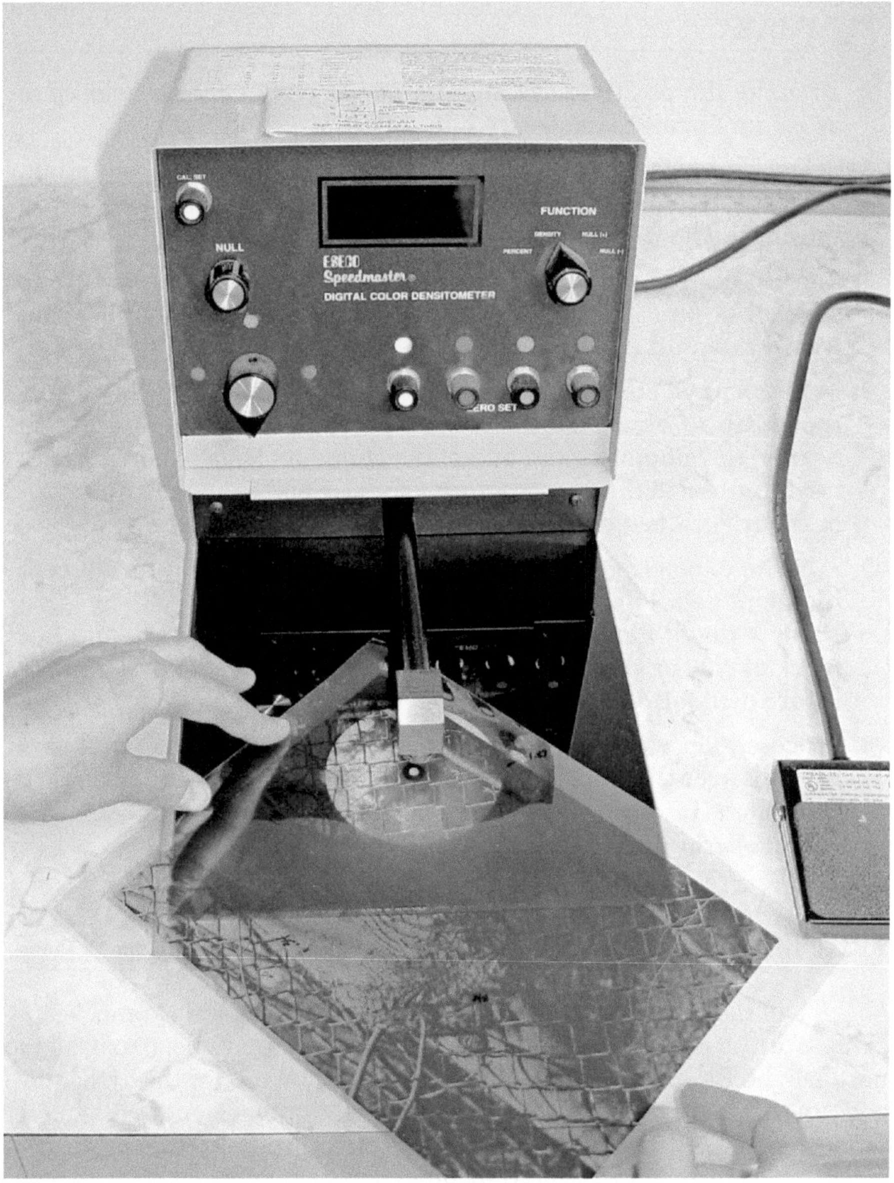

Figure 2-4 Transmission densitometer in use.

informative, will make it look brighter than it really is. The positive viewed by reflected light will give you a much more accurate indication of what it will look like as a gravure print, but with highlights that appear somewhat darker or stronger. All shadow detail should be strong and detailed. Muddiness or color cast will result in a poor print (Mertle and Monsen 1957, p. 334).

Note that all references to the densities on the Stouffer Step Scale and comparative areas on the film positive are based on the assumption that the film's base plus fog density is similar to the Stouffer Scale's base plus fog density (0.04). If the density exceeds 0.09 the comparison will not be meaningful. If the film's base plus fog density is much greater, test the safelights and process the film in total darkness to avoid the problem.

SUMMARY

1. To make the positive, use an enlarger or copy camera to enlarge or contact print the negative onto lith film.

2. Select and clean the negative and determine correct orientation. Carefully clean all surfaces of dust, hairs, and lint, and avoid fingerprints.

3. For lith film, have ready Kodak Polymax Paper Developer (dilute at 1:19) at 20°C (68°F), stop bath, rapid fix diluted for film, and wash water, all at or near 20°C (68°F).

4. Make stepped test exposures onto a sheet of film to determine correct ratio of exposure time and development time. Start with normal development and adjust development time to correct contrast. Agitate in the tray continuously. Use just enough fresh developer for each test or full sheet then discard.

5. Test the exposure and development times in order to produce a highlight detail density of 0.40 to 0.50 (Stouffer steps #3+ to #4), a shadow detail density of 1.65 to 1.85 (steps <#12 to #13), and full black at 2.00 or slightly higher. Dry the test strips with a hair dryer before evaluation.

6. Process then wash final film positives for 5 to 10 minutes. Use a wetting agent in the final rinse (optional). Squeegee (optional), then hang to dry from one corner in a dust-free environment. Handle carefully.

TROUBLESHOOTING

Fogging of the Film If the film has an overall fogged appearance, especially in areas protected from the enlarged image, it has been exposed to unsafe light. Be sure that the safelights are 1A light red (or the darker red) and not the green-orange OC safelights meant for black and white paper processing. If the film is very old or has been improperly stored near heat it can also show signs of overall fogging. If using a continuous tone blue-sensitive *commercial* film, it may be that the 1A safelights are too close or too bright. Try working with one light facing away from the work area and turn it off during development.

The Full-Sized Film Positive Does Not Look the Same as the Test Strip If the full sheet of film looks different than the test strip there could be several reasons: 1) the volume of the developer in relation to the area of film was not kept at the same proportion; 2) there were variations or inconsistencies in developer temperature, developing time, or developer agitation; 3) the developer was exhausted through use or time; or 4) exposure variations.

Uneven Development If working with highly dilute developers and/or short development times, the film may show signs of mottling, streaking, unevenness, or a color cast. Agitation technique can correct this or make it worse. Be sure to be consistent and careful that the film is gently and evenly agitated throughout the development time. Change the direction of the tray tilt every 15 to 30 seconds or so. Too little agitation can cause mottling; too much can increase contrast but add streaking. It can also add density to the outside edges and leave the middle of the image thinner. It is also important to use a large enough tray. Don't try to develop an 8″ × 10″ film in an 8″ × 10″ tray, for example. Also be sure that there is enough developer to completely cover the film at all times.

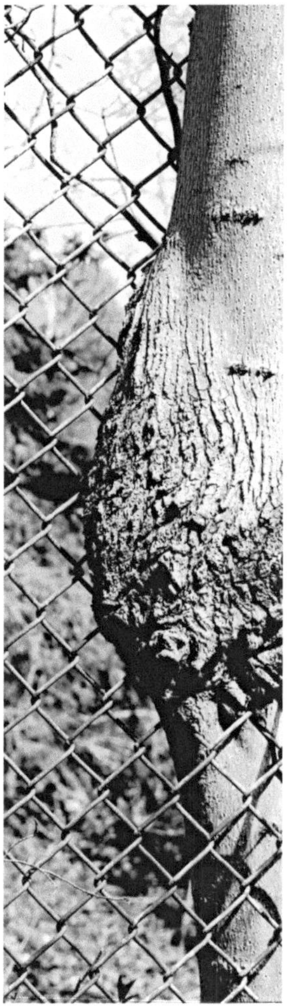

Figure 2-5 Example of an overly contrasty positive image.

Contrast Too High (See Figure 2-5.) To remedy this, decrease the development time if the shadows are too dense. You may have to increase the exposure slightly to maintain the highlight density. First check that you are not over-agitating or working at too high a temperature. If the developer is overly active, it may have to be diluted further. Be aware that overdilution or a very short development time can cause uneven development or mottle (see "Uneven Development" above).

Contrast Too Low If the image is too flat—but the highlight detail is good—increase the development time or the strength of the developer slightly. If the highlight becomes too veiled or dense, then decrease exposure a touch and retest the development. Use fresh developer for both the test and the full sheet.

Low Density If the positive is not neutrally black and has thin densities with a greenish or brown color cast, it is probably due to exhausted, overly dilute, or cold developer.

Grainy Image The grain of an enlarged 35 mm negative may result in "positive mottle" in the gravure tissue resist and ultimately in the plate (Cartwright 1939, p. 96). Use a diffused head enlarger to make the positive or use less diffused light for the gelatin tissue exposure.

Dust Specks Use canned air to clean the negative, the easel, and the film itself. Do not tilt or shake the can or it will discharge a liquid spray onto the film. If the relative humidity is low or if plastic surfaces are rubbed, a static charge can cause dust to cling to the film and negative. Use an anti-static cloth or brush if need be. When cutting the film, be sure the cutting surface is dry and dust free. Do not slide film over a surface. This can scratch the emulsion and may cause a static charge to build up.

Splotches or Fingerprints Be sure that your hands are dry when handling film. Wash them so there is no greasy residue. After working in the chemical sink, be sure to rinse all chemicals off your hands and dry them well. Avoid gripping the film tightly. Handle by the safe edges outside of the image area. Do not splash chemicals or water in the direction of the enlarger or dry work counter.

Scratches When the film is undeveloped or during development, it is especially susceptible to scratches that will mar the image. Be very careful not to slide the emulsion side across anything or vice versa.

3 Sensitizing the Gelatin Tissue

The gelatin tissue used for photogravure—also called pigment paper—consists of a thin layer of colored gelatin on a paper backing. It is very similar to the tissue used for carbon printing. Similar to other colloid-based photography such as carbon printing and gum bichromate, photogravure's gelatin tissue is sensitized using a solution containing a dichromate salt. Exposure to ultraviolet light causes the sensitized gelatin to become less soluble (raising its melting point) relative to the degree of exposure. The tanning or hardening effect is called *insolubilization*. This photochemical process produces a layer of gelatin that contains a hardened contour representation of varying densities, which correspond to the tonalities of the positive. The more light the gelatin receives, the greater the depth of hardening. The less light that passes through the positive, the shallower the depth of hardening. The exposed side of the gelatin is then bonded to the copper plate. Finally, the paper backing and remaining soluble gelatin is removed with warm water. The resulting layer is a three-dimensional contour image that will act as a permeable membrane between the copper plate and the etching mordant (Figure 3-1).

EQUIPMENT AND SUPPLIES

The Gelatin Tissue

The gelatin tissue we use is Autotype Pigment Paper G35. Autotype is the only remaining supplier of gravure tissue we have found to date. Another alternative is to make your own gelatin/carbon tissue, as outlined in Chapter 11. (See Color Plate 8.) The care and storage of gelatin tissue is important if the roll is to remain in optimum condition. The manufacturer's data sheets recommend that the tissue be stored at 20 to 22°C (68–71.5°F), and ideally at 60% relative humidity away from any possible contact with water, steam, or heat sources. If conditions are drier than the ideal, the tissue will curl more tightly as the relative humidity drops. As it reaches extremes—less than 25% relative humidity—there

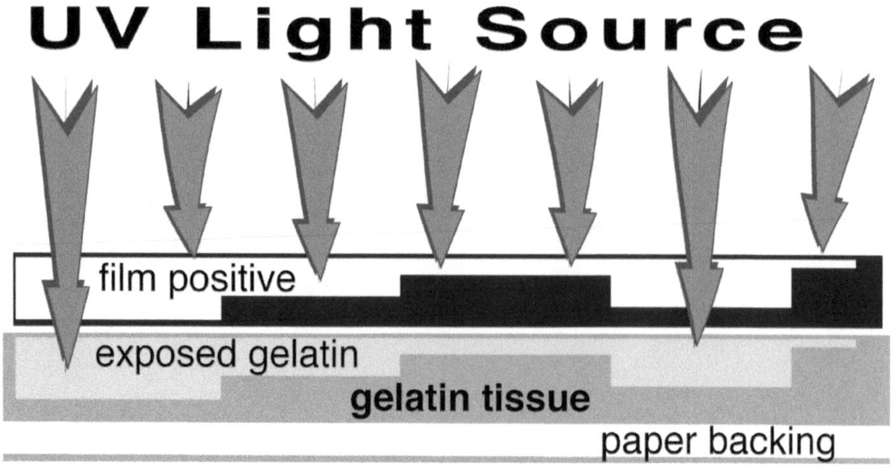

Figure 3-1 Cross section showing how light passes through the positive film and penetrates into the gelatin tissue in inverse proportion to positive density.

is a chance that the tissue may become too brittle to work with. Excessively high humidity will make the tissue tacky and prone to finger marks. It is important to keep the gelatin tissue away from chemicals it may react with including alum, photographic chemicals, free formaldehyde, and fumes from such sources as plywood, chipboard, and certain adhesive tapes.

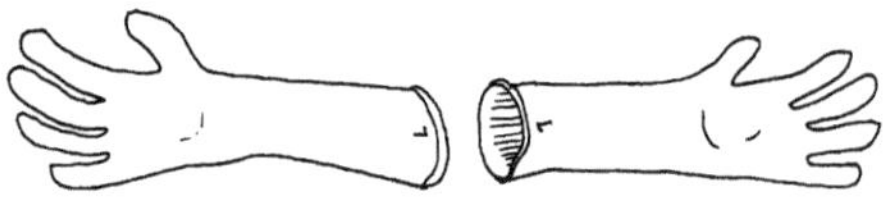

The Dichromate Sensitizer

Be aware that all dichromates are toxic and must be handled with extreme caution. The tissue is made light sensitive by soaking it in a solution of distilled water and potassium dichromate. The solution concentrations can be varied for contrast control. Normal contrast using G35 tissue is obtained with a 3.5% solution. The useful range of solutions is from 2.5 to 5%; the tissue manufacturer recommends 3%. The contrast will increase and the speed will decrease with lower solution concentrations, whereas the contrast will decrease and the speed will increase with higher concentrations.

Potassium dichromate is the sensitizer used most often for photogravure. Ammonium dichromate is also usable according to some sources, but requires a different dilution. "Identical concentrations of the two dichromates do not produce identical printing characteristics. A 3.5% potassium dichromate sensitizer actually has the same sensitivity, contrast and keeping qualities as a 2.5% ammonium dichromate sensitizer" (Crawford 1979, p. 184).

The dichromate sensitizer solution can be reused, but it is very important to keep a record of its age once used and how much tissue has gone through a given batch. It is recommended that the capacity should not exceed one square meter of tissue per liter of solution (de Zoete 1988, p. 37). If this level is exceeded, adverse effects include: 1) some tanning of the gelatin, which then peels off the plate during development or causes difficulty in adhering the tissue to the plate in the first place, 2) a seemingly fogged image due to overall hardening of the gelatin, which makes it difficult to peel the paper off the gelatin during development. These effects accelerate with time and use. As the solution is used, it picks up gelatin particles and becomes unusable over a prolonged storage time, even if it was not used to capacity. The solution should be stored in a cool environment, preferably a refrigerator. DO NOT STORE WITH FOOD. Be sure to keep the chemical secure and out of the reach of children because this toxic solution looks disturbingly like orange Kool-Aid. To avoid confusion, never store sensitizer in bottles once used for edibles. The maximum storage time for a partially used refrigerated solution is one month. Unused solutions should last indefinitely. Label clearly and store in a brown or light-proof container to avoid exposure to light. Do not pour used dichromate sensitizer down the drain. Bring it to a toxic waste disposal center.

The Drying Surface

After sensitizing, the wet gelatin tissue must be dried to an absolutely smooth and flawless surface. This is achieved by having the tissue dry in contact with a smooth methacrylate plastic surface such as Plexiglas, Perspex, Lucite, or chrome (as on a ferrotype plate). Do not use glass or Lexan because the tissue will stick and will be ruined. The Plexiglas support must be scratch free, cleaned, and degreased. Peel the protective paper off one side only, keeping the other side protected for later use. The Plexiglas must be several centimeters larger than the tissue. Have several sheets of a few standard sizes. Do not use these sheets for any other purpose.

Other Equipment

The sensitizing process should be done in a darkroom sink for chemical safety, water and light control, and access to basic darkroom supplies. Standard photographic supplies such as a timer, trays, funnel, stirring stick, graduated cylinder, and glass thermometer are required. You will also need a flawless stiff squeegee such as a screen printing squeegee or a good stiff photographic print squeegee or, alternatively, a wide soft rubber roller. Paper towels, paper coffee filters, a soft Hake-style brush, canned air, and sometimes ice packs are also needed. An accurate balance or electronic scale is needed to measure out the potassium dichromate. A bright safelight can be made using a 60-watt yellow bug light. A flat, nonmetallic surface is needed in the sink to support the Plexiglas while squeegeeing. The sink should be large enough to hold the trays and the squeegeeing surface to ensure that all dichromate splashes can be contained and washed away.

PREPARATORY STEPS

Mixing Potassium Dichromate Sensitizer

Prepare a 3.5% sensitizing solution by dissolving 35 grams of high grade potassium dichromate in 750 ml of distilled water at about 27.5°C (80°F). Stir constantly and then bring up the total volume to one liter after the dichromate salts are dissolved. Filter the solution through a coffee filter into a storage bottle to remove undissolved particles and impurities. Use *distilled*, rather than *purified* or deionized water. If minute air bubbles are a problem during subsequent steps, preboil the water in a clean pot and allow it to cool. This removes much of the air that is invariably dissolved in the water. The distilled water can be boiled in any metal container, but transfer the water to a glass or plastic darkroom graduated cylinder when mixing in the potassium dichromate. Dichromates should not come in contact with metal.

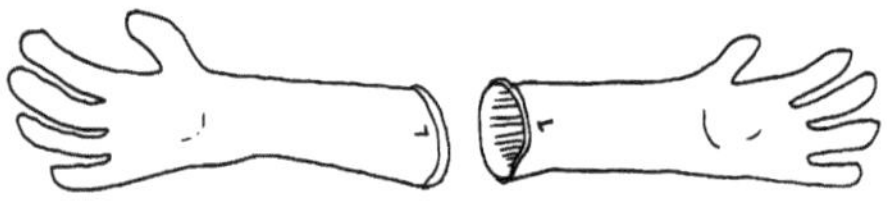

Dichromates are toxic and must be handled with caution. The inhalation of dichromate dust can be *fatal*. Wear a respirator or good dust mask, face shield or eye protection, gloves, and an apron when preparing the sensitizer solution. Skin and mucus membrane reactions can be long term and serious. Always wear gloves when working with the sensitizer solution or handling the sensitized tissue.

Cutting the Unsensitized Tissue

All cutting and handling of the tissue must be done carefully to prevent damage to its delicate surface. Cover all tabletops with newsprint or a clean self-healing cutting board. Be sure your hands are clean and dry when handling the tissue. You may even wish to wear cotton gloves. Make sure all straight-edges or rulers do not have burrs or tacky sections from old tape. Clean them beforehand with alcohol to remove any grease. Cut the tissue with a sharp blade cutter rather than scissors in order to prevent cracking and damage to the edges. Be careful not to crease or fold the tissue because all marks will become flaws in the resist and will ruin the plate. A second pair of hands is very useful to prevent the tissue from rolling up and getting away from you. You can also make little nonmarring weights to hold down the corners and edges as you cut. We use small bricks of zinc etching plate, each made up of four to six pieces of 5 cm × 10 cm × 0.2 cm (2″ × 4″ × 1/16″) zinc plate, stacked and covered with several layers of masking tape. We have about eight of them, and when it is dry out we could use even more (Figure 3-2).

After the tissue is rolled out, examine it carefully for creases, tears, or fingerprints. Cut around these areas when you lay out the pieces you need and discard the flawed tissue. It is counterproductive to work with damaged tissue. Use a clean, soft brush or compressed air to get rid of any visible dust or particles.

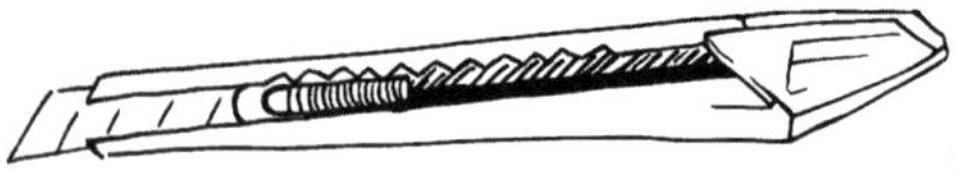

When working with the tissue, be careful not to touch the gelatin surface because body heat can melt it and skin oil or moisture will ruin it. Even prolonged pressure from the back can cause a blemish. If you have warm or moist hands, wear cotton gloves. After the tissue is cut it will have a tendency to curl, especially in a dry environment. If you are sensitizing immediately, the pieces can simply be moved to the work area as rolled

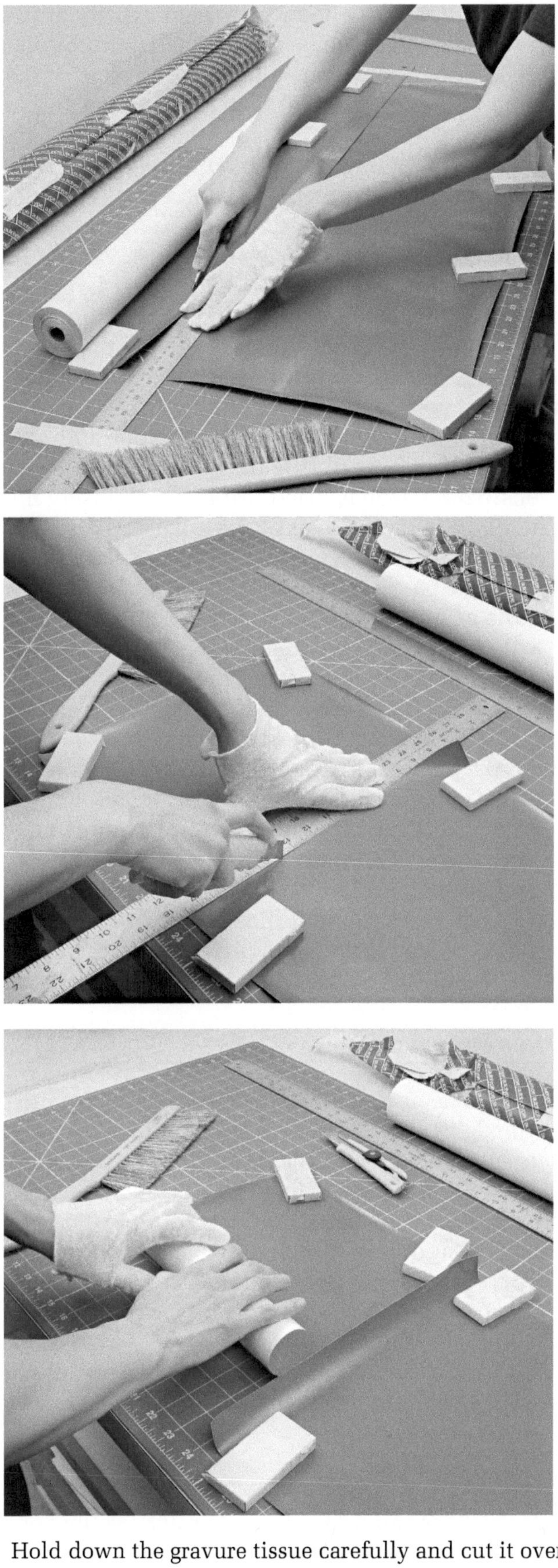

Figure 3-2 Hold down the gravure tissue carefully and cut it oversized for sensitizing.

up tubes. Be careful that they do not get wet or dusty. It is advisable to give the newly cut tissue a chance to acclimatize to the room's relative humidity. This is important because the leading edge of the tissue usually has a different moisture content than a few inches into the roll. This difference can affect absorption and cause the tissue to be unevenly sensitized (de Zoete 1988, pp. 32–33). If you unroll the tissue and hold it down with blocks or a metal straight-edge for an extended period of time, the area of contact should not be used because it will be of a different moisture content than the rest of the tissue and will sensitize differently.

Cut the tissue at least 3 to 5 cm (1.5–2 inches) larger than the final image. Include an additional allowance in one direction for the Stouffer Step Scale. It is important to cut the tissue larger than required to provide a handling edge and to allow for drying problems when sensitizing. Use only the central area of the tissue, trimming off any offending edges before exposure. This eliminates the frilly border and edge flaws and lessens the chance of creases during the lay-down.

After the sensitized tissue is dried, trim it to precisely fit the positive assembly. The final, trimmed size of the gelatin tissue should end up halfway between the image size and the edge of the copper plate. The copper plate should be at least 1 centimeter (0.5 inch) larger in each direction than the positive image chosen with a 2.5 cm (1 inch) allowance for the step scale. (For size relationships see Figure 5-3.)

For long-term storage, carefully wrap the cut pieces around the original roll. Rewrap the extra pieces around the remainder of the roll and hermetically seal it. Store in a cool place. We are forced to store ours at a lower humidity and are finding that the roll is becoming increasingly stiff and brittle as time goes on. So far, it still performs well, but it is hard to handle on dry days. Rehumidifying dry tissue is possible. Store it unwrapped and loosely rolled in a relatively humid environment for a period of time. This can be time consuming and tedious because you must readjust the roll constantly to ensure even rehumidification. Some sources recommend hanging the roll horizontally with weights attached to the bottom edge as it is unrolled. The exposed surface should be able to acclimatize evenly.

Plexiglas Plate Preparation

After the tissue has been sensitized, it must be dried against a Plexiglas support. The Plexiglas must be free of scratches or grease. Do not use oily cleaners such as Plexiglas cleaner. New Plexiglas should be initially degreased with TSP and water. See "Degreasing and Brightening the Plate" in Chapter 4 for the degreasing procedure. After that, and before each sensitizing session, it is sufficient to clean the Plexiglas with alcohol and dry with a paper towel. Just before use, position the Plexiglas on a firm elevated support in the darkroom sink next to the sensitizing tray. The support should be sturdy, level, and slightly higher than the sensitizing tray. This is done so that when you squeegee the tissue, the excess sensitizer is wiped back into the sensitizing tray.

SENSITIZING THE TISSUE

The sensitizing process should take place under a yellow or red safelight or subdued incandescent light of low wattage that is bounced off a wall or ceiling. A bug light is excellent and provides enough light to see well.

Put on rubber or nitrile gloves, a face shield, and a rubber apron for the entire process.

Find a clean, standard-sized photographic tray larger than the cut tissue. Ensure that the 3.5% solution of potassium dichromate sensitizer solution is stabilized at 10 to 13°C (50–55°F). This is easiest if you can store the solution in a *non-food* fridge. Alternatively, if the darkroom and water source are warm, float the tray in a larger tray of cold water with ice. You can even float a clean ice-pack in the solution to speed cooling. Remove the ice-pack when the temperature stabilizes. Never use it again for cooling food!

> Note: The gelatin should never be submerged into any solution above 15°C (60°F) except during the final development stage.

Before sensitizing, inspect the tissue for flaws or creases and be sure that there is no dust or debris stuck to the surface. Be careful not to get the tissue wet. If it is not too curled, try to carefully dust with a clean sable brush or blow with dry compressed air. Set the timer for slightly more than 3 minutes and start it. Immerse the tissue face up, sliding it smoothly and quickly under the solution at the 3-minute mark. If it is tightly curled up, unroll and reroll it back and forth like a scroll until it begins to go limp and lay flat (Figure 3-3 and Color Plate 9). Be careful not to splash the solution.

After the initial submersion be sure the whole surface is flooded with sensitizer as soon as possible. Use a wide soft brush with *no* metal ferrule—such as a wide Japanese Hake brush—to hold down and push the tissue flat and to gently clear the surface of any clinging bubbles or particles. Hold down the corners with the brush and/or your gloved fingers until the backing and gelatin absorb enough water to allow the tissue to lie flat on its own. This should happen after about one minute; it may take more time if the tissue is very dry due to low relative humidity. Turn the tissue over in the sensitizing bath and allow it to soak, fully submerged, for the remainder of the time—a total of 3 minutes as recommended by the manufacturer—while gently rocking the tray. Brush the back of the tissue to release air bubbles as well. The temperature and time need to be kept constant for consistent results.

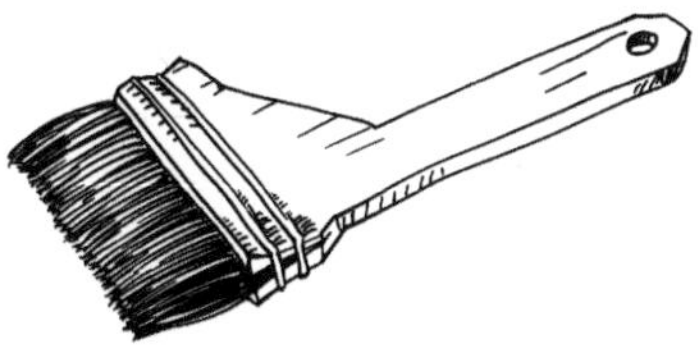

Adhering the Tissue to the Plexiglas

Gently lift the tissue out of the sensitizer and allow it to drain while holding it by one corner. Dribble a puddle of sensitizer on the center of the Plexiglas. Slide the tissue face down onto the surface of the Plexiglas in a continuous motion in the middle of the puddle of sensitizer. Be careful not to trap any air or dust particles under the tissue. Position (slide) the tissue to the center of the Plexiglas (Figure 3-4). Pin it in place along one edge and use a rubber squeegee to lightly stroke the back of the tissue, gently pushing out the excess liquid. Position the squeegee in the center of the tissue and stroke more firmly once in each of all four directions until there are no signs of lifting along the edges of the tissue. Do not use excessive pressure when squeegeeing. A screen printing squeegee is good because the edges are usually straight and sharp. A stiff photographic squeegee in new condition is also very good, but some photo

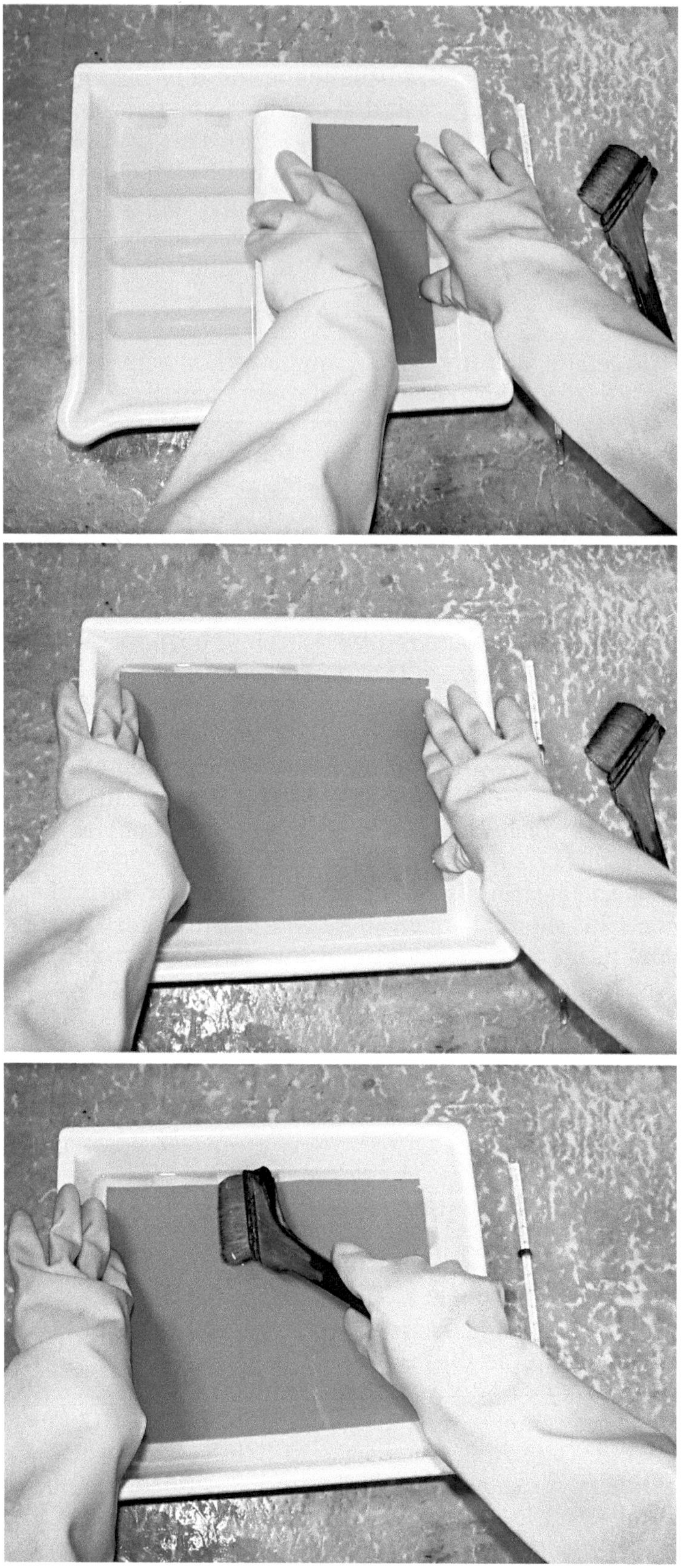

Figure 3-3 Hold the tissue under the solution, unrolling and rerolling until it begins to flatten out. Once almost flat, brush it to remove adhered bubbles and hold it gently under the solution until it is flat enough to turn over.

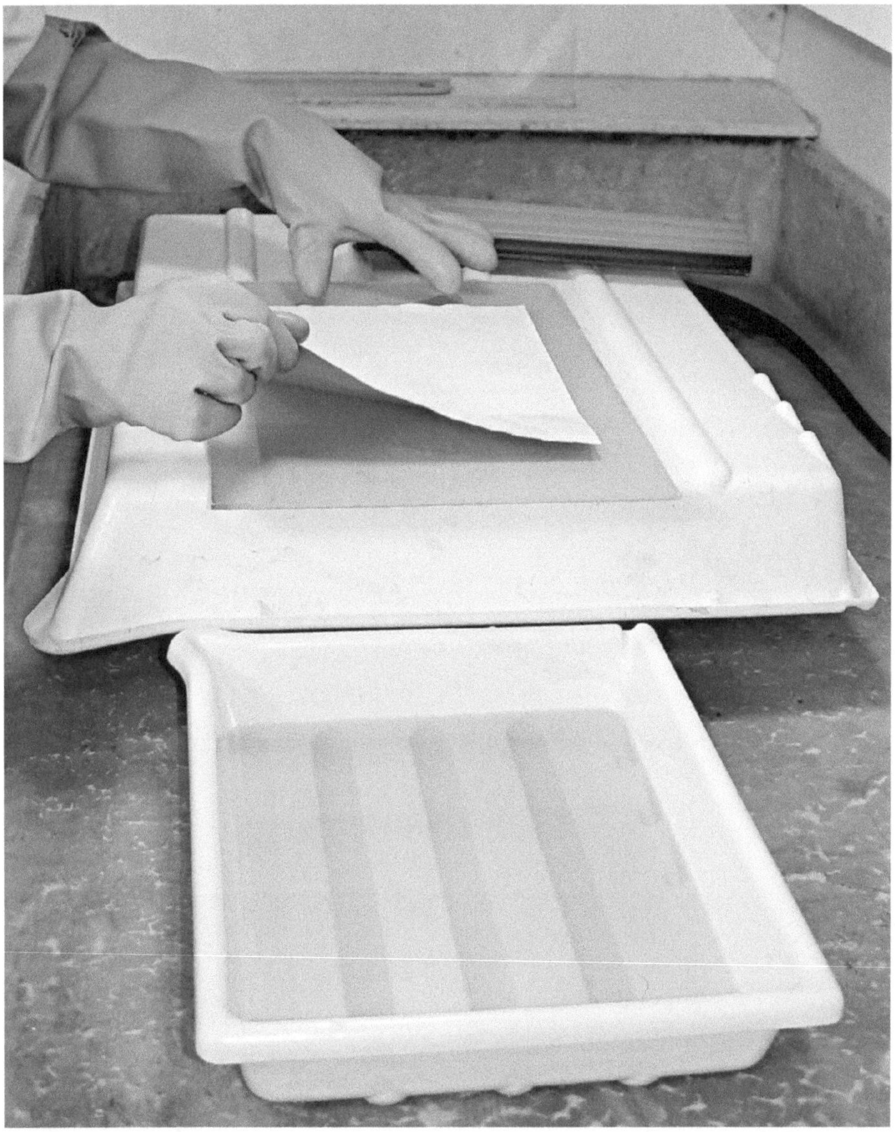

Figure 3-4 Center the tissue on a pool of sensitizer in the middle of the Plexiglas.

squeegees may be too soft or flexible (Figure 3-5). If any serious problems occur, such as buckling or creasing, discard the tissue and begin again. A creased gelatin tissue cannot be salvaged. We recommend that an extra tissue be sensitized in the event that one is damaged or flawed. If it is not required it can be stored frozen in a hermetically sealed envelope or box and used at a later date.

Immediately blot the back of the tissue with a wad of paper towels to remove any excess sensitizer. Be sure the edges and surface are free of any shiny wet areas. Wipe the edges of the Plexiglas and the back to remove any sensitizer solution (Figure 3-6).

All solutions that are to be stored and reused—such as the sensitizer—must be filtered after each session to remove hairs, dirt, or gelatin fragments. You do not want any gelatin particles to remain in the solution during storage because they will shorten its storage life. A No. 4 coffee filter and large funnel are ideal strainers (Figure 3-7).

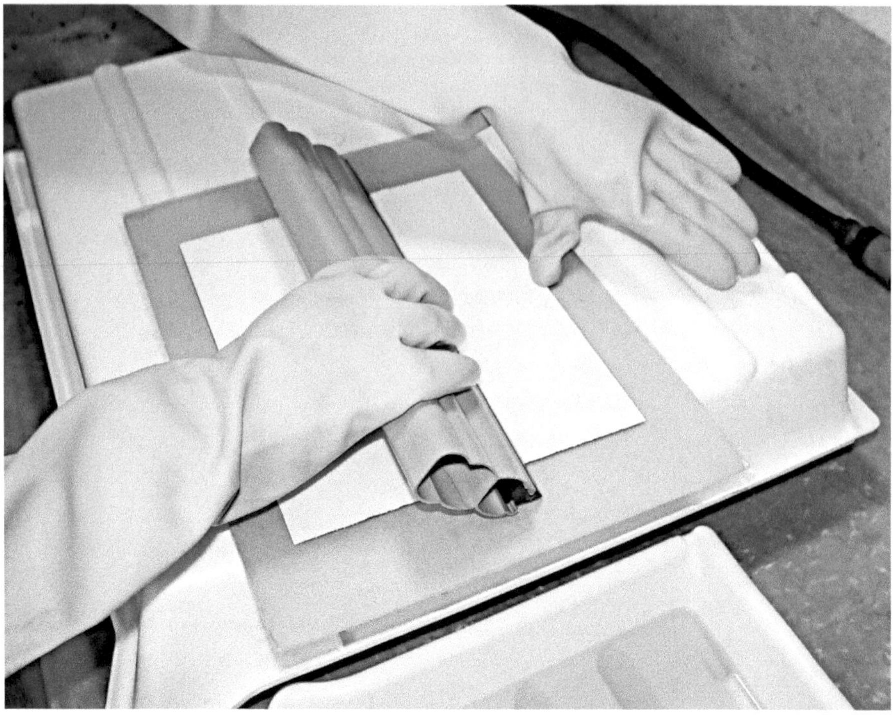

Figure 3-5 Squeegee the tissue from the center to each of the four sides.

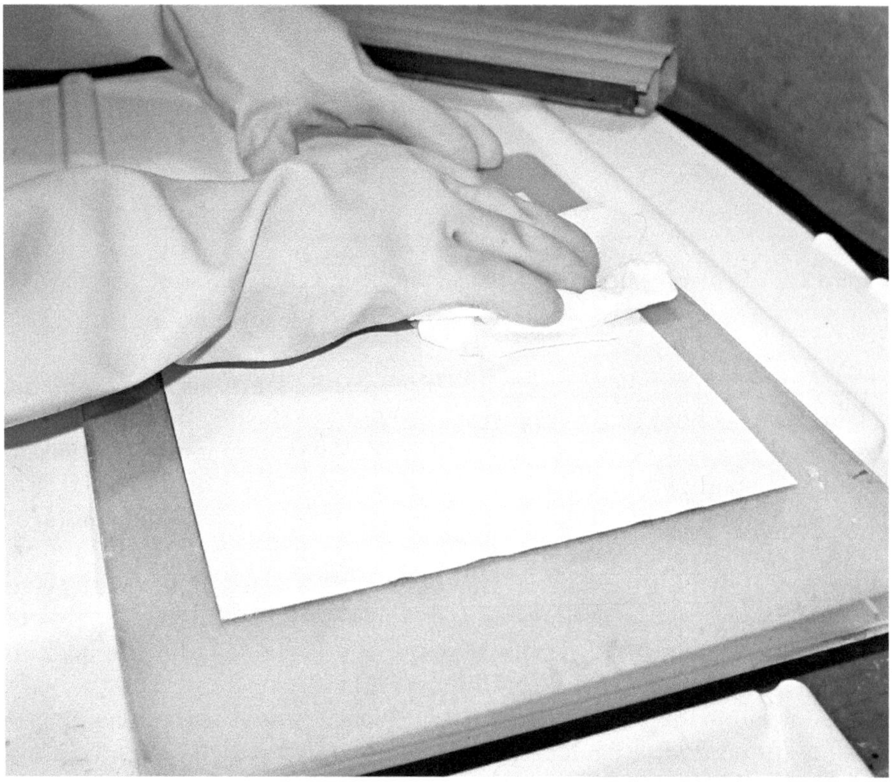

Figure 3-6 Soak up all excess solution by gently rubbing with a wad of paper towels. Wipe the edges and the back as well.

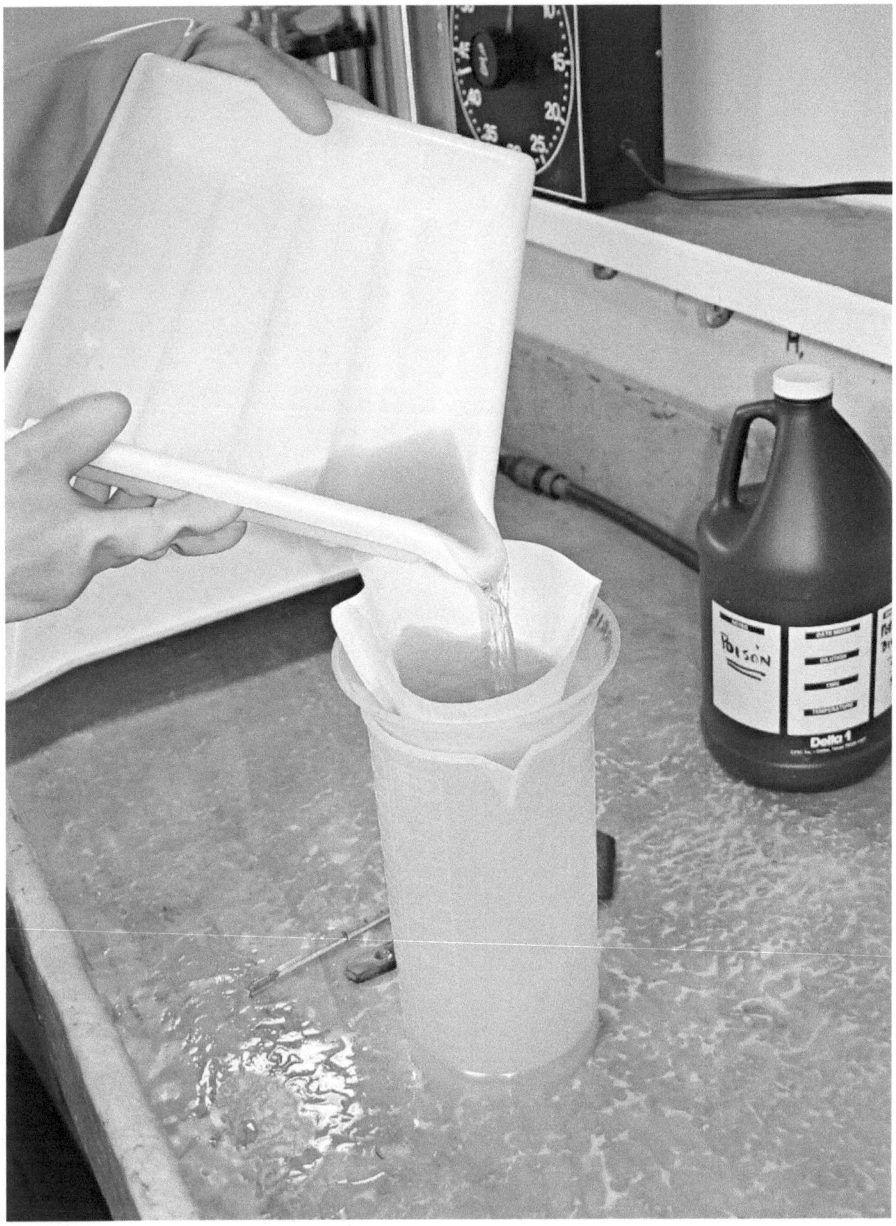

Figure 3-7 Filter the dichromate sensitizer solution through a coffee filter using a wide-mouthed funnel.

Drying the Sensitized Tissue

The tissue must dry evenly in order to prevent fractures and concentric rings from forming on the surface. If the tissue receives uneven air movements across its surface, the corners will dry before the center and begin to lift while the center is still stuck firmly to the Plexiglas. This will cause the tissue to form concentric fractures as it pops off the surface in small increments (see Figure 3-10). Ideally the tissue should pop or peel off the surface of the Plexiglas in one instant. Most sources mention the use of a fan to dry the tissues quickly. If the fan is large and the room is not too dry or hot, this method can work well. We have also experimented with stacking the wet tissues on Plexiglas between blotters. We have found that all of these methods presented various problems, some serious.

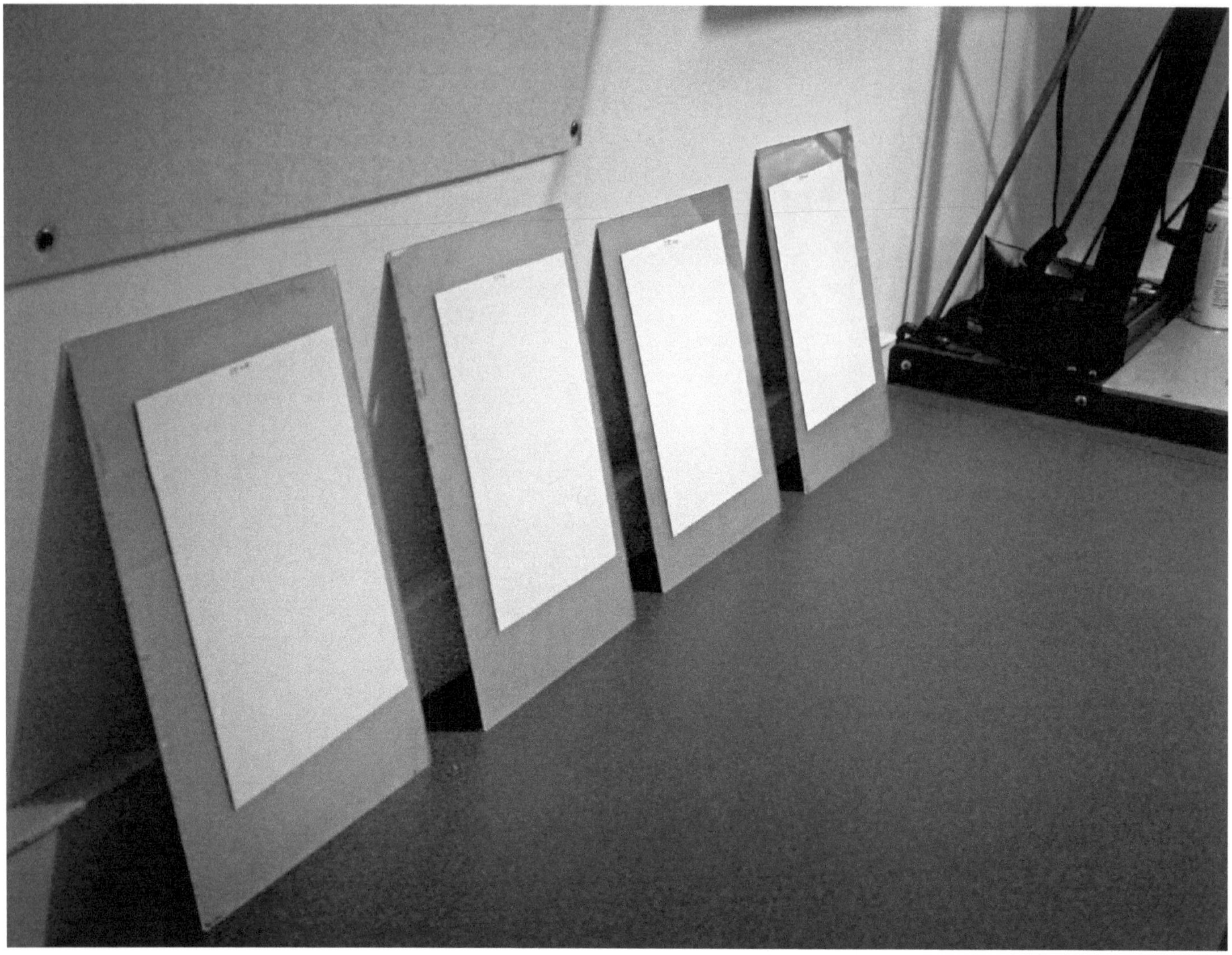

Figure 3-8 Plexiglas standing in a darkroom as tissue dries overnight. Be sure all lights are off.

Our solution, which seems to give us almost perfect results, is to allow the tissue to dry slowly and evenly while exposed to the still slightly humid air of an enclosed space. This prevents the edges from drying at a different rate than the middle of the sheet. To do this, it is best to leave the tissue uncovered while inside a dark cabinet or room for 8 to 14 hours. Stand the Plexiglas almost vertically on one edge (Figure 3-8). Because the tissue is now light sensitive, a dark cupboard is necessary if the room lights are to be on at any time while it is drying. If it is left in a darkroom to dry, examine it only by safelight (no white light) because it is now much more light sensitive than while it was wet. Do not leave any safelights on overnight while it is drying. Rotate the Plexiglas by 180° at least once while it is drying. Although most sources recommend 1.5 to 2 hours as the ideal time for drying by fan, Blaney (1895, p. 23) says to let it dry overnight, and the GTA Guide (Smeil 1975, p. 69) suggests a subsequent storage of 8 to 10 hours at room temperature to achieve uniformity.

The tissue should not over-dry to the point where you find it on the countertop in a tight little roll. It should be dried just to the point where you have to manually strip it from the Plexiglas (Figure 3-9). It will be

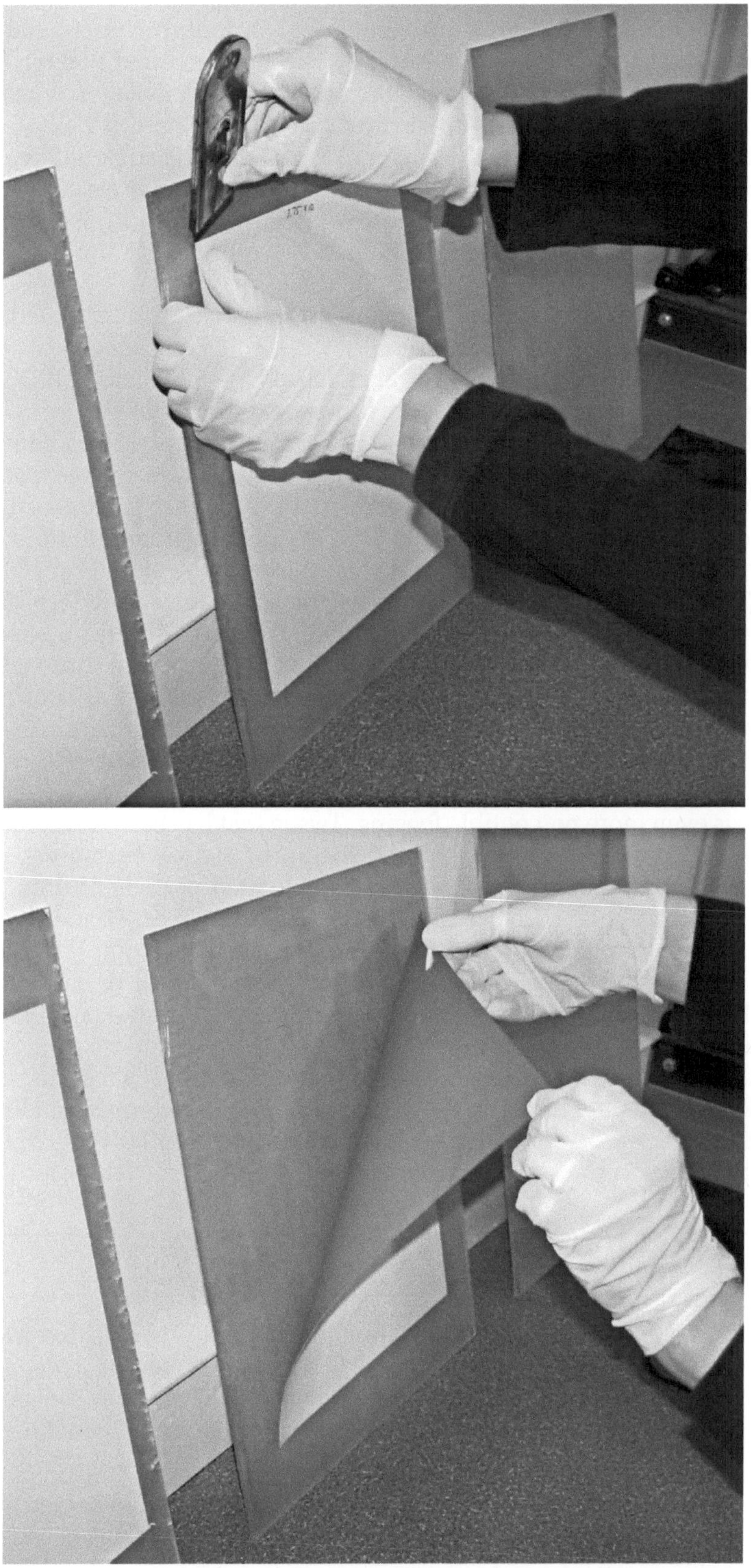

Figure 3-9 The tissue should peel off in one clean "woosh."

flat and have a better moisture content at that point. Under safelight, inspect the tissue for flaws caused by scratched or pitted Plexiglas, air bubbles, dust, or uneven drying. If the flaws are extensive do not use that portion of the tissue. Place the sensitized tissue in a hermetically sealed light- safe bag until used or frozen for storage. Black plastic photographic paper bags are ideal. Overly dry tissue will cause contact problems later during exposure. It is possible to correct this by rehumidifying through storage in a confined space of higher relative humidity until the curl goes limp.

The sensitized tissue can be kept at room temperature for a limited time before exposure. The dark reaction, or *dark effect*, causes the tissue to progressively harden as if it were being fogged by light. During this time, there is an increase in its *speed* because of the loss of the *threshold*. Speed and threshold are photographic terms used to describe how film and paper initially react to light. Refer to photography resources for more information on this. De Zoete (1988, p. 35) says that light sensitivity increases during the first 24 hours after drying, at which point the dark effect begins. After a day or two it is supposed to be at its best (Blaney 1895, p. 18), but after about seven days the tissue may be so fogged and unpredictable that it must be discarded. We normally use a sheet of tissue during the day after it was sensitized but have found through testing that after three days freshly sensitized tissue stored at cool room temperature shows no visible evidence of the dark effect. After a week, however, there is some slight fogging and evidence of dark effect. This change will accelerate if the temperature increases. After two weeks, there is even more perceptible fogging. Tissue that has been frozen immediately after sensitizing and stored for up to eight weeks gives good results. Tissue that has been frozen for extended periods of time (four months) can show evidence of the dreaded dark effect. Tissue that has been frozen for one year failed to adhere because it had completely hardened. It is best to freeze it as soon as it is dry before it over-dries and curls up. Put the extra tissue in a light-safe, hermetically sealed bag. Place in a chest freezer, preferably not a frost-free refrigerator freezer. When needed, remove it from the freezer ahead of time and leave the sealed package unopened during a thawing out period. After opening, allow it to achieve equilibrium with the room before exposure. This will prevent irregularities in the resist and final image.

SUMMARY

1. **Mix a 3.5% solution of potassium dichromate and distilled water for the normal contrast Autotype G35 tissue (5% for the lower contrast Autotype G25 tissue). One liter of mixed solution can safely sensitize 1 square meter of tissue. Keep refrigerated if possible (not with food!!).**
2. **Cut a piece of gelatin tissue at least 3 cm (1¼ inch) larger than the image area (including room for the 21-Step Scale).**
3. **Immerse the tissue in sensitizer at 10 to 13°C (50–55°F).**
4. **After the tissue relaxes and begins to lie flat, brush lightly in all directions with a very soft, wide brush to remove air bubbles.**

Use the brush to hold the tissue down under the solution until the curl completely relaxes and it flattens out.

5. **Leave the tissue face up for one and a half minutes and then flip over. Keep in the solution for a total of 3 minutes.**
6. **Have the degreased Plexiglas ready on a nearby firm support (cleaned in advance with 100% alcohol and then dusted with canned air).**
7. **Drain the tissue onto the Plexiglas to form a puddle.**
8. **Slide onto the Plexiglas and squeegee the tissue in place while holding one extreme edge. Repeat in all four directions from the center.**
9. **Pat the backing dry with paper towels to remove the excess sensitizer from the front and back of the Plexiglas.**
10. **Stand the Plexiglas vertically on a countertop in a not-too-dry, closed, light-safe (dark) area.**
11. **Rotate the Plexiglas after a few hours by 180 degrees, then allow to dry undisturbed overnight.**
12. **After the tissue pops off or peels off without resistance, it is ready for storage or use. Wrap well in a light-proof plastic bag or envelope like that which photographic paper comes in.**

TROUBLESHOOTING

Concentric Fracture Lines Radiating Toward the Corners When the tissue dries too quickly, the edges and corners dry faster than the center and tend to release first. After the tissue is removed, there are concentric rings visible on the surface of the tissue, especially near the corners (Figure 3-10). Contrary to many older texts, we advise that you avoid a direct fan or warm air current on the tissue during the drying time. A still and slightly humid room is best for a slow overnight drying period. We use our darkroom, and when the relative humidity is very low we add water to the sink to raise the relative humidity and slow the drying. This is also important in order to prevent the tissue from over-drying and curling up.

Cupping and Edge Frilling When the tissue dries unevenly, it will not be flat. It may be cupped or domed, or it may have frilled or wavy edges. If there are problems with the edges frilling or drying faster than the center, sensitize a larger sheet than needed and simply trim off the wavy edges. This is particularly useful in dry climates (≤ 35% relative humidity). This problem can make the dry lay-down technique next to impossible.

Air Bubbles and Pinholes If you are getting a lot of tiny air bubbles from using the puddle method of adhesion, use water that has been preboiled and then cooled. If the problem persists, you can adhere the tissue to a submerged Plexiglas sheet using the adhesion to copper procedure described in Chapter 6 ("Adhering the Tissue to the Copper Plate: The Wet Lay-Down Method").

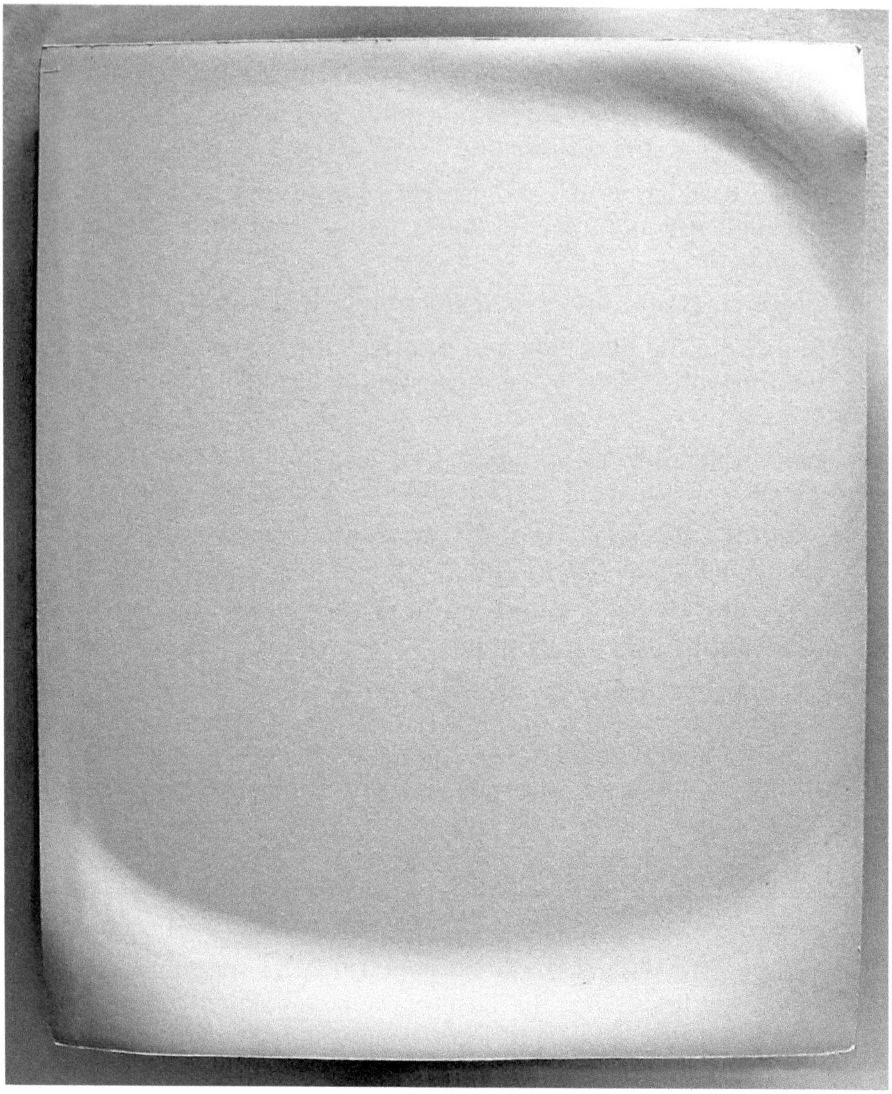

Figure 3-10 Concentric fractures resulted when the tissue let go in small increments, because the corners dried at a faster rate than the middle of the tissue.

Blemishes and Spots (water, air, grease) Areas on the Plexiglas that had a greasy smudge will cause a blemish or blister in the tissue during lay-down or become apparent during the etch. Clean the Plexiglas with 100% alcohol and paper towels immediately before adhesion. If you trap air under the tissue during lay-down, the blister formed prevents the gelatin from bonding to the Plexiglas. This area dries differently and causes a blemish.

Pits and Bumps, Scratches and Flaws If a scratch or pit mars the surface of the Plexiglas, it will leave a tiny bump on the surface of the newly sensitized gelatin tissue. This will cause a *sunspot* during exposure by holding the positive slightly away from direct contact with the tissue at this point, allowing the light to spread underneath. Be sure the surface of the Plexiglas is flawless (Figure 3-11).

Uneven Sensitizing (not apparent until the plate is etched) If the moisture content of the gelatin tissue is not even when immersed into the

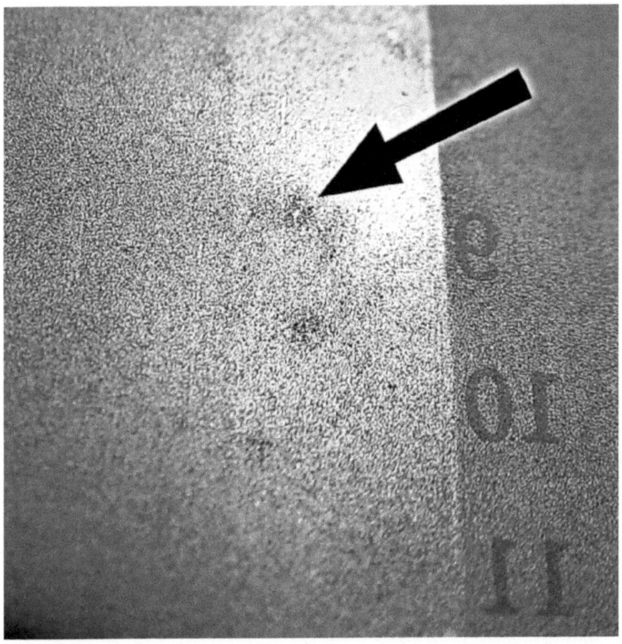

Figure 3-11 A bump in the gelatin resist caused by a pit in the Plexiglas used to sensitize the tissue. If etched, this will result in a *sunspot.* (See Figure 5-9.) This is a good indication that it is time to replace the Plexiglas.

sensitizer, it may cause uneven absorption. Give the tissue time to acclimatize first. When the sensitizer is used beyond its capacity or the tissue is not submerged fully or agitated evenly during sensitizing, the tissue may acquire unpredictable areas of variable sensitivity. This will show in the etched plate as splotchy tones or coarse mottle. Take precautions to ensure that the sensitizer is not over-used or too old and that you keep the tissue fully submerged during sensitizing.

Fingerprints The warmth of your fingers can actually melt the gelatin resist, especially when pressure is applied. Wear cotton gloves when cutting and trimming tissue. Be sure to wear rubber or nitrile gloves at all times when sensitizing the tissue and cool your gloved fingers in the bath as you are working. Don't use a lot of pressure when handling the tissue. Avoid touching the surface of the gelatin.

4 Preparing the Copper

Copper has been used for traditional intaglio for centuries because of its malleability and resilience. Its ability to bond to the gelatin and react with ferric chloride in subtle and controllable ways also serve to make it ideal for photogravure. This chapter describes the preparation of the copper for photogravure, a process that, due to its fine detail and finicky nature, is less forgiving than other intaglio processes.

EQUIPMENT AND SUPPLIES

Copper Sheets

The copper necessary for photogravure should be highly polished and without flaws. You can buy it ready-to-use or you can polish standard sheets of roofing copper. Either type can be purchased in a variety of *gauges* or thicknesses; we recommend between 16 and 20 gauge. Copper that is too thin will be hard to print due to the difficulty of obtaining high pressure in the press. Copper that is too thick is extravagant and may give a plate mark that is too deep or abrupt. It is also hard on the blankets.

We prefer to use a high-quality, 18-gauge copper that is sold with a mirror-polished finish, making it consistent and convenient. It usually comes with a plastic anti-tarnish film adhered to the surface. Roofing copper is a less expensive alternative that is purchased as-is—search for the least scratched sheet. It must be cut into manageable sizes and hand-polished to a mirror finish. We find that the labor needed to achieve this is hardly worth the savings. The hardness of the copper can vary as well. This may have an effect on the etching characteristics and the durability of the plate when printing. When using copper from a new source, be prepared to make test plates in order to be sure that it is of appropriate quality. Once you find a reliable source, stick with it. (See Appendix I, Suppliers.)

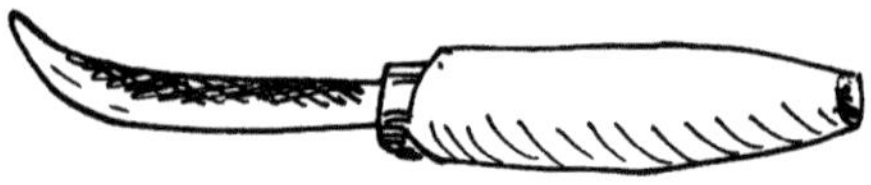

Tools and Polishes

Regular printmaker's tools such as a burnisher and a scraper are needed to remove pits and fine or deep scratches. If you use roofing copper, polishing materials include jeweler's rouge, rottenstone, superfine wet/dry sandpapers, Brasso, Twinkle, Putz Pomade, automotive polishing compound, soft rags, felt pads, and solvents. Mirror-finished copper will rarely require the use of these polishing compounds and needs only a light polish with Brasso. Good quality new metal files—coarse and fine—are needed to bevel the edges of the plate. Wet/dry silicon carbide paper of 320 to 400 grit is also useful (see Figure 4-1).

Degreasers and Brighteners

Trisodium phosphate (TSP) in its crystal form is used to degrease the copper plate, and is available from most hardware stores. TSP is a strong alkali and gloves are required when using it, especially in the concentrations we recommend. Avoid inhaling TSP dust.

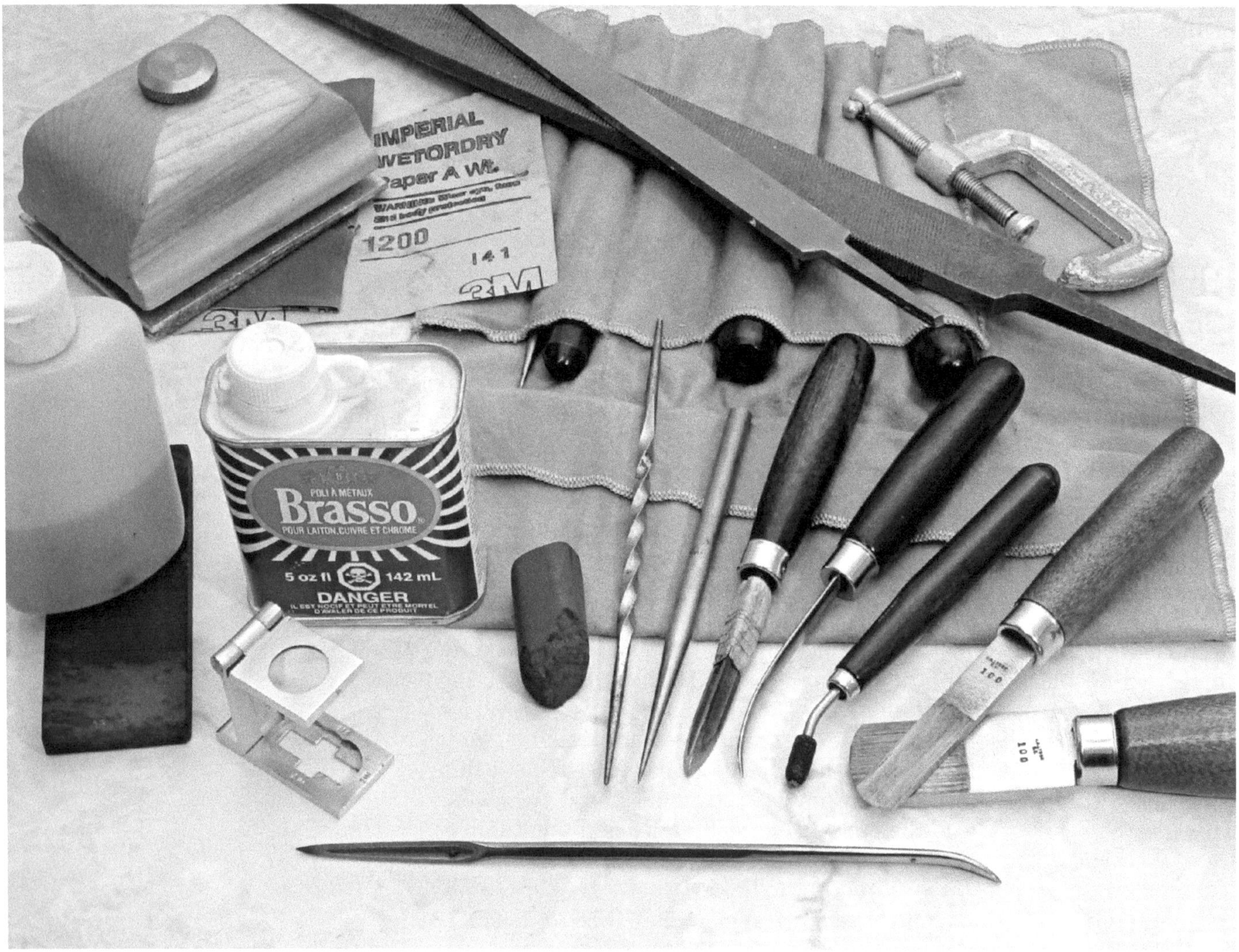

Figure 4-1 A selection of printmaker's tools used to rework and polish copper.

After degreasing, a brightener is used to give the plate a final and complete cleaning. This causes a perceptible change in the "brightness" of the copper. A simple brightener is made from one part glacial acetic acid, one part common table salt (by volume), and eight to ten parts distilled water. Use extreme caution with acetic acid in its glacial form because it is a strong oxidizer and can burn you severely. Be sure to add the acid to the water (not vice versa). An alternative brightener is a solution of one part muriatic acid (hydrochloric acid at 20° Bé/31.45% industrial strength) and ten parts water. Do not leave the plate immersed in this solution for more than a few seconds because it will eventually etch the surface and dull it.

PROCEDURE

Cutting the Copper Sheet

Mirror-finished copper is available from our supplier in 36″ × 96″ sheets. We have it cut into four pieces (24″ × 36″) to make shipping easier. Roofing copper can usually be found locally in full 4′ × 8′ sheets. Have the supplier cut it into more manageable sizes. To cut copper into working plate sizes you need access to a plate cutter with a throat of at least the width of your copper (Figure 4-2). Most printmaking studios have these, or a sheet metal fabrication shop should be able to do it for a fee. The latter facility would probably use a mechanized cutter, which could leave indentations on the face of your copper.

Be very careful when having someone else cut it for you. In any case, protect the surface of your copper from the hold-down of the cutter with heavy paper or even matboard if it has firm grippers. Do not try to use a hacksaw or metal cutting blade on a table saw. It will deform the edge of the copper enough to give you problems later on. The table saw blade can also overheat the copper and change its properties along the saw cut. You will need to cut your copper into plate sizes that allow a minimum of 1.2 cm (1/2 inch) around the image on each side plus an additional 2.5 cm to 4 cm (1–1.75 inches) on one side for the 21-Step Scale. (See Figure 5-3 in Chapter 5.)

Edging and Polishing the Plate

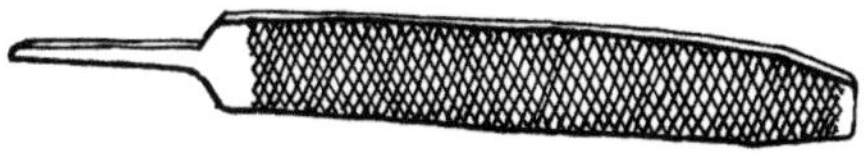

All copper plates must have the edges beveled or rounded over to prevent the sharp, freshly cut edges of the copper from cutting the paper and the printing blankets when run through the press. After the plate has been cut to size, use metal files to bevel the face edges and slightly round the corners. A common method is to create a 45° angle, but the bevel can also have two facets or a smooth bull-nose to give a more gradual shift (Figure 4-3). Mirror-finished copper comes with a protective plastic film adhered to it. Leave this on the copper at all times until you are ready for the final polishing. For roofing copper, it is a good idea to protect the good side with a piece of self-adhesive plastic shelf liner (Mactac) or a piece of paper held in place with masking tape just shy of all edges.

If you cannot hold the plate steady while filing, overhang the countertop 1 to 2 cm (1/2–1″) and clamp it down with a little C-clamp. Be sure to use a small square of thick matboard between the clamp and the copper and do *not* over-tighten. Sometimes a C-clamp attached directly

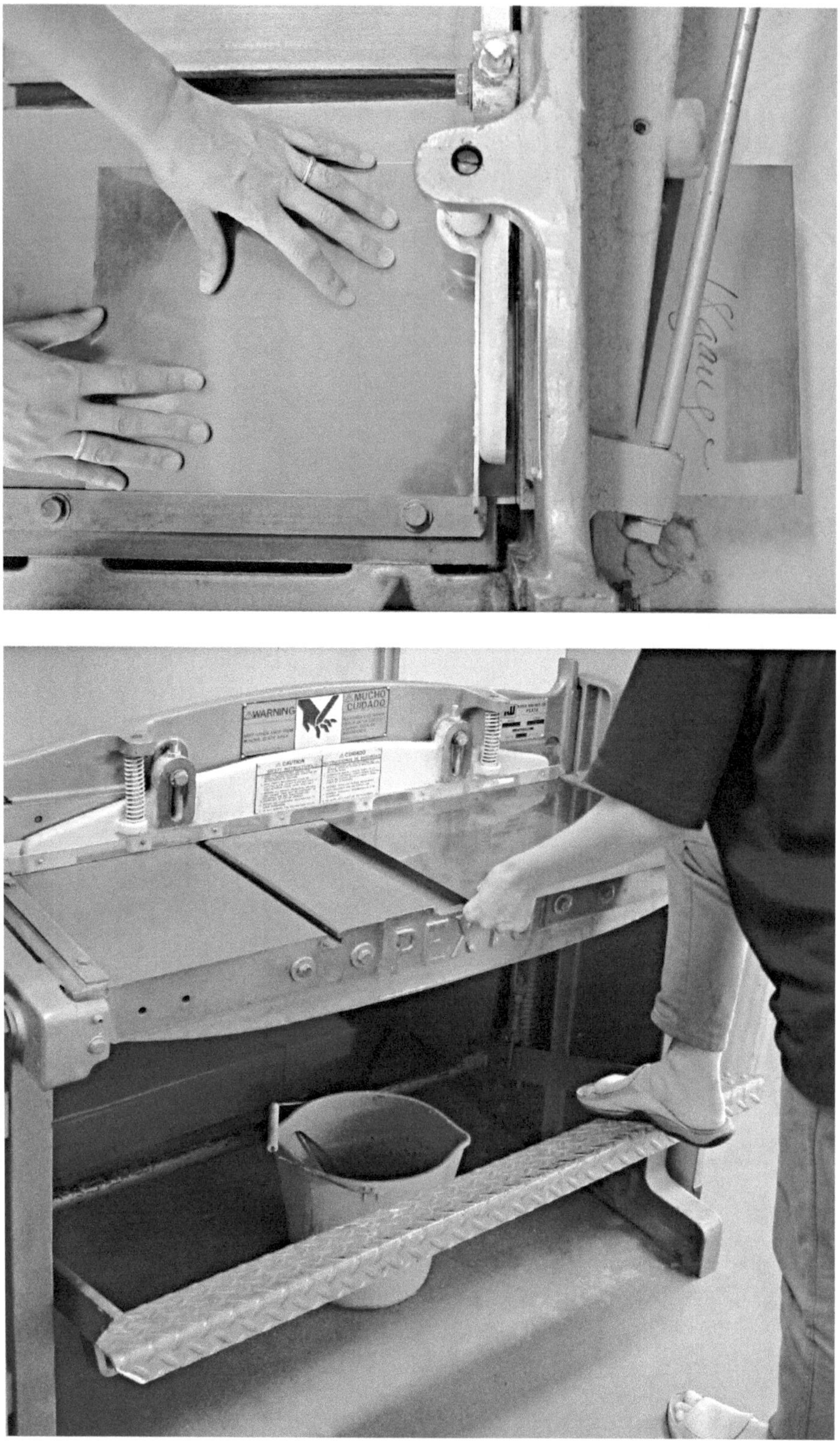

Figure 4-2 A small plate cutter in use.

to the counter edge is all you need to butt the plate against without actually clamping the copper plate down.

Start filing with a coarse file to shape the profile quickly. File off the face edge to 45°. Be sure to use the file in the right cutting direction, moving from the tapered end toward the handle end. Lift and repeat (Figure 4-4).

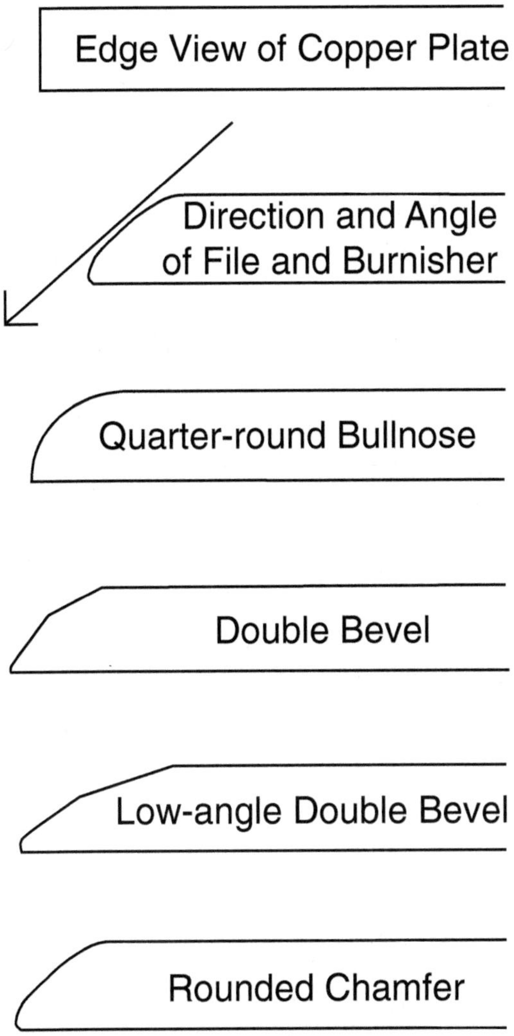

Figure 4-3 Diagram of possible profiles for filing a beveled edge on a copper plate.

After the profile is shaped on all four sides of the plate, use a finer file to smooth out the marks, corners, ridges, and burrs left by the coarse file and to add a bull-nose to the profile of the bevel. Brush away the filings periodically to prevent copper filings from cutting through the protective layer to the copper's face. If you retighten the C-clamp, be absolutely sure that no stray filings are caught between the mat board and the plate. Frequently clean the file with a wire brush to keep it cutting well and to prevent nicks on the edge of your plate.

The burr that is often formed on the bottom edge must be removed carefully with the fine file. It is very sharp and can cut you easily. Be careful not to add a second bevel on the back, though (Figure 4-5). Also, to protect the darkroom trays used for processing, you should round off the pointed corners slightly. Be sure the bevel extends over these as well, and again, remove the burrs. Rub the freshly filed edge with fine wet/dry sandpaper on a sanding block to remove all the fine file marks. The final burnish will then be easier and quicker. A final burnishing of the bevel can be done now, but is usually left to when the plate is trimmed for the final printing. For a description of burnishing the plate's beveled edge,

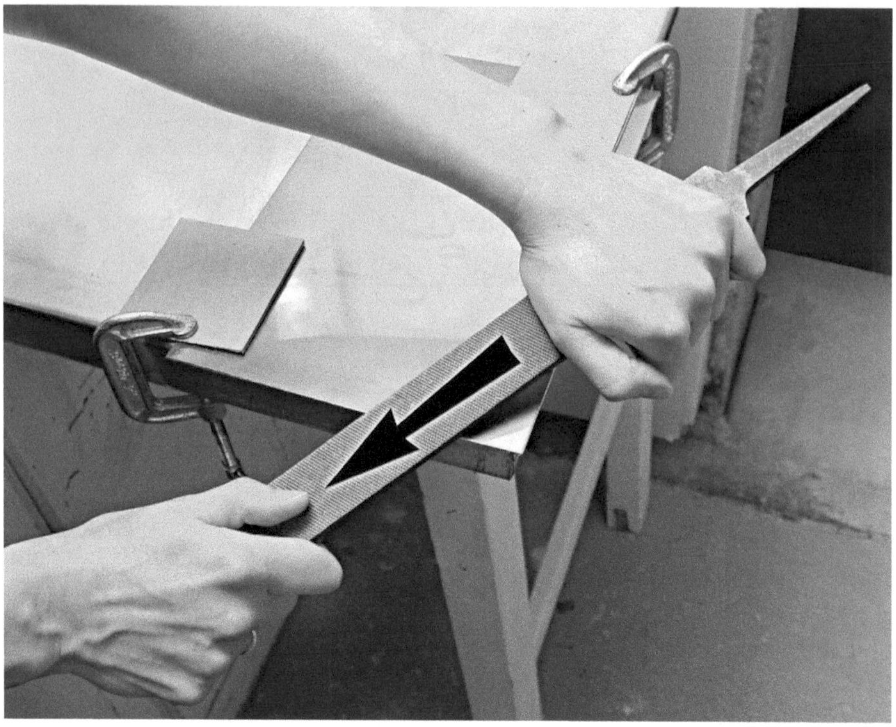

Figure 4-4 Rough filing the clamped plate. Direction of stroke is down in this case.

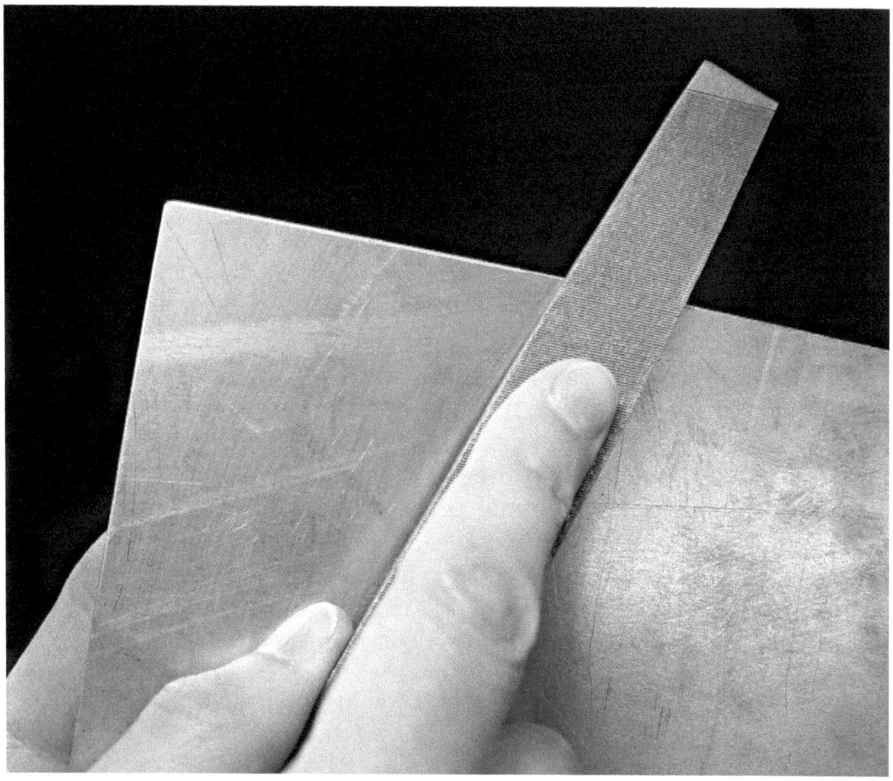

Figure 4-5 Remove the burr that forms on the back edge of the plate. Do not overdo it.

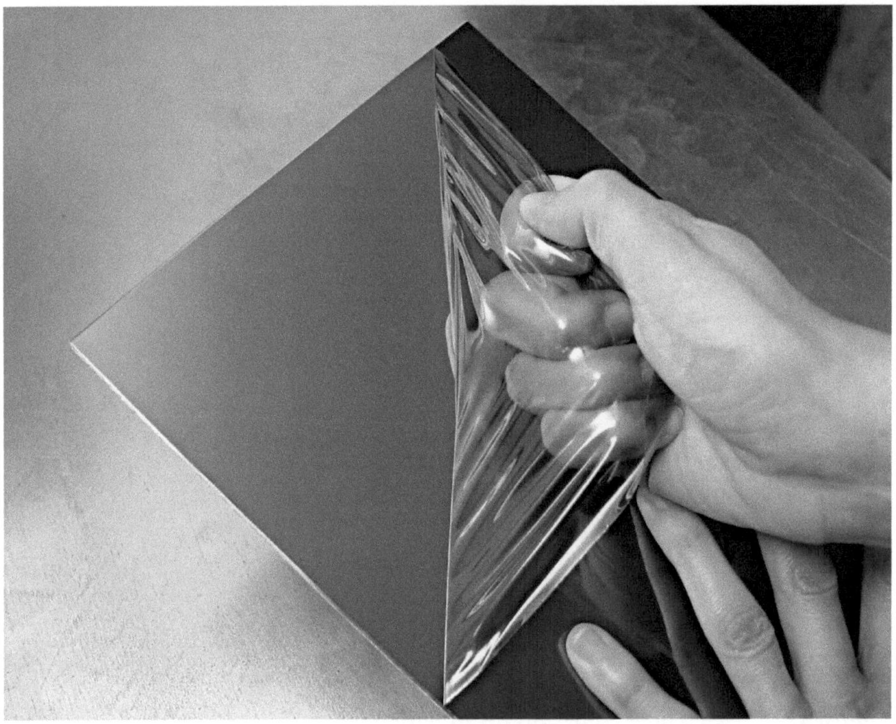

Figure 4-6 Peel the plastic coating off the mirror surface and inspect for flaws.

see the section on Reworking the Plate in Chapter 9. With mirror-finished copper, the protective plastic film can now be removed in preparation for final polishing (Figure 4-6).

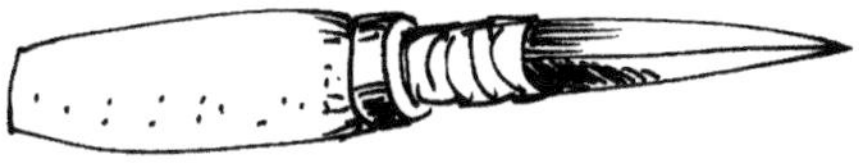

When using roofing copper, ensure that the plate surface has minimal lines, gouges, and scratches. To remove deep scratched lines, use a very sharp printmaker's scraper to level the area around the scratch and then burnish using a printmaker's burnisher and oil. Small shallow scratches can sometimes be burnished directly (Figure 4-7). This technique is well described in books on printmaking. To remove many overall fine scratches, first use fine wet/dry sandpaper. Use water or oil as a lubricant, starting with 400 grit if the condition of the copper is not great. Progress to 600 grit and then finish with 1500 grit—or 2000 grit if available. Fine waterproof sandpapers are available from automotive supply stores. Alternatively, one can use fine pumice and oil on a felt block. Follow this with rottenstone and oil on another felt pad (each backed by a flat piece of wood). This step will remove the small scratches left by the pumice. Next, use a polish such as a fine automotive polishing compound or Putz Pomade, and apply with a soft cloth. Rub and buff to a shine. Jeweler's rouge and oil will almost mirror-polish the surface. (New mirror-finished copper should not require these aforementioned treatments, other than the beveling and burnishing of the edges and corners.) Next, for all plates, including mirror-finished copper, clean the back of the plate with naphtha, then use Brasso and a fresh soft cloth for a final polish on the face of the plate (Figure 4-8). It is very important to make sure that the soft cloth is free of stray grit or other foreign bodies or old hardened polish. Use a clean work area where all abrasives and grit have been thoroughly and carefully removed. The plate should end up with an unflawed mirror finish with no tarnish marks.

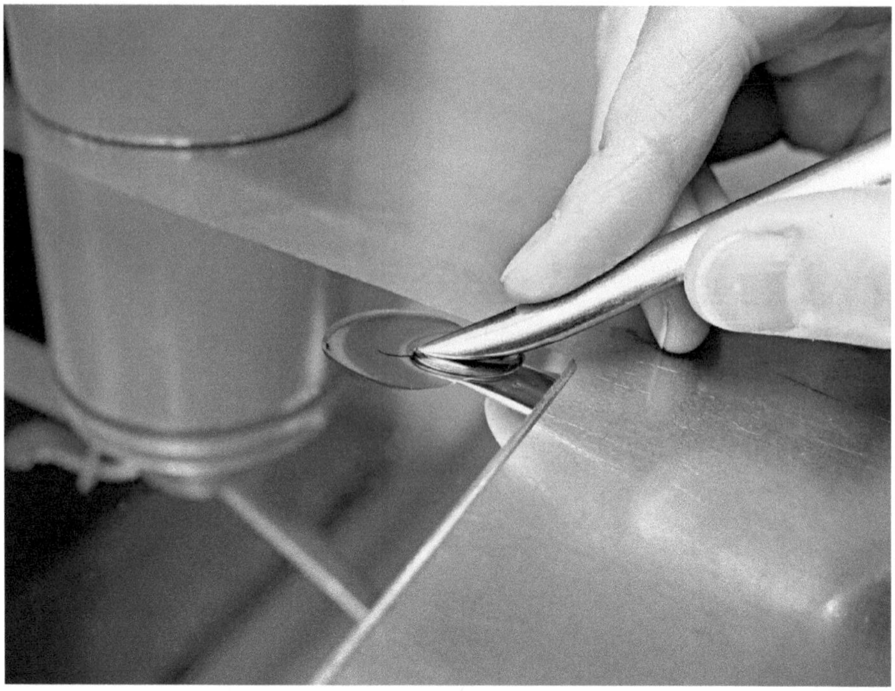

Figure 4-7 Use a burnisher and oil to flatten a small scratch.

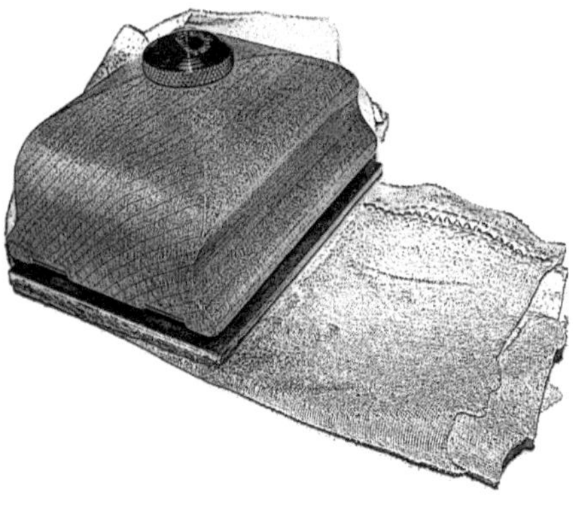

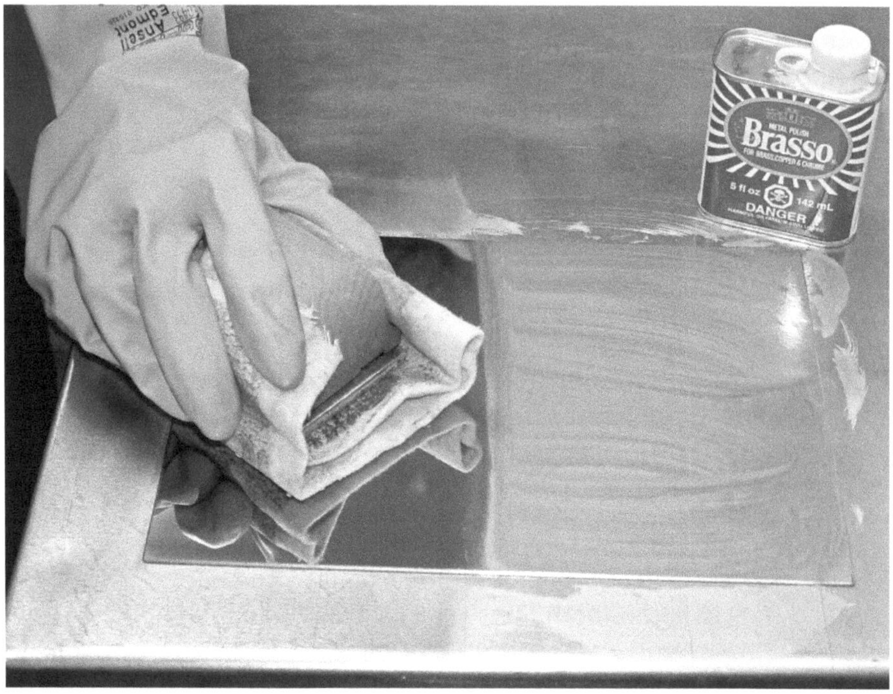

Figure 4-8 Finally, polish the copper using Brasso, a block, and a soft cloth.

Degreasing and Brightening the Plate

Degreasing and brightening should be done just before applying an aquatint or adhering the exposed resist tissue. In the latter case, this can be done while the tissue is being exposed if there is enough time (6 to 10 minutes). Set up a sequence of photographic trays in a large darkroom sink. You will need one tray in which to degrease the plate and another filled with preboiled tap water cooled to 10 to 15°C (50–60°F) in which to store the plate before you adhere the exposed gelatin tissue. Many sources suggest drying the plate before the adhering stage. We found that the dry plate was prone to retarnishing. Submerging it virtually eliminated this problem.

First, lay the prepared copper plate face up in the empty tray. Put on rubber gloves and an apron because the TSP is very corrosive to skin. Sprinkle a small amount of water in the center of the plate. Add TSP to the water on the plate. A heaping teaspoon should do for a small plate under 525 cm^2 (80 sq. inches). Be careful not to inhale any of the TSP dust. Wear a mask if you are vigorous when you work. Use a cotton ball to push the TSP crystals and the water around the surface of the plate for at least one minute (Figures 4-9 and 4-10). Be sure to cover all areas, especially along the edges and corners. The water will initially bead up. As the TSP takes effect the water will sheet and no longer bead or pull away from the edges. Make little circular motions with the cotton ball and do not use too much pressure.

Be sure to pay special attention to the edges and areas where beading was most prevalent. After the front is done, wipe down the back to remove the greasy film. Do not worry if it still beads up on the back, though. Rinse the plate and tray well with cool tap water after the TSP

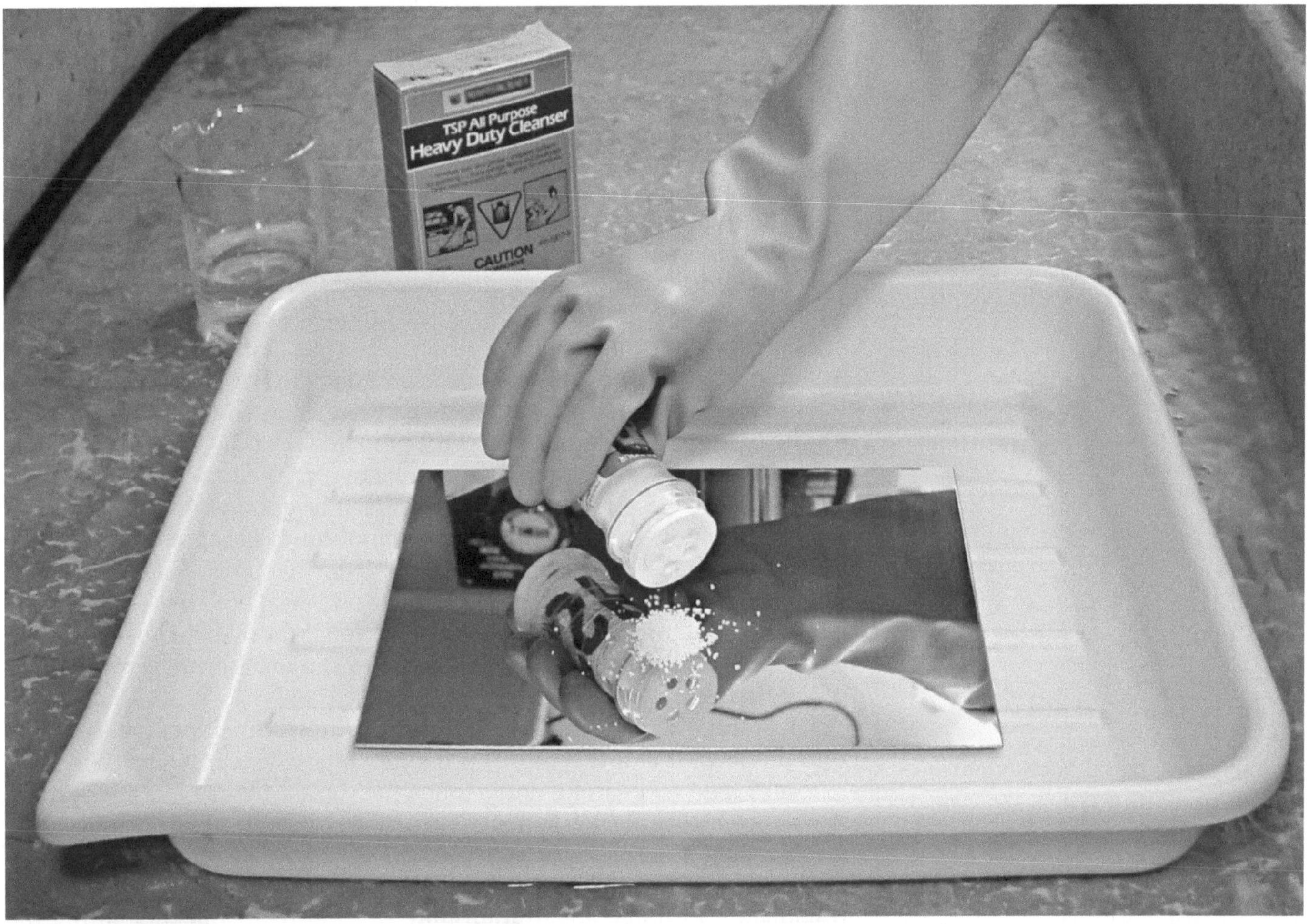

Figure 4-9 Sprinkle a teaspoon of TSP on the center of the plate.

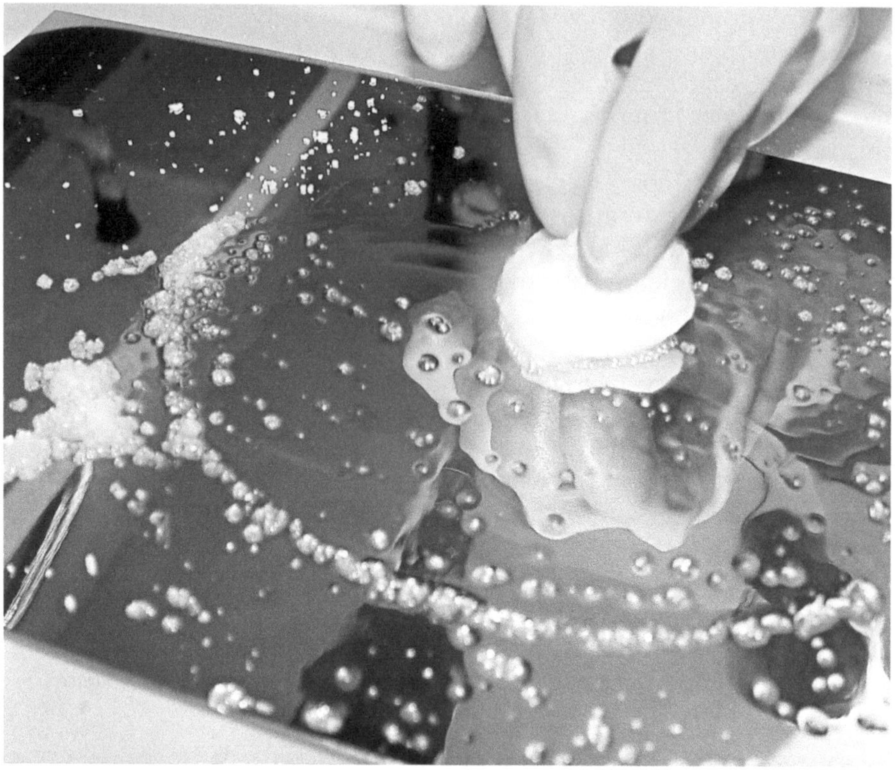

Figure 4-10 Gently rub in a circular motion. Note how the water is still beading up.

has done its job, and leave the plate resting in the half-full tray. Make sure the plate is constantly covered with water between stages because air drying may result in spontaneous tarnishing (see Figure 4-12).

The next step is to flood the surface of the rinsed plate with the brightener. You will see the plate noticeably brighten immediately on contact with the brightener. There are two techniques for submerging or flooding the plate with brightener. You can slide the plate into a tilted tray of brightener so that it covers the surface in one quick smooth wave as you lower the tray and plate together (Figure 4-11). Or you can rest the plate on the bottom of an empty tray and pour the brightener over its entire surface in one smooth motion from a large-mouthed container or measuring cylinder. In either case, do not splash brightener on the surface of the plate or allow the plate to be flooded too slowly because this can cause the sudden appearance of tarnish streaks. If this happens, rinse and dry the plate and return to the Brasso stage, and then repeat both the TSP degreasing and brightening stages. Save the brightener for repeated use. Replace it when its color becomes too blue-green or if there are any signs of floating particles or cloudiness.

BRIGHTNER

- one part Glacial acetic acid
- one part table salt
- 8 to 10 parts distilled water

CAUTION: Add acid to water. Glacial acetic acid can cause severe burns.

Rinse the plate and tray well and immediately resubmerge the plate into the second half-full tray of preboiled tap water cooled to 10 to 15°C (50–60°F). Be very careful not to touch the surface of the plate at any time. If you observe any spontaneous tarnishing, dry the plate and return to the Brasso polishing stage. Degrease and brighten again.

The plate is now ready for the lay-down of the exposed gelatin tissue. If using a traditional aquatint ground, dry the plate and then apply the aquatint ground. See Chapter 10 for more information on aquatint grounds.

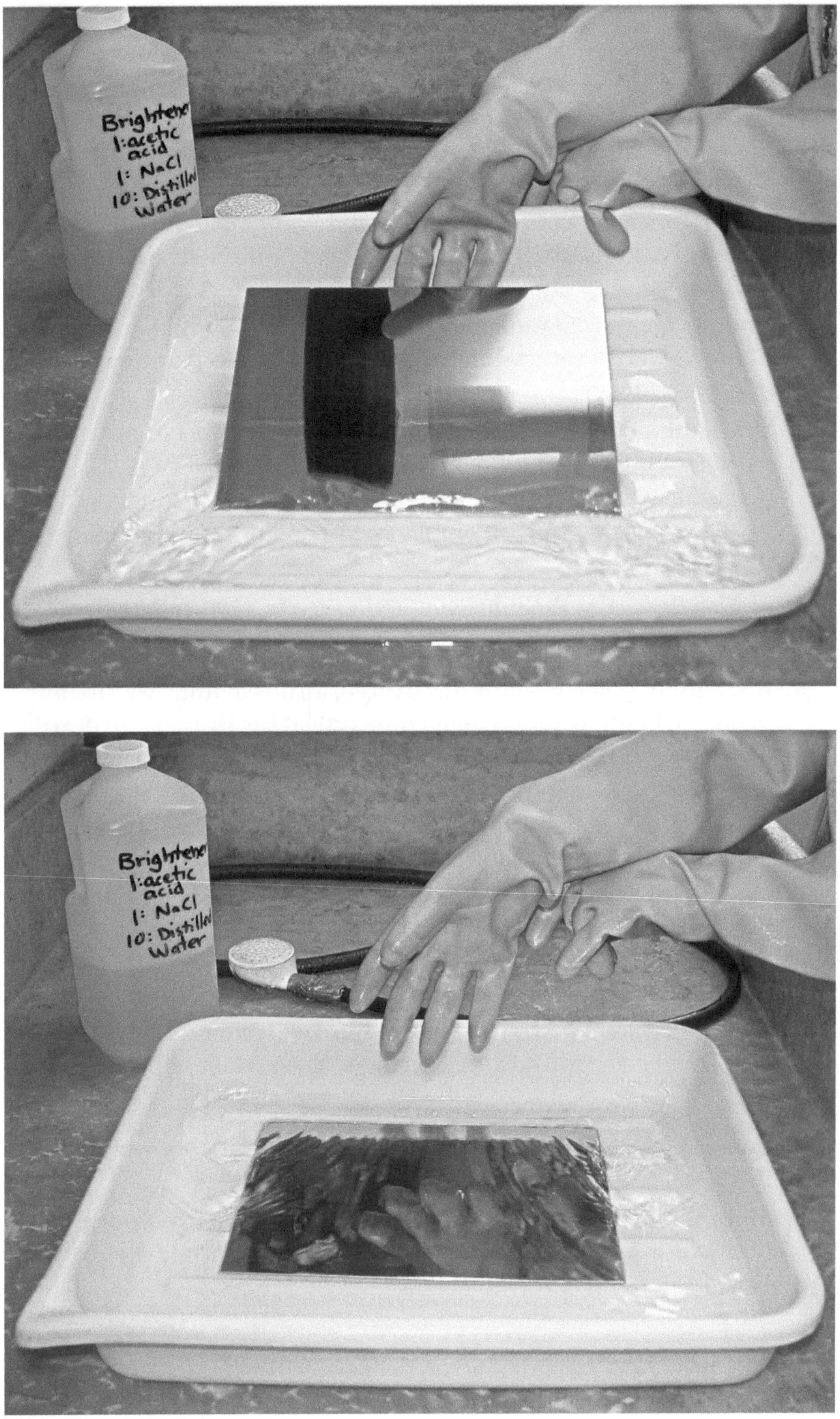

Figure 4-11 Note how the brightener solution floods over the plate in a single wave.

SUMMARY

1. **Use a file to bevel the edges of a 16- to 20-gauge mirror-finished gravure copper or good quality roofing copper. Sand and burnish the bevel, round the corners, and remove the burr on the back. Clean the back of the plate with naphtha.**
2. **Mirror-surfaced copper should simply be polished with Brasso to remove all tarnish in order to be ready for the degreasing and brightening step (#6).**
3. **Roofing copper: Scrape and burnish any major scratches on the plate. Rub with rottenstone and oil on a felt pad if there are a lot of fine scratches. If none, proceed to the next step.**
4. **Unless the copper plate is already mirror-surfaced, polish the plate to a high gloss following the order listed below:**
   a) **Use wet/dry automotive sandpaper starting at 400, then 600, then 1000 or 1200 grit with water (depending on the condition of the copper).**
   b) **Use Brasso as the lubricant and move to 1500 grit, then 2000.**
   c) **Polish with Brasso and a soft cloth.**
5. **If singular scratches are discovered at this point, use the burnisher with oil to flatten them, then repeat the Brasso polish with emphasis on the corrected areas.**
6. **Move the plate to a tray in the sink and degrease for a minute or two with a teaspoon of trisodium phosphate (TSP) and a few tablespoons of water. Rub gently with a cotton ball until the water no longer beads. Finally, be sure to rub down the edges and back of the plate.**
7. **Rinse well with water.**
8. **Immerse quickly into the brightener (acetic acid/salt* brightener recipe or muriatic acid—hydrochloric acid at 20° Bé/31.45% industrial strength—diluted 1:10). (* The former is preferable.)**
9. **Rinse well with water again.**
10. **Immediately submerge into the second tray of cool water at 10 to 15°C (50–60°F).**
11. **Store the plate under water while awaiting the exposed gelatin tissue. If there are tarnish marks, repeat from the last polishing with Brasso, Step 4c.**

TROUBLESHOOTING

Scratches and Swirls When concentric swirls made up of fine scratches appear on the surface of the plate, they are usually caused by some grit or hard particles on the polishing rag. Change to a new rag and be sure it is clean and soft. Also be sure the plate is positioned on clean newsprint or adhered to a clean countertop with loops of masking tape on the back. Previous abrasive steps may leave debris on the counter or along the edges and back of the plate. Clean everything carefully when advancing to the next polishing step.

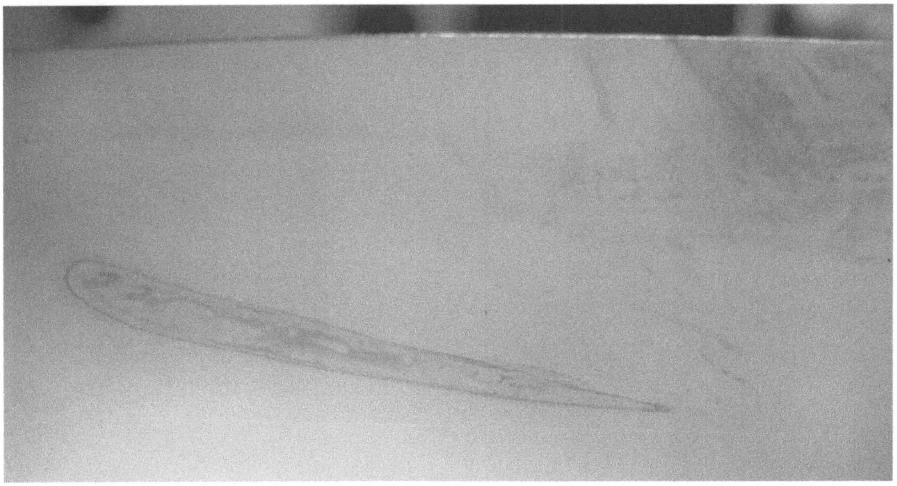

Figure 4-12 Reddish tarnish streaks can spontaneously occur at any time. Be sure to check the plate before adhering the tissue. Repolish and degrease if streaks appear.

If the burnisher itself is not highly polished and without flaws it, too, may create scratches instead of burnishing them out. Polish its convex surface with jeweler's rouge and then Brasso on a rag, being sure to use a lot of pressure.

Cuts to Your Hands The edges of a newly trimmed copper plate are very sharp. Applying a bevel to the top edge can make the bottom edge even sharper, like a knife blade. Be sure to lightly file these edges as well and even use the burnisher to smooth them slightly. The burnishing itself may raise a burr, which is also very sharp. Refile lightly to remove the burr and burnish again. Avoid running your fingers along the edges of the plate. Do not forget to file off the sharp corners as well.

Tarnish and Streaks After polishing and degreasing, the plate will sometimes react to the air or any contact with a rag or your fingers, causing the sudden appearance of a darker reddish streak or smudge (Figure 4-12). This is tarnish, and must be removed. It will interfere with the adhesion of the gelatin resist or will show in the etched tone. Repolish lightly with Brasso, then degrease and brighten again. Be especially careful not to touch the surface of the plate. Submerge it in a half-full tray of cool tap water immediately after the brightening stage to await the adhesion step. Prepare the plate during the final exposure so that the plate is not sitting too long.

Splash Pattern Is Visible on the Plate after Brightening The brightener has been splashed on the plate instead of covering it in an even flood. The streaks that sometimes form must be polished out and the plate must be re-degreased. Review the methods for submerging the plate and modify your technique. If the brightener is fresh or too strong it may be the source of the problem. Season the brightener with a strip of copper or remix with ten parts water instead of eight.

5 Exposing the Gelatin Tissue

In order to create a resist matrix for the copper, the positive film image must be exposed onto the sensitized gelatin tissue. A grain or screen texture also must be applied over the image to create ink holding wells in the etched areas of the plate. A traditional aquatint grain can be applied directly to the copper plate, or more easily, a screen pattern can be exposed onto the gelatin tissue prior to the positive exposure.

When the sensitized gelatin is exposed to light, the resulting hardening effect, called tanning or *insolubilization*, is directly proportional to the amount of light hitting the gelatin. Exposure to light will affect the gelatin by raising its melting point from 32°C (90°F) to as high as 93°C (200°F). *Continuing action* will occur for up to an hour after exposure, causing increased insolubilization. It may render the gelatin tissue useless. It is therefore important to progress immediately to the development stage. The order of separate screen and positive exposures makes little difference to the final plate. It is interesting to note, however, that although the continuing action is likely to be beneficial to a screen exposure by reinforcing its density, it is not good for the positive image. Therefore it is theoretically advantageous to expose the screen pattern first, followed by the positive, and then to immediately adhere and develop the resist. To prepare for these exposures, the positive and step scale must be stripped into a sheet of light-proof paper and the gelatin tissue must be trimmed to the correct size.

EQUIPMENT AND SUPPLIES

Goldenrod, yellow, red, or thin black paper is used for stripping the positive. Alternatively, the positive can be edged with Mylar tape (silver or red) to establish an unexposed *safe edge* surrounding the resist image. When the gelatin tissue is exposed, the positive should always be accompanied with a step scale such as the Kodak No. 2 Step Scale or a Stouffer 21-Step Scale No.T2115. (See Figure 2-1 in Chapter 2.) These scales are basically the same, but we use the latter because each step is slightly larger and clearly numbered. The scale should be stripped into the positive assembly off to one side of the positive, with its emulsion surface oriented in the same way as the positive. Leave a space between the scale and the image to allow for any clear border you will want on the final plate. (See Color Plate 10.)

In order for the copper plate to maintain detail and hold tone during the etching and printing process, the plate must contain a structure of high points or islands, or a linear grid pattern. These will act as a web of retaining walls or peaks that hold ink in the recesses of the etched surface. This can be achieved through an aquatint that is adhered to the plate before the application of the resist (or even after in the case of rosin). See "Applying an Aquatint" in Chapter 10. This can also be done by exposing a pattern of high density ridges or dots onto the resist itself with a commercial gravure screen or a homemade, random-patterned, hard-dot screen (Figure 5-1). Several degrees of fineness are possible, usually 250 to 300 lines per inch. Overly fine screens will make it harder to prevent foul-biting problems during the etch. See Appendix B for information on how to make a screen.

Whether using an aquatint or a screen pattern, the desired coverage is 50%, an equal balance of solid and open areas. The degree of fineness is critical in determining the final image character. The screen films required for photogravure are hard-dot screens, as compared to commercial halftone or mezzotint screens, which are soft-dot (each dot is vignetted). A soft-dot screen can make it difficult to determine the standard exposure and does not result in the required hard-edged pattern for the lands or high points on the etched plate. With this in mind, it is evident that photogravure is not truly a continuous tone process because it breaks down the image into a pattern of regular (screen) or random (aquatint) wells or pits. These wells and pits, however, vary in depth due to the etching process. They deposit spots of ink of different densities at a very fine resolution, which tend to blend together to hide the intermittent high points. In this way the plate is able to reproduce tonal variations much more smoothly than the dots of ink in a regular halftone reproduction.

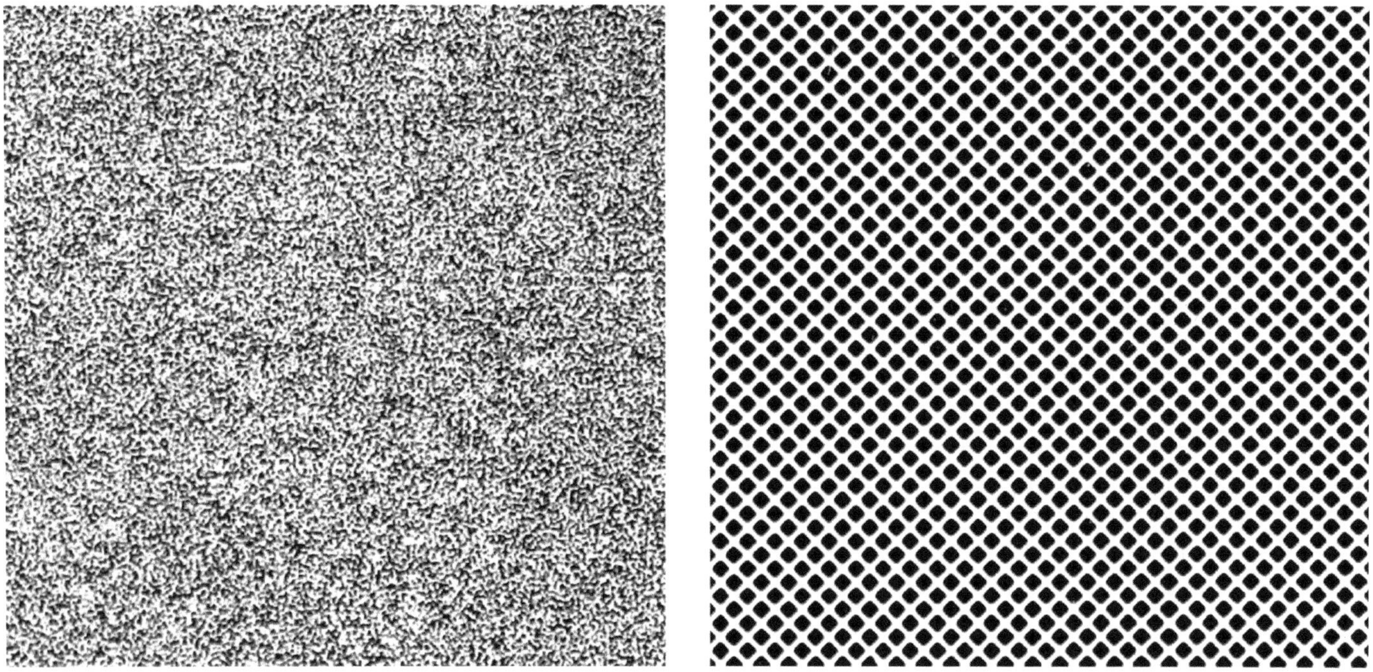

Figure 5-1 Pattern sample of a second-generation, random-patterned, hard-dot screen (left) and a commercial gravure screen of 260 lines per inch, at a ratio of 1:3.4 (right), each highly magnified.

For exposure, a vacuum frame is required to provide the tight overall contact needed to preserve the detail of the positive and the sharpness of the screen. Graphic arts vacuum frames are becoming more scarce but can still be found as used equipment. Small sizes (16″ × 20″) are premium, but larger ones are great if you have the space. Be sure the glass is unscratched, the seals are airtight, and the vacuum pump and PSI gauge are in good working order.

A bright, actinic light source is required, high in the UV and blue-green portion of the spectrum. The lights should ideally cover the spectral range between 3600 and 4200 Ångström (360–420 nm) because the peak sensitivity of the sensitized gelatin tissue is at about 3800 Å (Mertle and Monson 1957, p. 336). The spectral range can affect the contrast of the resist. Within the range of 360 to 420 nm (which falls within UV-A), the shorter wavelength UV will produce a flatter image than the longer wavelength UV (Cartwright 1939, p. 64). See Appendix A for UV safety considerations. The light should be consistent and easily timed. A light integrator would be great, too. We have used a 4000-watt metal halide lamp, but currently work with a bank of eight 24-inch BL fluorescent bulbs (F20T10-350BL).

Do not work with BBL bulbs. We work with a diffused light system as it is easier to build for a home studio and because of the advantages for the positive exposure. See Appendix C for information sources about building your own light exposure unit.

PROCEDURE

Stripping the Positive

The first step is to make a safety border of opaque or semi-opaque material for the positive. This safe edge blocks the light on the tissue surrounding the image during exposure. This underexposed area provides a border of thin resist between the actual edge of the resist and the image, and prevents the tissue from frilling or peeling off during development. When working with an aquatint, it is better to use stripping that is not entirely opaque. Cut a hole in a piece of goldenrod, yellow, orange, red, or black paper the size of the positive image. Make a second rectangular hole to fit the Stouffer 21-Step Scale, usually at one end or side so it wastes the least copper. Then strip the positive into the colored or black paper along with the step scale. A hole may also have to be cut through the film edge about 7 mm (1/4″) from the image along one side so as not to obstruct the step scale. (See Color Plate 10.) Be sure that the scale fits neatly into the space flush with the emulsion side of the positive. Tape the ends into place from the top. During exposure, the emulsion side of the positive and the scale must be in tight contact with the gelatin tissue with absolutely no interruptions by tape or paper along the edges. Make sure all stripping materials and tapes are laid over the top (shiny) surface of the positive rather than under it (Figure 5-2). Finally, use a fine-tipped permanent pen to outline the position of the edges and corners of the trimmed gelatin tissue on the underside of the colored or black paper so that it is visible when looking at the emulsion side of the positive. This will help you position the tissue prior to exposure.

When working with the sensitized gelatin tissue, you should have a clean, dust-free work surface under it at all times. Spread out clean paper and use thin vinyl or latex gloves because the dichromates that are dried

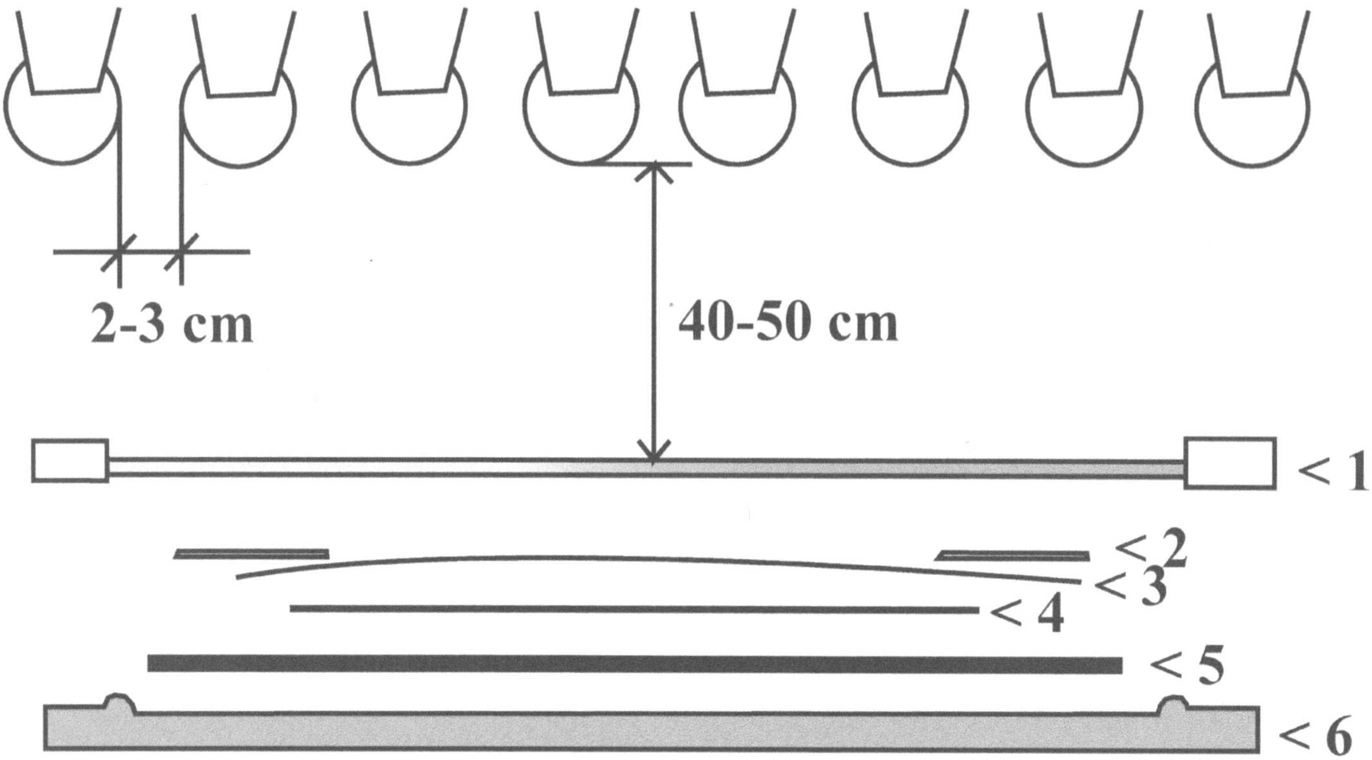

Figure 5-2 The order of materials in the vacuum frame: 1) the glass of the vacuum frame, 2) the stripping paper and tape, 3) the positive, emulsion side down, 4) the gelatin tissue, face up, 5) a smooth black or red backing card, 6) the bottom of the vacuum frame.

into the backing and gelatin are toxic. Do not touch the surface of the tissue at any time. Do not press the backing for any length of time because the heat from your fingers will melt a flaw into the tissue. You can use a folded piece of cloth as an insulating pressure pad when you have to hold the tissue in place from the back. (See Color Plate 11.) Avoid breathing or especially coughing or sneezing onto the tissue. This will ruin it. Wear a mask or face shield if you have a cold. Sensitized tissue must not be exposed to any light except red or yellow safelight. A 40–60 watt bug-light works well as a bright yellow safelight.

If you are working with frozen tissue, allow the sealed package to stand at room temperature for an hour or more before opening it to avoid condensation on the gelatin surface. The GTA guide also recommends that you allow it to stand for a half hour after opening to stabilize the gelatin to the room humidity (Smeil 1975, p. 69). Be careful of light fogging or over-drying if the relative humidity is low.

Cutting the Gelatin Tissue to Size

The sensitized gelatin tissue must now be trimmed to the final size required, extending slightly beyond the positive image and step scale. It must, however, still be smaller than the copper plate, but large enough to include the safe edge established around the positive image. The size should match the outline previously drawn on the emulsion side of the stripped positive (Figure 5-3). Use a clean ruler to measure the size of the tissue and a plastic triangle to ensure that the corners are square. Make sure any straight edge or triangle used while cutting the tissue is clean

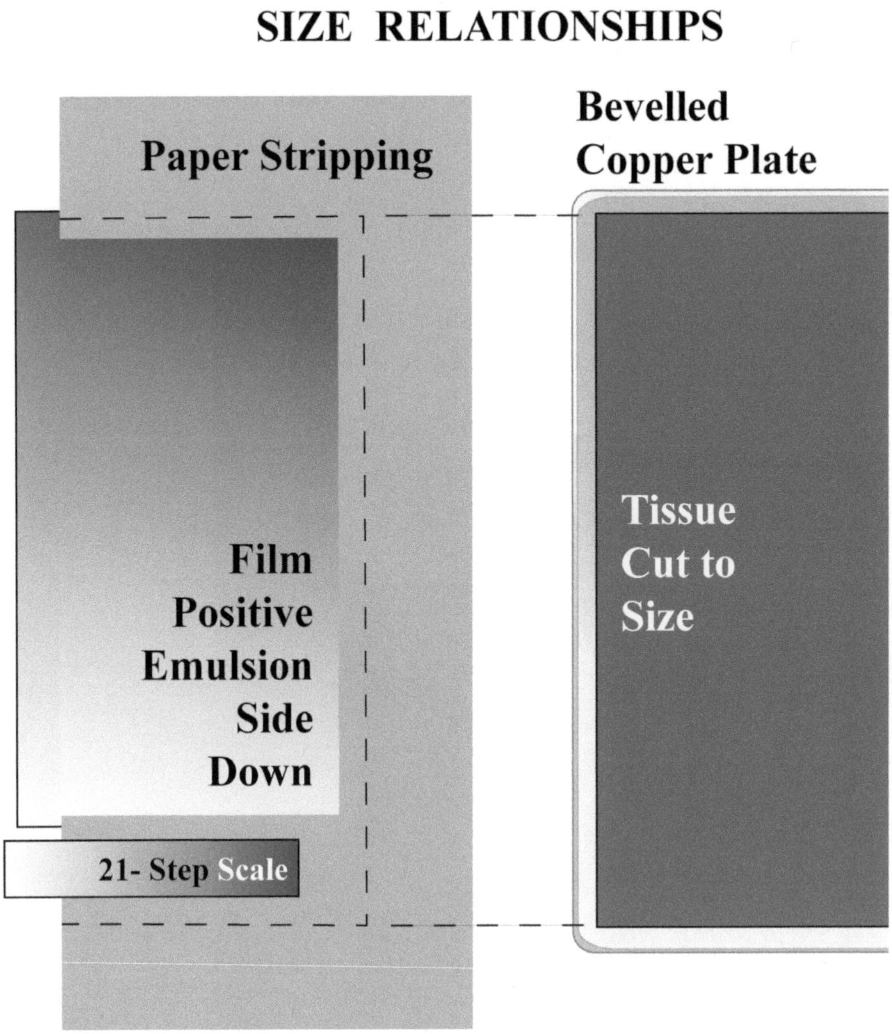

Figure 5-3 Diagram showing relative sizes of positive, tissue, and plate, all trimmed to their working dimensions.

and grease free (Figure 5-4). Cut the tissue with a sharp segmented blade or X-acto knife. Use a thin, clean paper folder to surround the tissue if you insert it in a rotary paper cutter.

Another method of trimming the tissue is to use the copper plate itself as a cutting guide. First, cover a cutting board with fresh clean paper or blank newsprint. Place the tissue face down on this surface. Clean both sides of the copper plate with alcohol to be sure that there are no oily residues on it. Place the plate face down onto the center of the back of the gelatin tissue. Hold it in place and use a pencil to trace its outline on the top edge and one side. Use a sharp blade to trim the tissue on the opposite two sides. Shift the plate so that the pencil lines are parallel to the copper plate and are about 9 mm to 13 mm (3/8″ to 1/2″) from it. Now trim these edges. The remaining tissue is now cut to exactly the shape of the copper with a 5 mm to 7 mm (3/16″ to 1/4″) space all around. These allowances are small in order to minimize the waste of copper. When you are learning to use these materials, you might want to use greater allowances so that positioning the tissue on the copper is easier. Be sure the copper plate is large enough to allow for this increase.

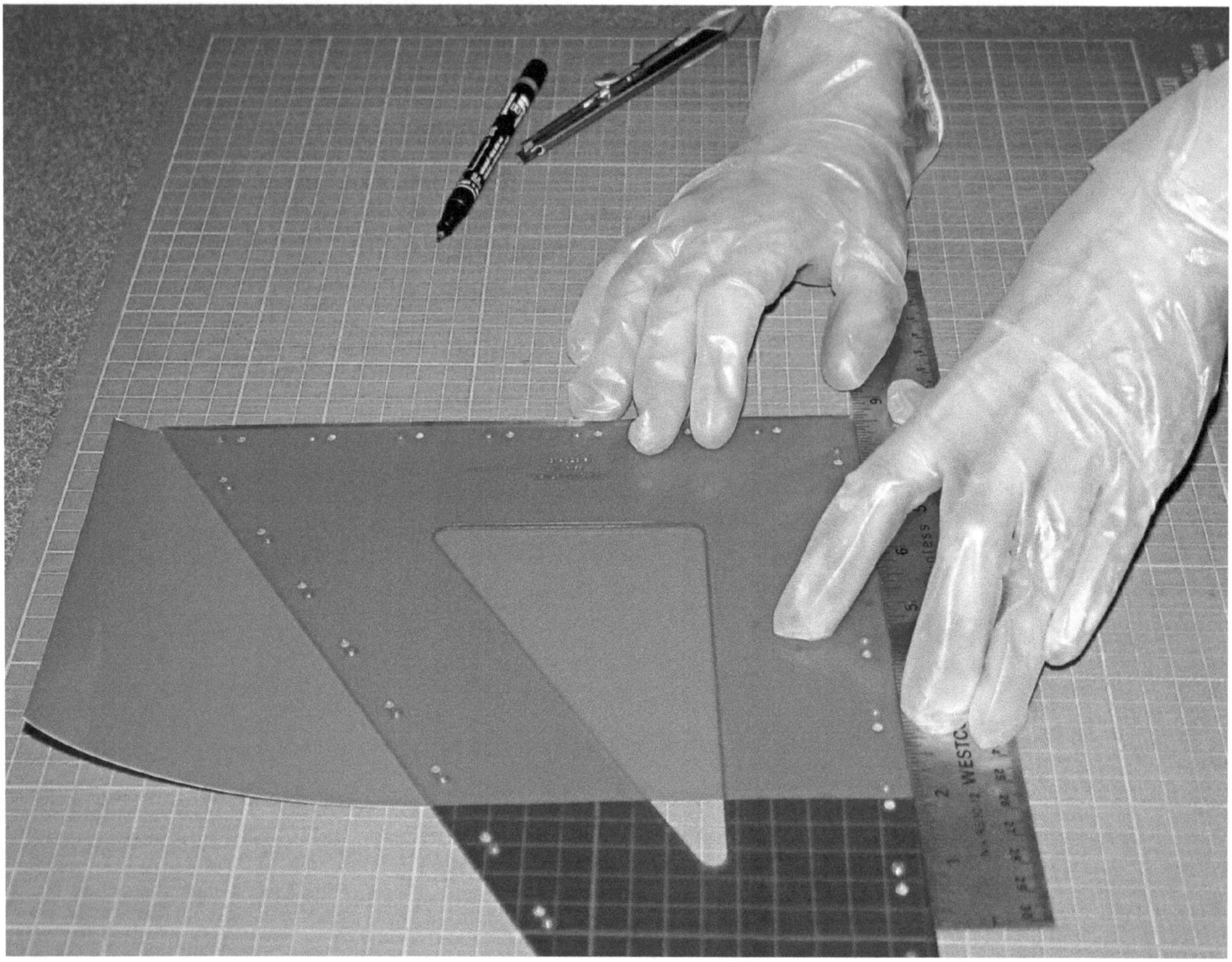

Figure 5-4 Note the use of latex gloves for handling the sensitized tissue. Do not press on the surface of the gelatin or the heat from your fingers can cause a blemish that will show up during the etch.

Exposure Times

In our darkroom we use a bank of eight 24-inch BL fluorescent bulbs (F20T10-350BL) spaced at less than 2.5 cm (1″) apart and at a distance of 43 cm (17″) from the vacuum frame (Figure 5-5). We expose the tissue through positives of standard density for 8 to 12 minutes. Regardless of the light source used, it is very important that the light does not heat up the glass and tissue beneath and ruin it. Fluorescent (BL) bulbs are a good choice because they do not produce heat. A good way to determine a standard exposure with your own lighting setup is to expose a series of tests through Stouffer Step Scales and adhere them to glass instead of copper. Details of this procedure are explained in Appendix C.

When determining the standard exposure time for a positive that conforms to ideal densities, consider the following: the thinner, highlight portions of the positive (with a transmission density of 0.40 to 0.50) allow up to 50% of the transmitted light energy to reach the gelatin tissue and produce a thick resist. The middle tones on the positive (a density of 1.0)

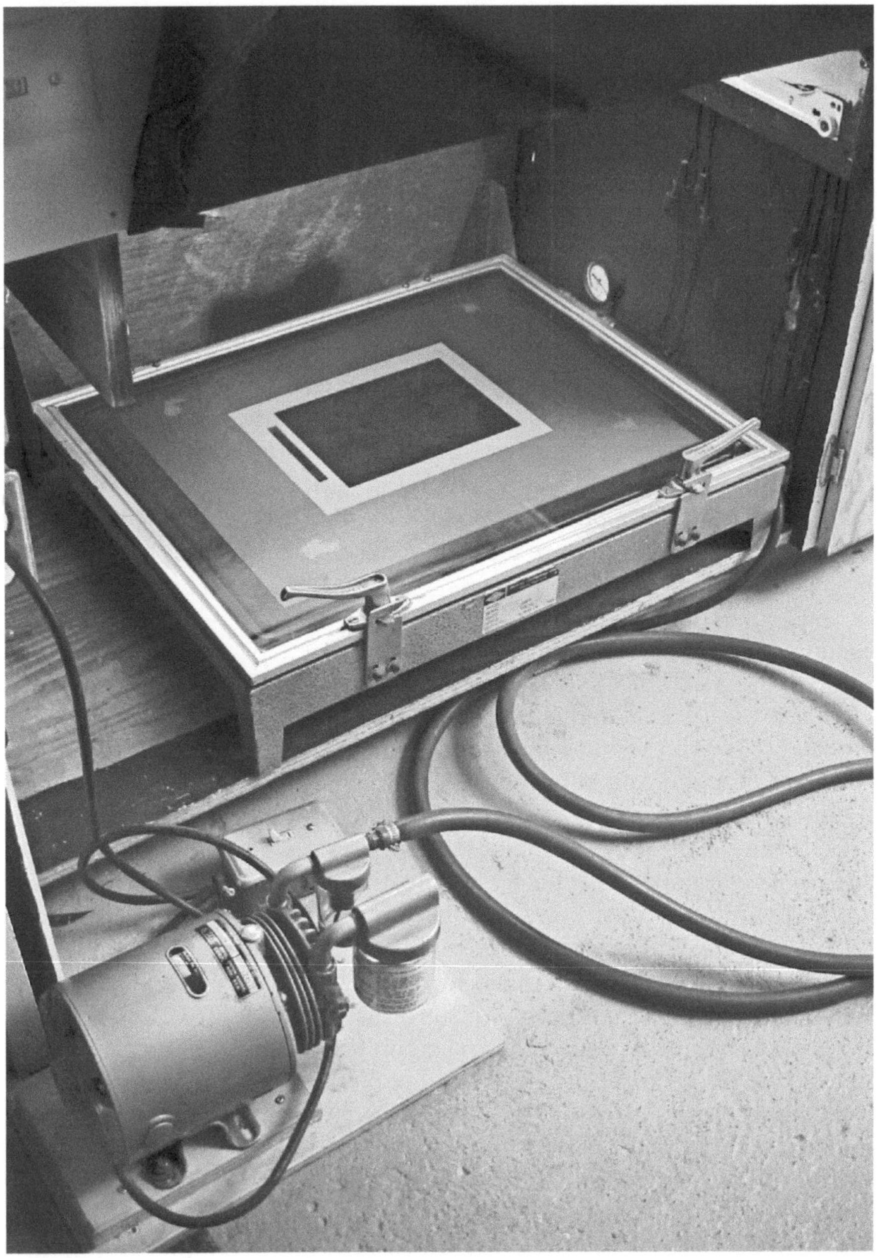

Figure 5-5 An under-counter lighting setup spaced 43 cm (17″) from the bulb array to the glass. The light bank is shielded with a short opaque curtain (raised in this view).

pass only 10% of the light energy and produce medium-thick resist areas. The shadow areas of the positive (with a density of 1.65 to 1.85) pass as little as 2.5% or less of the light energy and result in a very thin resist (Smiel 1975, pp. 69–70). Many different factors affect the exposure time so it cannot be stressed enough that a consistent working procedure is absolutely necessary to determine future adjustments and make repeatable results possible. On a resist exposed to a Stouffer 21-Step Scale, the steps higher than #14 may not retain detail or separation on the dried resist. They can be so thin that they may show a slight iridescence—like gasoline on water—when viewing the dry resist on the copper. Steps #12 to #14, however, should never be so thin as to show this iridescence.

Iridescence is much less likely when the screen exposure is by slightly diffused rather than collimated light. The iridescence on these thin steps will be more obvious when a screen exposure is not used.

Remember, all surfaces, including the glass of the vacuum frame, the positive, the screen, and the gelatin tissue, must be absolutely free of fingerprints, dust, or loose material. Use a can of compressed air to blow each layer clear before assembly. The glass should be clean and dry on both sides. The pebbly rubber surface often found in vacuum frames as well as the long amount of time spent under high vacuum can impart a texture on the tissue, either physically or by mottling the exposed tones. Be sure that the surface under the tissue is smooth. Place a sheet of stiff card stock or smooth mat board larger than the tissue on the bottom of the vacuum frame. It should be red or black and thick enough to prevent any bottom texture from coming through. Allow the negative pressure of the vacuum frame to reach at least 25 PSI. Run the vacuum for a minute or two before turning on the exposure light(s). Go through this process for both the screen exposure and the positive exposure.

Screen Exposure

After the gelatin tissue has been trimmed to fit the positive array, it is exposed through the hard-dot screen. Some sources say that the screen exposure can be done before or after the positive exposure; we choose to do it first. The tissue usually has a tendency to curl, especially on dry days. If the tissue is curling excessively, it is probably too dry. A curled stiff tissue can make it difficult to achieve tight contact during the screen exposure and can result in mottling. Rehumidify the tissue if necessary by storing in a humid atmosphere—closer to 60% relative humidity—until it uncurls a bit and is more supple.

The screen exposure must cover the entire surface of the tissue. Be sure to have a hard-dot random pattern screen that is large enough. Use compressed air to dust off the tissue and lay it face up in the vacuum frame. Remember to use a black or red sheet of smooth, thick card stock, slightly larger than the screen film, so that the surface under it is smooth. Dust off the screen film and lay it over the tissue, emulsion side down (Figure 5-6). Close the frame, turn on the vacuum pump, and leave it running for a couple of minutes to be sure there is absolutely tight contact before turning on the exposure lights.

The screen exposure can vary as much as 10 to 20% from the positive exposure. We have found that when using a slightly diffuse light system, a screen exposure at 100% of the normal positive exposure gives a good screen pattern in the resist, which will give strong high points on the plate. The percentage chosen for the screen exposure will depend on the density of the screen. If the screen density is the ideal 50%, then the screen exposure should be the same or slightly more than the positive exposure (100 to 110%). The tonality of the image may determine shifts in screen exposure. A low-key (dark) image may require additional screen exposure to ensure that the peaks do not over-etch. A high-key (light) image may benefit from a shorter screen exposure so that the pale tones are smooth. Working with collimated light may require longer screen exposures. A direct gravure positive made of fine lines would work best with a screen exposure of slightly less than 100%.

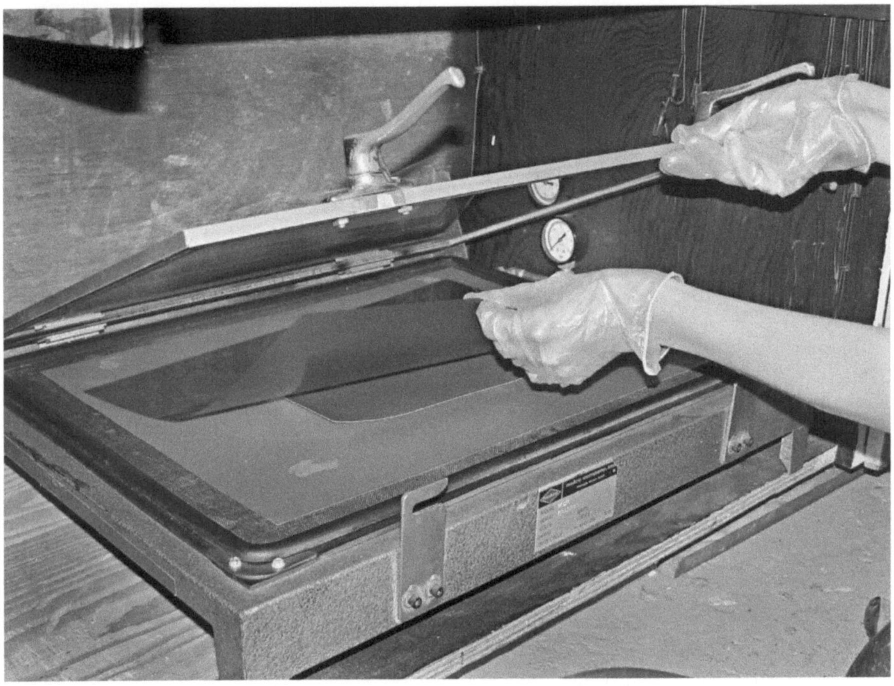

Figure 5-6 The dust-free screen is placed emulsion side down onto the tissue. Note the card stock on the bottom. This protects the tissue and film from mottling due to the textured rubber backing in the vacuum frame.

An overly long screen exposure will result in the thickening of the screen pattern and excessive hardening of the gelatin. This surface hardening may affect the proper adhesion of the tissue onto the copper. It can also slow down the etching and may affect the tonal scale or quality of the image. This can also be caused by making the screen exposure with light that is too diffused (i.e., placed too close to the UV bulbs). Using collimated light for the screen exposure will, in theory, give a screen pattern with steeper or sharper edges. We have found, however, that this does not noticeably improve or change the sharpness, resolution, or tonal smoothness of the etched image. If anything, it may contribute to weaker high points prone to undercutting during the etch. Unfortunately, overly diffused screen exposures tend to cause screen mottle, especially if the contact between the screen and the tissue is not perfectly tight (Cartwright 1939, p. 95).

If the screen exposure is too short, the screen pattern will be too thin in the resist. This will allow undercutting or side etching of the well walls and thus eliminate the peaks in the dark areas before the highlight details can be etched. The risk of the shadows being foul bitten will increase as a result. The screen exposure must be long enough that the screen pattern is more deeply established than the image highlights.

Positive Exposure

The sensitized gelatin tissue is now placed in contact with the positive. The positive's emulsion also faces the gelatin side of the tissue. The tissue's tendency to curl requires it to be attached in some way. It is easy to center the tissue on the positive assembly if you place the tissue face

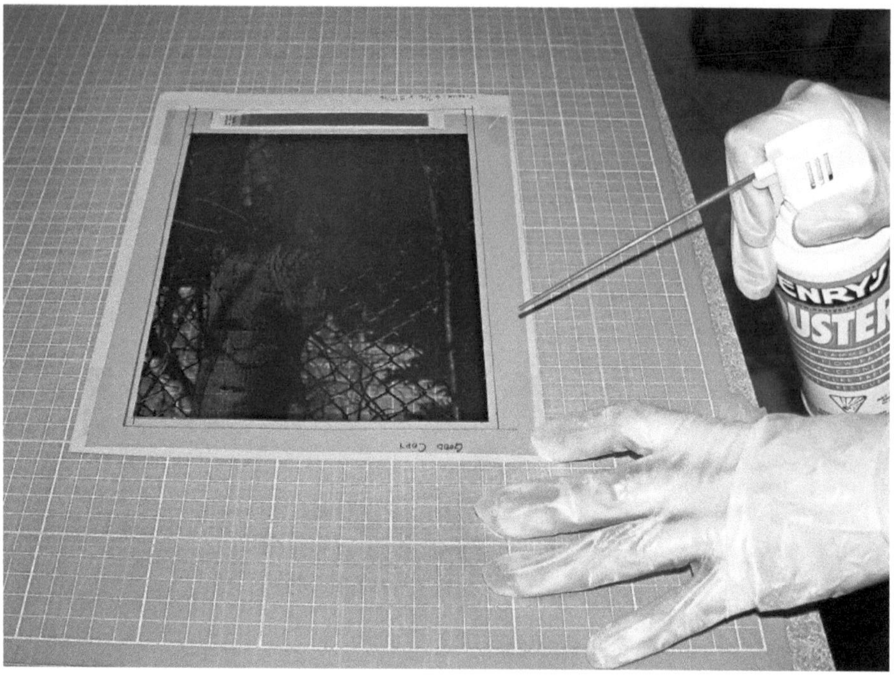

Figure 5-7 Canned air is used on all surfaces to remove loose dust. Don't tip the can or allow it to spray its liquid contents.

down onto the positive—which is emulsion side up—and use two or four very small pieces of tape on opposite ends or corners of the tissue to hold it flat and in place. (See Color Plate 11.) Be sure this assembly is absolutely free of dust or debris and that the exposure or vacuum frame glass is absolutely clean. Clean with glass cleaner and use compressed air to clear all surfaces of loose dust, including each layer of the positive assembly (Figure 5-7).

After the standard exposure time has been determined through testing and experience, you can adjust it to suit the varying densities of your positives. The standard exposure time will be appropriate for positives with shadow details falling within the range of 1.70 (step #12) to 1.85 (step #13). The most important consideration in determining the positive exposure time is that the overall exposure should not be lessened in an attempt to compensate for thin highlight densities; this will result in underexposed shadow detail. You can, however, use the fact that the change in density from step to step on the scale is equal to one-half a stop to select an exposure for a positive with thin or heavy shadow densities. For example, if the shadow detail density of the positive is 1.55 (step #11) and the plate is given the standard exposure, the resulting tones in the final print may be a bit light because the gelatin resist for the shadow detail will have the same thickness as normally given by step #11 instead of the ideal, step #12. By decreasing the exposure one-half of a stop, the thickness of the gelatin resist in the shadow details will now have an equivalent thickness to that of step #12 (for a standard exposure). The slightly thinner resist will now result in slightly darker tones. Conversely, a positive that is too heavy in the shadow detail, 1.99 or step #14, can be overexposed by one-half of a stop to add gelatin resist density to the shadow detail and bring it to the equivalent of step #13 on a standard exposure.

It is important to have the shadow detail fall within these steps so that separation in the tones can be maintained without becoming too light in tone (so that it reads as a mid tone) or too blocked in (so that it reads as a black). When you adjust the exposure in this way, all of the densities will move proportionally. Therefore, consider the densities of the highlight detail when determining your exposure and try not to have them fall in step #2. If this cannot be avoided, your positive has problems in the highlight detail density and the plate may be difficult to etch. In the event that shadow detail densities are higher than 1.85 and highlight detail densities are lower than 0.45, the contrast of the resist can be lowered by flashing the tissue prior to exposure. Flashing will add proportionally more gelatin resist density to the tones corresponding to the higher-numbered steps on the scale with little change to the lower-numbered step scale densities. To avoid this complication, the correct contrast and highlight density should be established when making the positive.

> The contrast of the print [gelatin tissue] can also be modified by lengthening the exposure to increase contrast or flashing the print to flatten or soften contrast. Since the .35 highlights pass about 50% of the light energy and the 1.65 shadows pass only 2.5%, the thickness of the resist is increased much more in the highlights than in the shadows. In flashing exposed carbon tissue, since no positives are used, the exposure is applied evenly all over the print and affects the shadows much more than the highlights. In a 5% flash situation the shadows receive three times the light energy as on a normal exposure, while the highlights receive only 110%—only 10% additional. This additional flash would reduce the 1.6 shadow to less than 1.2 density, with little or no effect on the highlights. Middle tones, obviously, would be flattened out proportionately, but not seriously. (Smeil 1975, p. 70)

Note: We have found that this flashing adds a fair amount of time at the beginning of the etch. It is slow to start, but once underway, things seem to progress normally.

After each exposure, continuing action effects the gelatin and must be minimized. If the gelatin hardens too much due to this effect, it will not adhere well to the copper and may be very difficult to wash clear (develop). Once the positive has been exposed onto the tissue, go immediately to the adhering and water development stage to halt the continuing action.

SUMMARY

1. **If using an aquatint, proceed to the positive exposure in Step 3. Otherwise, sandwich the screen of choice with the sensitized gelatin tissue, emulsion side to emulsion side, in the vacuum frame. Depending on the density and nature of the screen and image, the screen exposure can vary from 90% to 130% of the calculated positive exposure time. Be especially careful of dust. Clean all surfaces before each exposure. To ensure tight contact, allow the vacuum to run for a couple of minutes at 25 PSI before turning on the light. Don't forget to wear gloves when handling the tissue and do not touch its surface.**

2. Expose the tissue to a high UV light source for the optimum length of time. Take appropriate precautions to protect eyes from dangerous UV light levels.

3. Remove the screen and carefully align the tissue with the marks on the positive, including the Stouffer 21-Step Scale. Carefully remove all dust with compressed air. Allow the vacuum to run for a minute at 25 PSI in order to ensure tight contact before turning on the UV light for the second exposure.

TROUBLESHOOTING

Newton Rings Newton rings can sometimes be seen as an iridescent pattern of contour lines on the surface of the film or glass when the smooth surface of the film or screen is pressed in close contact with the glass of the vacuum frame. During the positive exposure, this phenomenon can be transferred as a pattern onto the resist and can ultimately show up in the lighter areas of the print. If the images are dark and detailed, the rings should not be a problem. Newton rings pose no real problem during the screen exposure. We have rarely had any problems with this phenomenon while using diffused light. Newton rings can be avoided by adding a layer of less smooth material between the glass and the shiny side of the film. Old sources often mention varnish, cellophane, or even talc. We wonder if the cure is worse than the affliction (Figure 5-8).

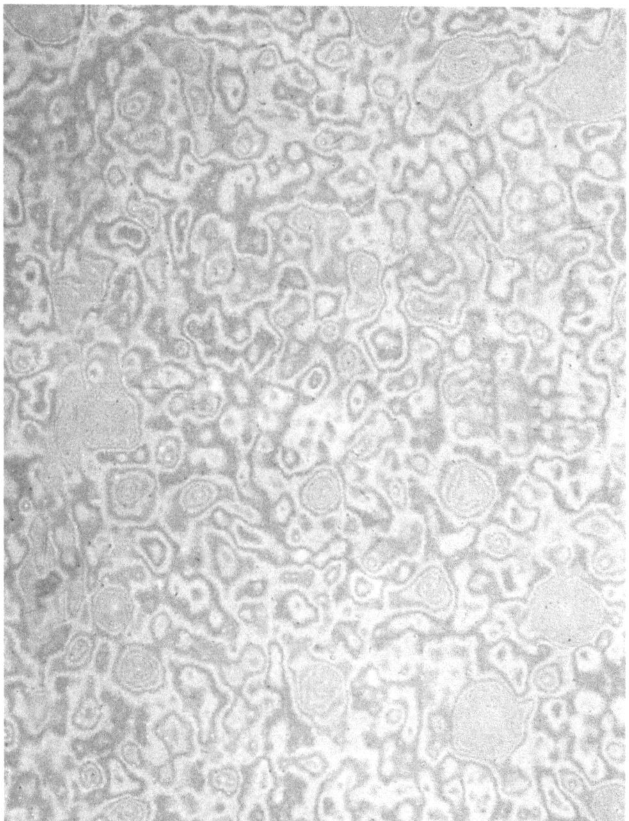

Figure 5-8 Newton rings have this distinctive pattern—sometimes tighter, sometimes looser—but with iridescent colors.

Dust A dust-free atmosphere and clean surfaces are paramount when working with the tissue resist. A large amount of dust could create a mottled texture in the resist. A single dust speck will result in a spot on the resist, either photographically transferred or physically under the wet resist. The former can also originate from dust during the exposure of the positive onto the tissue and would result in a black pit on the copper plate—a spot that received no light. If the dust was on the negative during the production of the positive, the positive would have a clear spot, which would then translate as a heavily exposed spot on the resist. This would print as a white spot, a high point on the copper plate. The white spot can easily be retouched with a needle. If the positive has a dense black spot, however, it will cause a hole in the resist, which will etch as a pit. This is especially problematic in a light area of the image, and is almost impossible to repair. It can be prevented in the first place by stopping out or sealing the spot on the dried resist with a fine-point permanent black Lumocolor pen prior to etching. See "Staging the Plate" in Chapter 7.

Sunspots Sunspots are also caused by dust particles. When a piece of grit is caught between the tissue and the positive, it separates the two surfaces enough that a halo is exposed around the speck where the film is not in tight contact with the tissue during exposure. This will show up in the etched plate as a soft light ring around a dark spot (Figure 5-9).

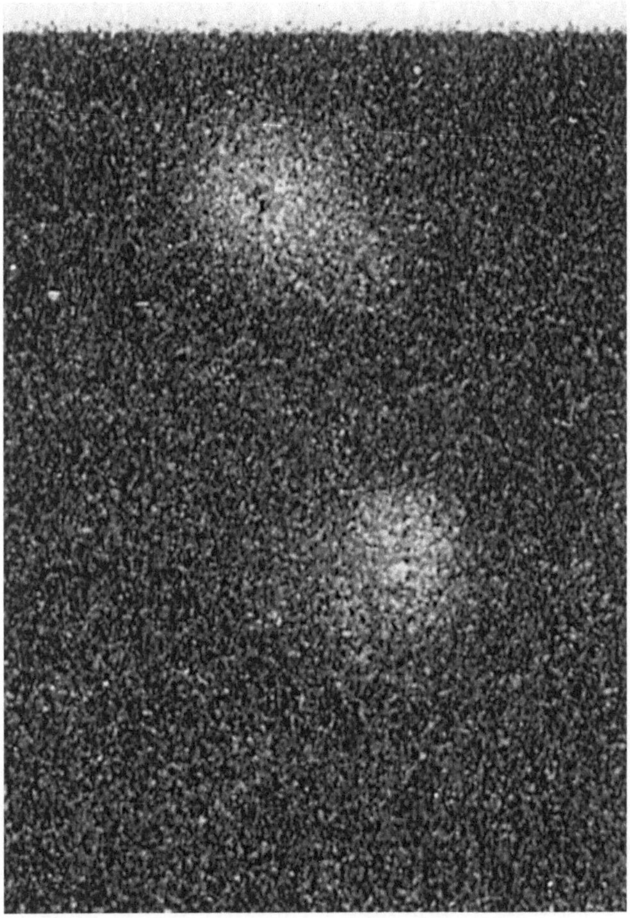

Figure 5-9 A closeup of a sunspot on a dark area of a print.

Mottle Splotchy or mottled areas can indicate problems with positive-to-tissue or, more commonly, screen-to-tissue contact. If the vacuum pressure was weak, there might be some soft-focus areas. If the vacuum frame's base surface below the tissue is pebbled or rough, this pattern can be transferred to the image because of the pressure points. Be sure that the surface under the tissue in the vacuum frame is smooth. You can add a black or red piece of mat board or stiff card stock. If the screen exposure from the UV bank was too diffused—with the light source being too close to the glass—the lateral direction of the light can cause mottling to appear in thinner areas of the resist. The solution is to move the vacuum frame further from the light source or use collimated light. Check to be sure that the screen itself is not mottled, especially if it was home-made.

Uniform positive-to-tissue contact is very important for sharp images and smooth, unmottled tones. Mottle can be even more apparent if the contact between the positive (or screen) and the tissue is not constant throughout the exposure. Be sure the vacuum frame does not lose its negative pressure. Grainy positives can also exaggerate mottle (Cartwright 1933, pp. 75–77). Mertle and Monson (1957, p. 337) recommend placing a sheet of ground glass over the positive in the vacuum frame during exposure. Mottle can also be caused by overly dry tissue that curls too tightly or is too stiff to smoothly come into contact with the positive. If the gelatin tissue is too dry or stiff from low relative humidity, it may not be possible to obtain tight contact during exposure. Humidify it slightly before exposure if necessary. This can be done by storing it in a sealed cabinet with dampened cotton or rags nearby. Tissue can become overdried if it is allowed to dry too long after sensitizing or if the room has a very low relative humidity or is too warm. One solution is to raise the humidity of the drying room (water in the sink, for example) or check the tissue sooner, before it drops off the Plexiglas and curls up. Always store sensitized tissue in a hermetically sealed bag or plastic photographic paper envelope as soon as possible.

Continuing Action The continuing action of light after exposure will cause the gelatin to continue hardening in the same way as if the exposure itself was lengthened. Excessive continuing action will cause the gelatin to harden so much that it prevents adhesion. The tissue should be adhered and developed as quickly as possible after exposure to avoid this effect.

A NOTE ON USING SCREENS OR APPLYING DUST-GRAIN AQUATINTS

The two types of screens available for gravure are the ruled commercial screen and a random pattern hard-dot screen. A commercial gravure positive screen is a positive image of a cross-hatched linear grid. It has clear lines and sharp-edged, dense squares in a diagonal orientation. This very regular grid pattern makes a tough plate with a smooth-toned image. It is designed for the rotogravure industry for large-scale publication runs. The screens themselves are very expensive. The random pattern hard-dot screen is easy to make and does not impart a noticeably regular pattern or mechanical quality to the image. (See Color Plate 12.)

Traditionally, a dust-grain aquatint resist can be applied to the plate before the exposed gelatin is adhered to it. It creates a resist of random-patterned tiny little spots. If it is too coarse where the grains are too large, it causes a noticeable grainy texture throughout the image. The asphaltum method is the finest, and if properly done provides a virtually invisible pattern, giving very smooth tones. Unfortunately, asphaltum powder is toxic, a possible carcinogen, and very messy. The rosin method is slightly less fine, but is safer and easier to obtain. Rosin is soluble in alcohol, so it is not practical to apply it under the resist because alcohol is used during the processing steps. A rosin aquatint seems to work quite well when applied on top of a dried resist, however. See Chapter 10 for more information on aquatints.

6 Adhering and Developing the Gelatin Tissue

In preparing an intaglio plate for etching, a resist is used to control the etch and create an image. In photogravure this resist is made from the photographic contour map of gelatin. The exposed gelatin tissue is immediately adhered to a copper plate, then developed in hot water. Hot water is used for the wash-out development process to effectively remove all of the unexposed or underexposed gelatin and leave the tanned gelatin image intact on the surface of the plate. This three-dimensionally contoured rendition of the original image will act as a graduated acid resist for the copper plate. The extremely narrow range of thickness within this resist layer is from virtually nil in the darkest blacks to about 15 microns (0.015 mm) in the densest highlights (Crawford 1979, pp. 103–104). Control of this process depends on stable temperatures and clean solutions.

REQUIRED SOLUTIONS

In the wet lay-down procedure, a solution of 25% denatured ethyl alcohol (our stock alcohol is 85% pure ethanol plus 15% methanol as the denaturing agent) and 75% distilled water (1:3) is used as a presoak bath. It should have a specific gravity of 0.970 at 20°C. The distilled water can be aged or preboiled to drive out any dissolved air and thus prevent the formation of air bubbles. The alcohol/water mixture can be reused if you filter it after each use to remove bits of dust, hair, and gelatin. After a time, when it has changed color from light yellow toward orange (from the leached dichromate), it should be changed. Remember to wear gloves whenever you handle anything involving dichromates, including this solution.

A second bath of 80% denatured ethyl alcohol and 20% distilled water (4:1) is required as a post-development bath. It should have a specific gravity of 0.861 at 20°C. It can be filtered and reused for quite some time. A few squirts of straight 100% alcohol after each use will help to maintain the alcohol concentration needed for this solution. The third solution you will need is distilled water or clean preboiled tap water, cooled to 15°C (60°F).

> Avoid the use of drugstore isopropyl rubbing alcohol or even pure isopropanol for any part of this process. It has created severe problems for us during and after development. (See "Alcohol Problems" in the Troubleshooting section at the end of this chapter.) Although both methanol and isopropyl *should* work for these processes, pure methanol is highly toxic and should be avoided, and the isopropyl alcohol we tried caused adhesion problems and mottling. If you encounter problems with ethyl alcohol, question how pure it is and what was used to denature it. Anything used to denature alcohol that is not totally water soluble will cause a problem.

EQUIPMENT AND SUPPLIES SETUP

Work in a roomy darkroom sink under ventilation. Use an inverted tray or other elevated flat nonmetallic support that can be easily rinsed off. Two more trays are needed, one for the soaking solution and one for the brightened plate waiting under water. Figure 6-1 illustrates the setup. A good stiff rubber squeegee or smooth soft rubber roller is required to attach the tissue to the copper plate. A screen printing squeegee or a photographic squeegee is fine if it is unscarred and straight. The rubber roller must be the full width of the gelatin tissue to work.

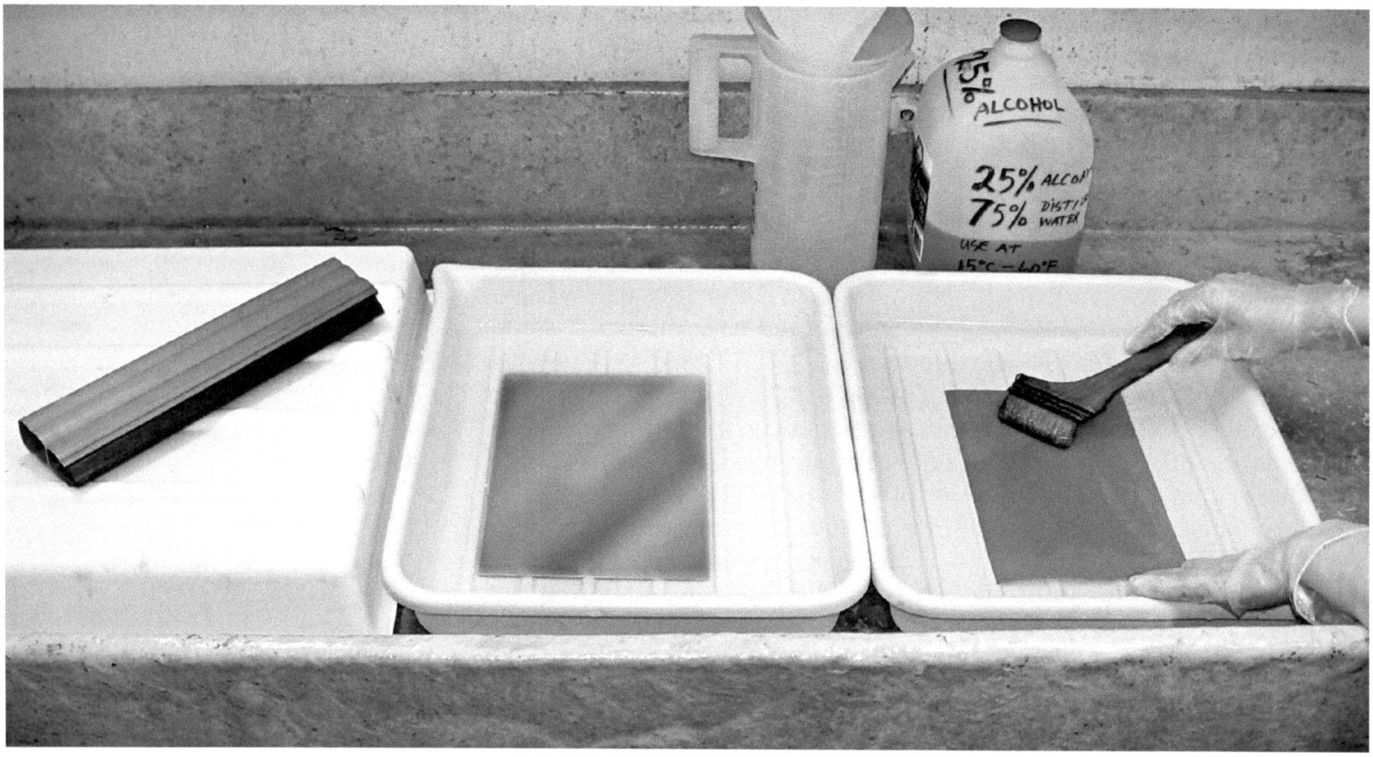

Figure 6-1 The tray setup for the adhesion process. Left to right: A stiff squeegee is on a firm support. A newly degreased and brightened plate is waiting in a tray of cool water. A soft metal-free brush is used to dislodge air bubbles from the surface of the exposed gelatin tissue as it soaks and flattens in the cool 25% alcohol bath.

PROCEDURE

Adhering the Tissue to the Copper Plate: The Wet Lay-Down Method

The wet lay-down method of transferring the tissue to the surface of the copper plate is preferred when working in low humidity conditions. This involves a complete soaking of the gelatin tissue before it comes into contact with the copper surface. This presoak removes the curl of the dry tissue, softens the gelatin, and makes the tissue manageable when positioning it on the plate. See Chapter 10 for the dry lay-down method, which can be useful for very large plates or multi-color multiple plate registrations.

Work under a yellow safelight and wear thin vinyl or latex gloves. Immediately after exposure place the exposed tissue in the 25% mix of alcohol and distilled water at a temperature of 12.8 to 15.5°C (55–60°F). Do not use alcohol if there is a rosin resist under the gelatin; use distilled water only. Slide the tissue carefully under the surface of the alcohol/water bath without allowing air to be trapped under its surface. Brush the surface gently with a soft Japanese Hake brush to remove air bubbles or dust, which may cling to the gelatin. Hold the tissue down by the corners until the backing and gelatin absorb enough alcohol/water solution to allow the tissue to lie flat on its own. This should happen after about one minute (Figure 6-2). At this point, the tissue is turned face down and allowed to soak fully submerged while you gently rock the tray. Continue until

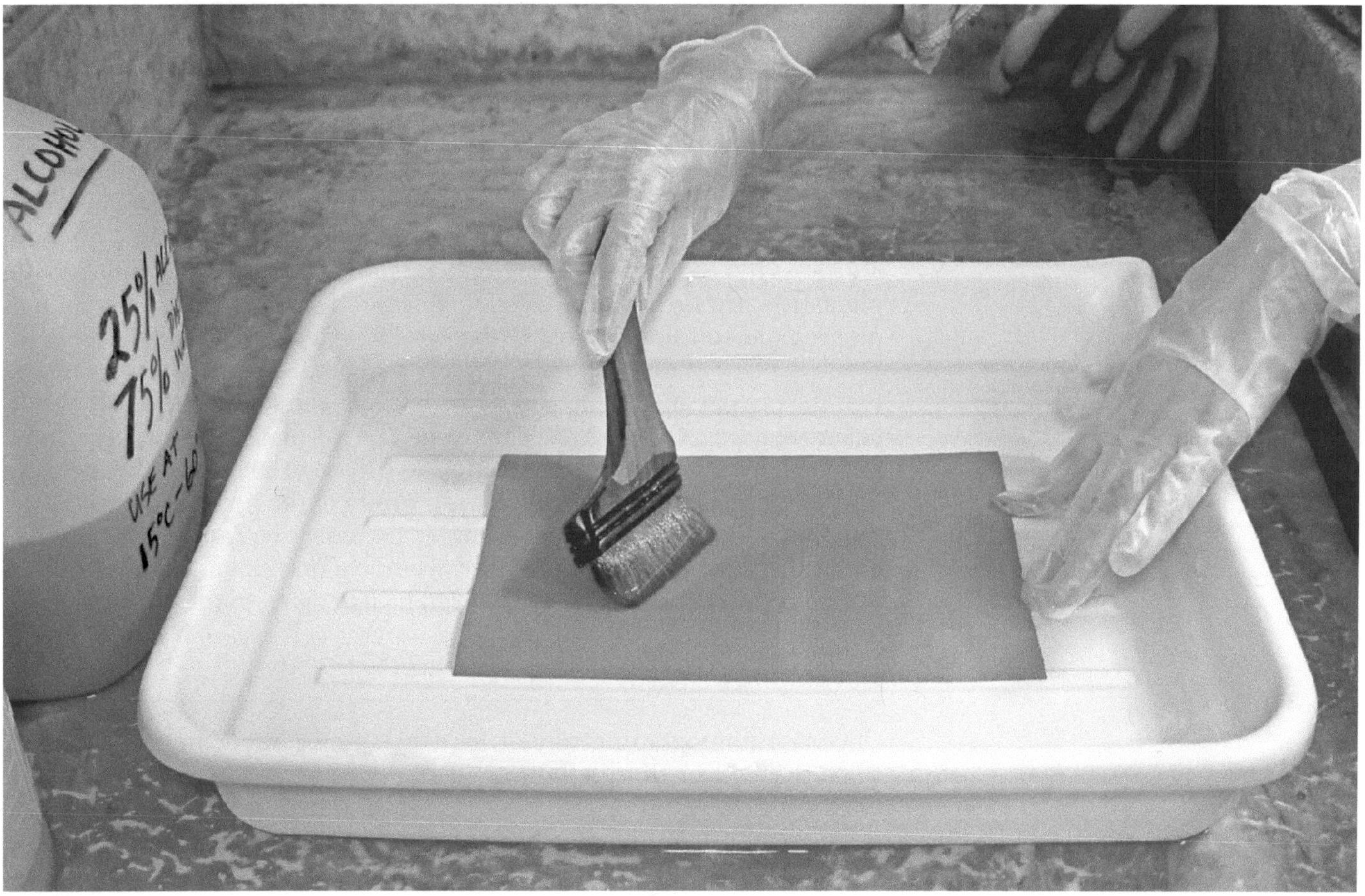

Figure 6-2 Use a soft Hake brush to dislodge bubbles while the tissue goes limp in the 25% alcohol presoak solution.

there are signs that the curl is just beginning to flatten or reverse. Do not leave submerged so long that the corners actually begin to curl back. Oversoaking, like undersoaking, will interfere with adhesion to the copper.

The degreased and brightened copper plate (Chapter 4) should be waiting submerged in a tray of preboiled tap water that has been cooled to 12.8 to 15.5°C (55–60°F). Remove the presoaked tissue from the alcohol solution, hold it up by one corner to drain briefly, and then slide it gelatin side down into the tray of water containing the plate. Slide it above the surface of the copper plate and position the tissue so it is centered on the plate. Be careful not to press the tissue against the copper until it is in place, then lightly press the extreme far edge of the tissue to bond it to the plate. Lift that end of the plate and tilt the copper—and even the tray—to allow the water to slowly escape from between the tissue and the surface of the copper. While holding this top end of the tissue against the copper, gently run the back of your gloved fingers down over the back of the tissue to push out the excess water (Figure 6-3). See also Color Plate 13.

Lift out the copper plate and place it on a firm, flat surface. Start the squeegee at the pinned edge and move softly across the tissue toward the loose end. Squeegee again, but this time start at the middle with slightly increased pressure. Rotate the plate 180° and repeat in the opposite direction with the squeegee placed in the middle of the tissue again. Be sure to avoid hard or prolonged finger pressure on any part of the tissue. Even vinyl or latex gloves can allow body heat to cause a blemish or distortion in the gelatin, which would ultimately show up in the etched plate. Squeegee in four directions from the middle with steadily increasing pressure until there are no signs of lifting along the edges of the tissue. Do not use excessive pressure because it may distort the details of the image or screen and may squeeze the soft gelatin out from under the backing paper. An alternative to the squeegee is a rubber roller.

Pat the paper backing dry with paper towels. Make sure that no excess moisture is left on the paper or surrounding plate. Hold the plate up to the safelight to see if there are any blisters or loose edges (Figure 6-4). If there are, press the edges with the towel to be sure that they have bonded to the plate. (Discard the contaminated towels safely, because they will contain dichromate solution.) Set the plate aside in a splash-free environment while the tray of development water is prepared. Several sources suggest that the copper and tissue should be left to stand for 10 to 15 minutes to ensure bonding. Others go so far as to suggest the plate be placed face down on a blotter and weighted for 15 minutes (Blaney 1895, p. 29). This allows the gelatin to expand and attach itself firmly to the plate. We have found that no waiting time (nor weighting) is necessary because the bonding is almost instantaneous and the plate can be developed immediately. Pressure points on the paper backing may even be damaging to the gelatin.

Development Wash

Choose a photographic tray at least two times larger than the copper plate. Ribbed or dimple-bottomed trays work best because they help with the flow of water and make it possible to quickly and easily pick up the plate off the bottom of the tray. It is important to have the edges and especially the corners of the plate well filed and rounded so that vigorous agitation in the tray will not damage or poke holes in the tray. Fill the tray with water stabilized at a temperature of 43°C (110°F). Set the timer

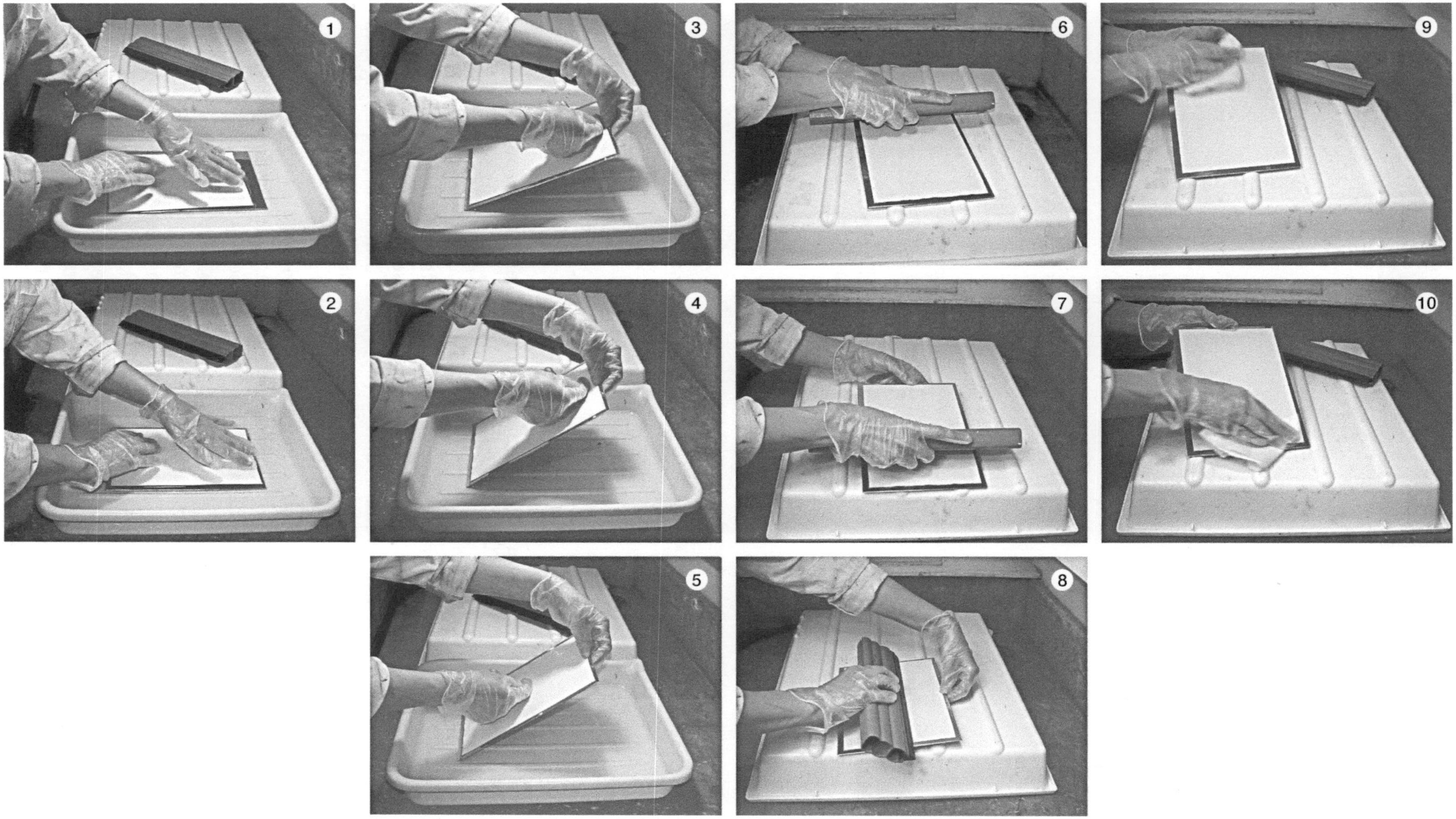

Figure 6-3 Position, then tilt the copper plate to allow the water to drain out from underneath and to lock the position of the tissue. Squeegee in several directions and use paper towels to soak up excess liquid and to press down the edges. The dichromates leach out, making the gloves necessary.

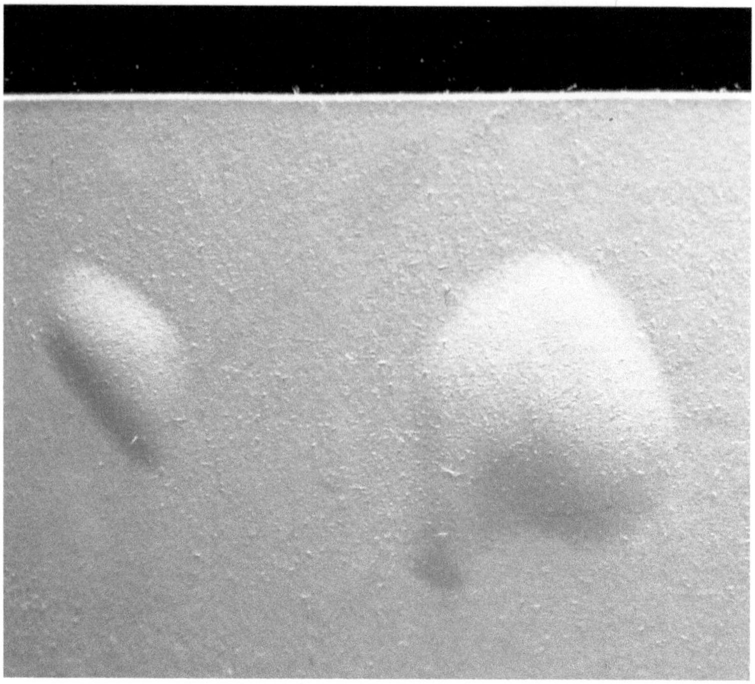

Figure 6-4 If there are blisters under the tissue, these will be voids in the resist. If they are near the edge it may be possible to press them down. If not, strip the resist off and make another one.

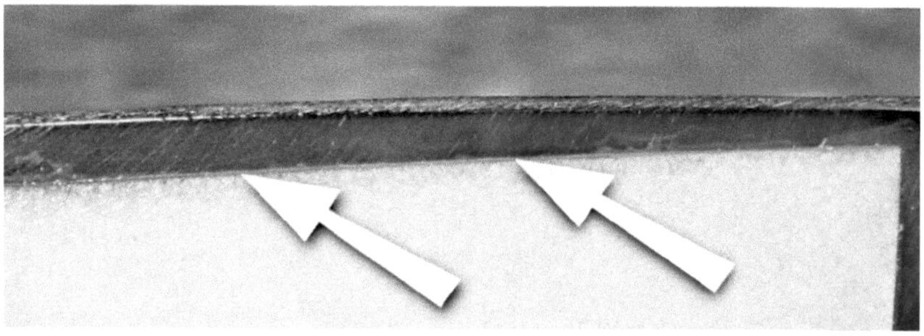

Figure 6-5 A thin margin of red gelatin and a row of tiny bubbles appear along the edge of the paper backing.

for 20 minutes. Immerse the plate, tissue side up, into the 43°C (110°F) water and allow to soak for 15 seconds. Some sources suggest starting with water at 26.7°C (80°F) and immediately raising it to 43°C (110°F). We find this unnecessary. Once the plate is immersed, gently rock the tray and occasionally add more water at 43°C (110°F) to maintain the temperature of the bath. Continue for 3–4 minutes or until there are minute signs that the gelatin is oozing out from under the paper backing (Figure 6-5). If this is excessive, you have waited too long.

Test whether the paper backing is ready to be removed from the resist by slowly and gently peeling from one corner. Keep the plate submerged the whole time. If there is only slight resistance, remove the paper in one smooth, steady pull without jerking or stopping (Figure 6-6). See also Color Plate 14. If there is strong resistance or any sign of the gelatin resist being lifted off the copper, stop immediately and continue to soak. After a moment try removing the backing paper from a different corner. Do not be concerned if you see irregular or oval-shaped "cauliflowers" or "cat's

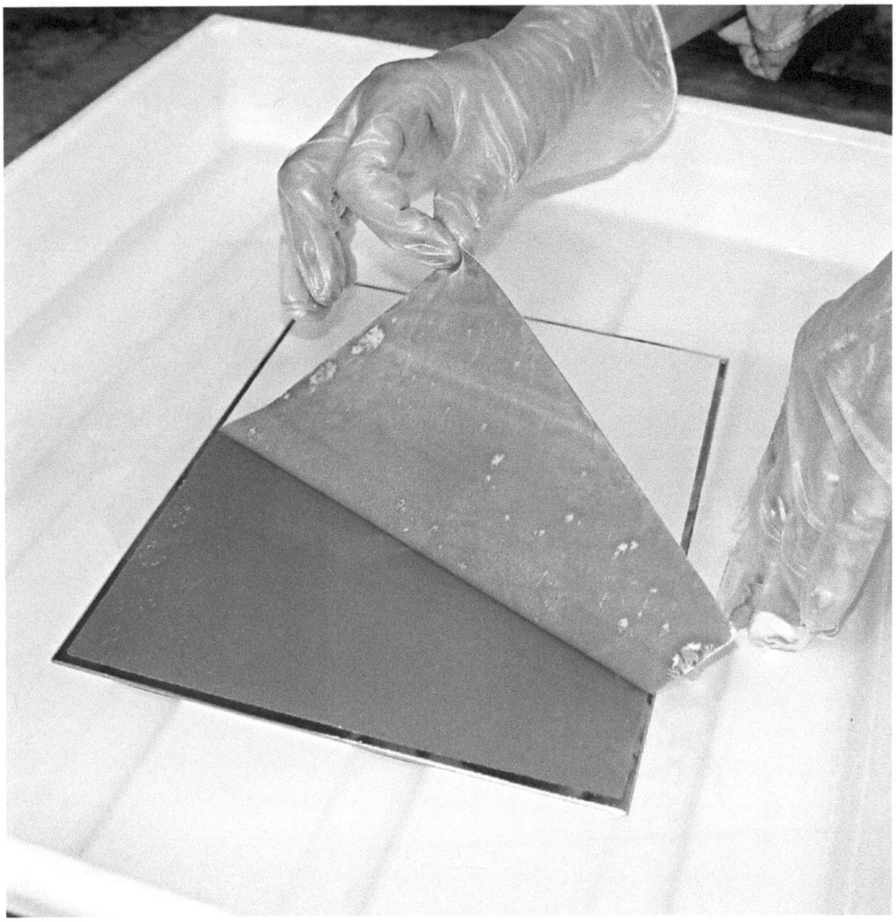

Figure 6-6 Peel the paper backing from the resist.

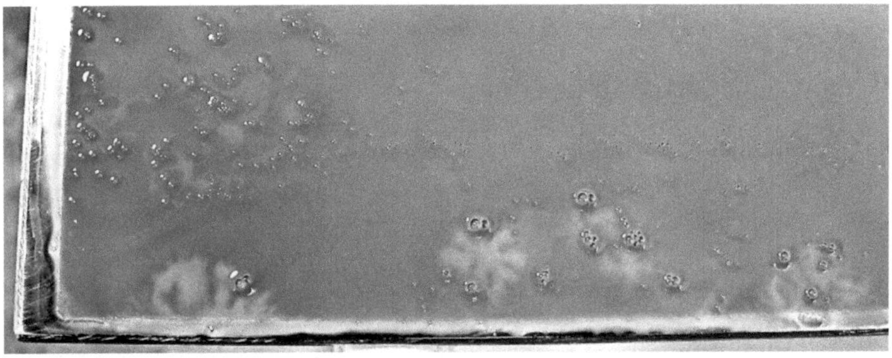

Figure 6-7 "Cauliflowers" or "cat's feet" in the gelatin resist.

feet" on the surface on the gelatin (Figure 6-7). They indicate where water has penetrated the paper backing through a pin-hole in the paper and shouldn't affect the outcome of the image. They should disappear during the wash out. If they are excessive, begin the peeling stage earlier next time.

After the backing is peeled off and discarded, periodically add fresh hot water to the tray, still at 43°C (110°F). Rock the tray to cause a surge of water to run over the plate. Continue this rocking for the rest of the processing time. Rotate the plate 90° every 30 seconds or so to produce an even agitation. Rocking back and forth and alternatively side to side will also allow the water surge to rush across the plate in opposing directions.

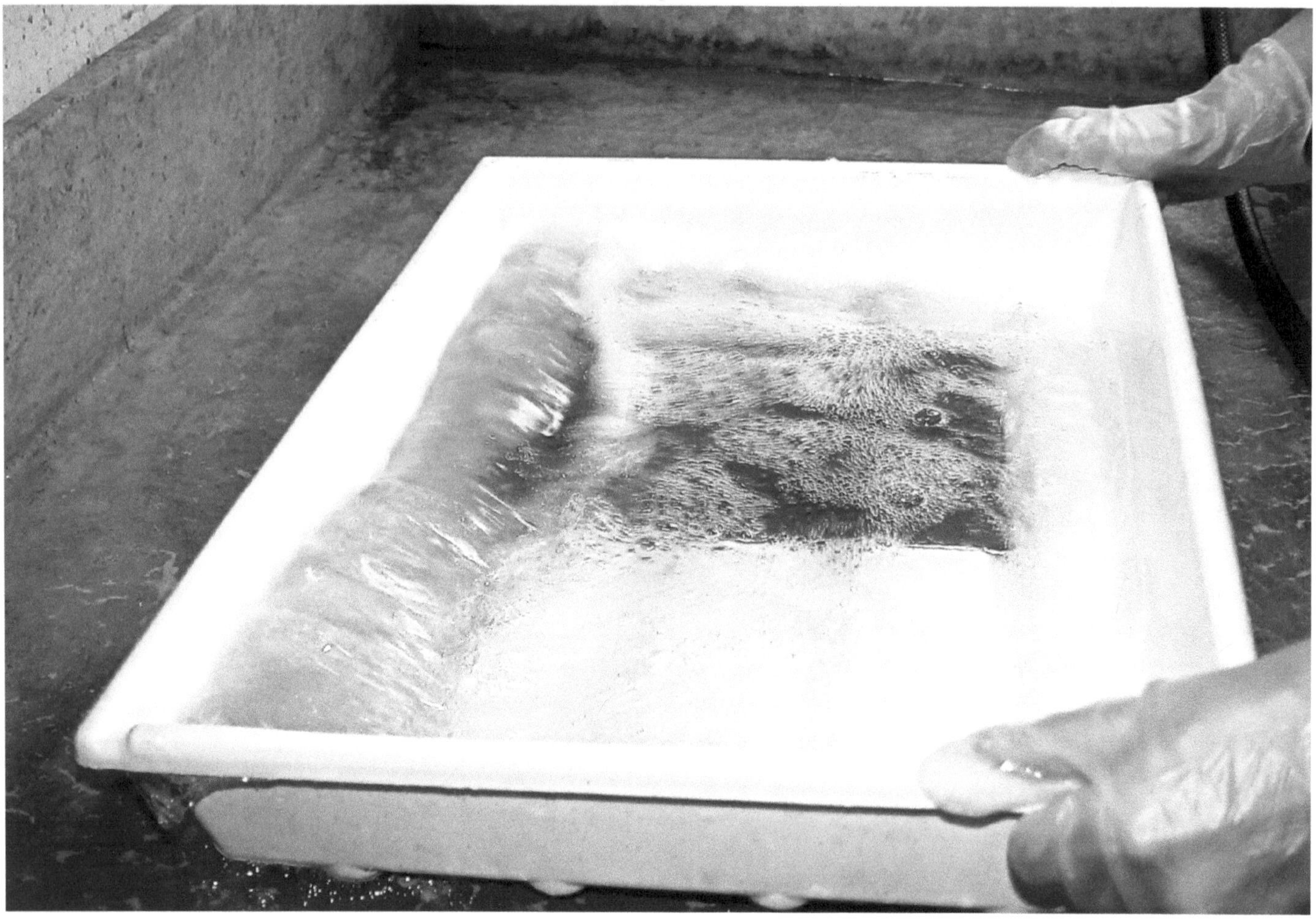

Figure 6-8 If you get the rhythm right, you can get the water to fall almost straight down onto the plate, while at the same time keeping the plate centered in the tray rather than bouncing off the sides.

As the tray is rocked, the unexposed gelatin should dissolve and wash away. Increase the speed and vigor as the time progresses, especially during the last few minutes. If you slide the tray, the water should surge back and forth across the plate (Figure 6-8). Continue to maintain the level and water temperature throughout development. This is when a temperature control panel is most useful (Figure 6-9). The hose spray, if not too harsh, can be used directly on the gelatin by this point, but be careful. If the water remains clear and bubble-free for a time, all remaining vestiges of soluble gelatin should be dissolved. Be very careful not to touch or bump the gelatin surface with anything, as even your fingertips will easily scar the gelatin. It is difficult to overdevelop the plate, so it is better to err on the side of a long developing period until you feel sure that you can identify when the resist has been fully cleared of soluble unexposed gelatin. If developed at too high a temperature or for an excessively long time, the resist may end up thin (de Zoete 1988, p. 75). Normally, 15 minutes after initial immersion will fully develop a properly exposed resist under the right wash-out conditions. Cartwright suggests 20 minutes from when the backing paper is stripped (1936, p. 92), but we find it happens faster than that. See "Development Problems" in Troubleshooting.

After the development is completed, continue to rock the tray while lowering the temperature of the water to room temperature. Dump the

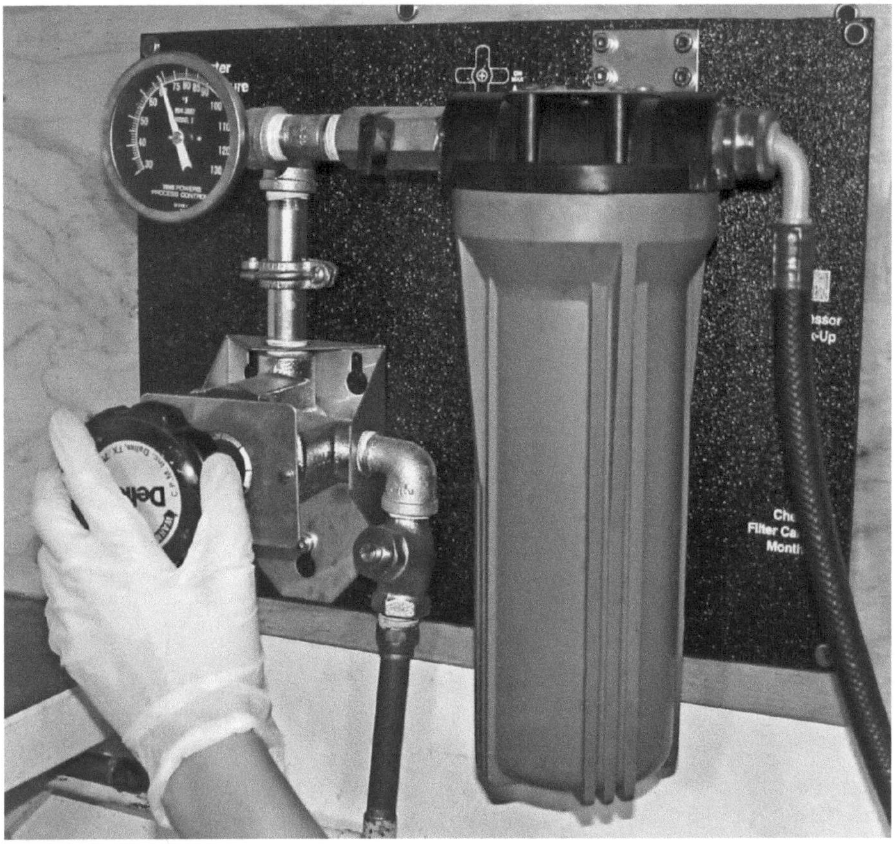

Figure 6-9 A commercially available temperature control panel designed for darkroom use is most useful for maintaining resist development temperatures.

warmer water as you add cooler water until the plate has been fully cooled to 20°C (68°F). This will stabilize the softened gelatin and make it less susceptible to damage during the following steps. Do not cool the plate too quickly and do not drop the temperature of the water below 18°C (65°F). A sudden, extreme cooling can cause reticulation.

Drying and Stabilizing the Resist

Remove the plate from the cool water and quickly immerse it into a mixture of 80% denatured ethyl alcohol and 20% distilled water (4:1) at 20°C (68°F). Do not do this step if there is a rosin aquatint under the gelatin resist. Leave it to soak for 5 minutes, agitating frequently. Add a splash of 100% alcohol when you filter the solution back into the storage bottle. An optional shift to a 100% solution of alcohol for the final minute can complete the process (Crawford 1979, p. 259). The alcohol displaces most of the water in the gelatin, allowing the resist to dry quickly and evenly so that no irregularities occur in the etch. After a total of 5 minutes, remove the plate and hold it at a slight angle from the back (Figure 6-10). For even drying, rotate the plate as it is draining. Tiny flecks and hairs can sometimes be dabbed off gently with the corner of a damp paper towel. Don't overdo it. After draining off the excess, use a folded wad of paper towels to soak up the bead of alcohol from the lower edge of the plate as you continue to rotate it from one edge to the next (Figure 6-11). Continue until the wet surface loses some of its sheen. Use a hair dryer

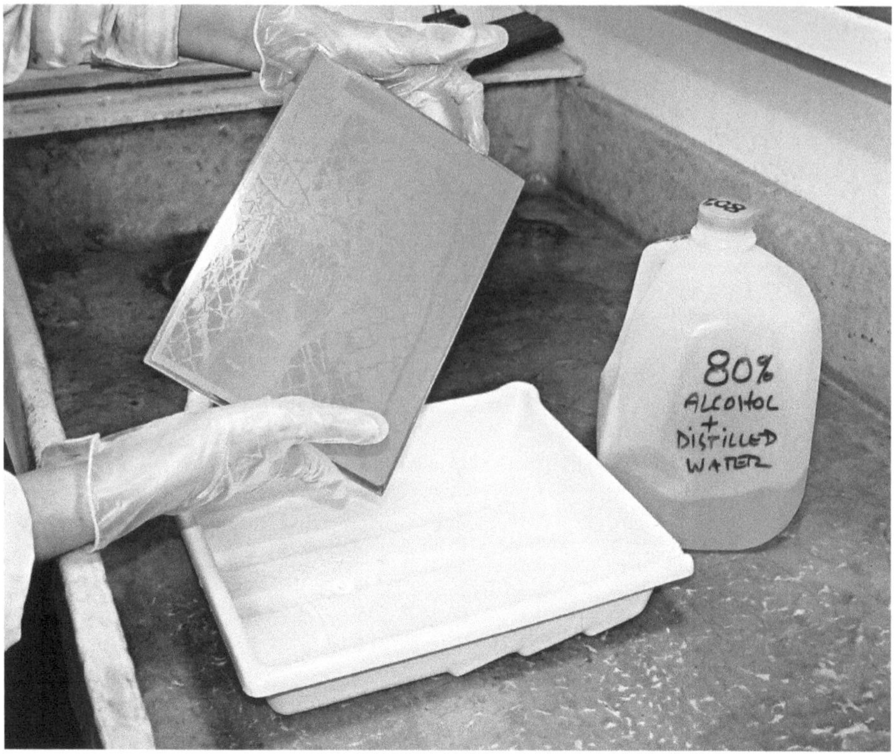

Figure 6-10 Drain the excess alcohol/water solution back into the tray from one corner.

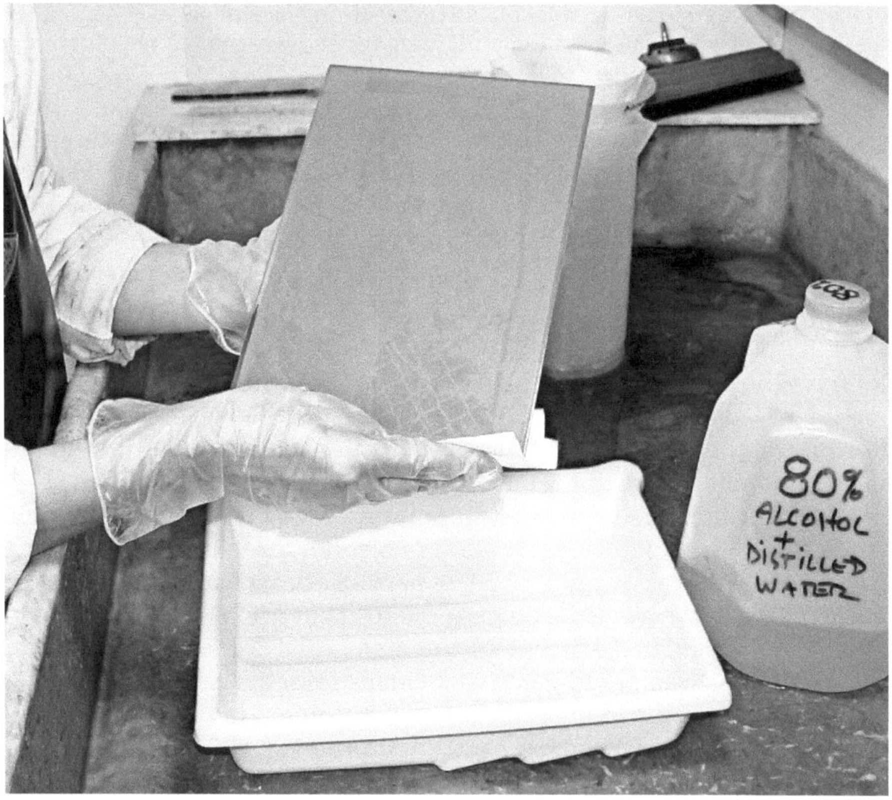

Figure 6-11 Rotate the plate and rub each edge with a folded paper towel until you can see the moisture pulling away from the edges. Be careful not to rub the gelatin resist; it is still very soft and fragile.

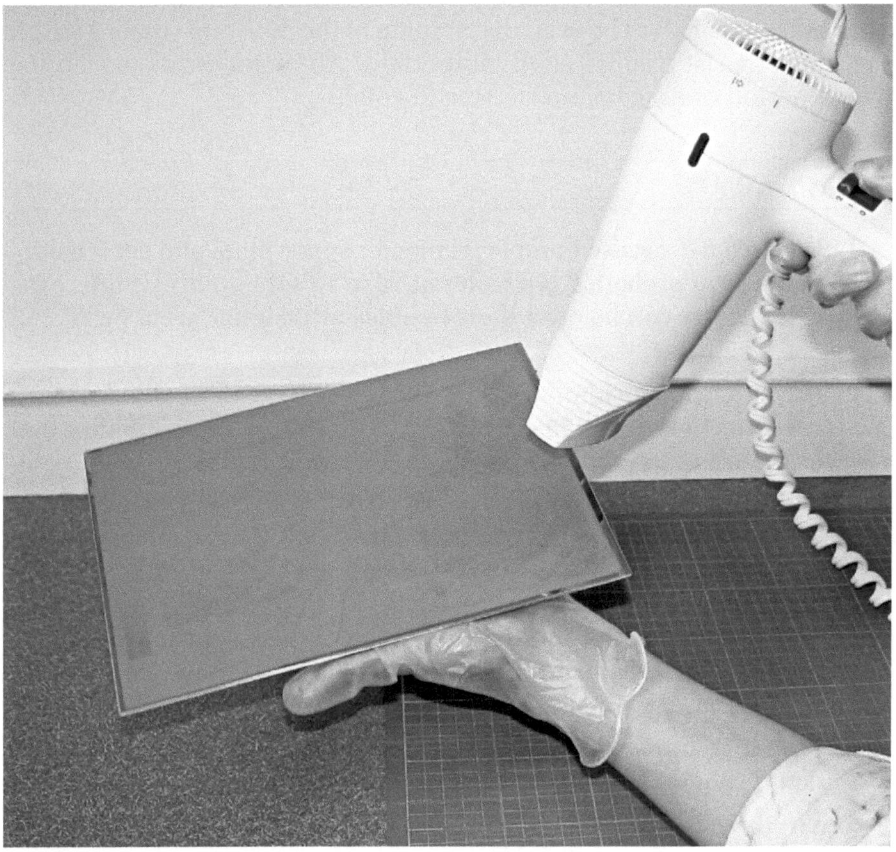

Figure 6-12 Immediately dry the surface with a hair dryer set to cool. Dry the gelatin until there are no signs of ripples or wet areas left in the resist. If there is an odd liquid quality to the gelatin as it is drying that results in a mottled surface, the alcohol is contaminated and should be changed. See also Figure 6-15.

to give the plate a final and quick drying (Figure 6-12). Denison says that plates dried with alcohol should be fan-dried immediately, whereas if air drying is done without alcohol, do not use a fan (1895/1974, pp. 67–69).

The plate must now be placed in a safe, dust-free area away from drafts and in a situation where the gelatin resist's moisture content can stabilize to the room's relative humidity. The time between the development and the etch should be at least 4 hours, but the resist should ideally be left overnight to come to equilibrium. Some sources allow for a shorter drying time between the development and the etch. We have found that an overnight wait is safer because it guarantees a more even and thorough drying. Other sources use a drying oven to drive out all remaining moisture from the gelatin. We have not tested this method and assume a rehumidifying time would be needed before the etch.

Relative Humidity Factor

Before etching the plate, it is important that the moisture content of the resist be uniform. According to the manufacturer, a relative humidity of 60% gives the best results. We have been working at a relative humidity as low as 30% and it may well be that many of the modifications we have made to standard practice are ways of compensating for the low humidity. A relative humidity higher than 70% may cause the etch to start or progress too quickly. The even distribution of the moisture content within the gelatin resist itself is vitally important. Allow the resist to stabilize over a period of time before starting the etch.

SUMMARY

1. **Prepare a degreased and brightened copper plate and set it aside in a tray of preboiled tap water at 12 to 15°C (55–60°F). Stale, distilled water can be used if air bubbles or residues are a problem. See Chapter 4.**
2. **Have the 25% alcohol adhering solution (1:3 with distilled water) ready at 10 to 15°C (55–60°F). Leave it in the bottle until after the exposure to prevent evaporation. If there is an underlying rosin aquatint, use a tray of chilled distilled water instead.**
3. **Wear gloves during this procedure. Place the freshly exposed tissue in the tray of 25% alcohol. Use a Soft Hake brush to remove air bubbles from the surface and to help hold it under the surface of the solution. Continue until just before a reverse curl begins. Note: Filter the alcohol bath before storage.**
4. **Remove the tissue from the solution and slide it gelatin side down onto the copper plate, which is submerged in a shallow tray of cool preboiled water.**
5. **Pin the middle of one edge against the plate.**
6. **Tilt the tray, with the pinned edge high, to allow the water to run out from under the tissue as you lightly brush the back of the tissue with the back of your gloved fingers.**
7. **Lift out the plate and tissue and place onto a firm support.**
8. **Squeegee from the pinned edge across the tissue using one smooth stroke. Repeat in each direction from the middle, being careful not to trap air beneath the tissue.**
9. **After squeegeeing, pat the back of the tissue dry with paper towels. Press the edges firmly and check for air blisters or edge frills.**
10. **Put the plate in a splash-free environment.**
11. **Fill a large tray with water at 43°C (110°F). Set the timer for 15 to 20 minutes.**
12. **Quickly immerse the plate, tissue side up. Periodically add water at 43°C (110°F) to maintain temperature. Play the water over the back of the plate or agitate the tray.**
13. **As soon as you see signs of gelatin oozing along the edges (3 to 4 minutes when the screen exposure is from a diffused light source, less if collimated), peel off the paper backing. There should be slight resistance—but no tearing or lifting—as you remove the paper.**
14. **Develop by vigorously and continuously rocking the tray and rotating the plate 90° every 20 to 30 seconds. Continue to maintain the water temperature throughout. Be very careful not to touch the gelatin surface.**

15. Once the water remains completely clear, development is done. Development time after the paper backing is removed should be between 8 and 12 minutes.

16. Cool by gradually lowering the water temperature to 20°C (68°F).

17. Remove the plate and soak in an 80% alcohol bath (4:1 with distilled water) at room temperature for 5 minutes—except if using an underlying rosin aquatint. Agitate briefly. Add a "squirt" of 100% alcohol to the solution as you filter it back into its storage bottle. Move the plate to a 100% alcohol bath for the last minute (optional).

18. Remove the plate from the alcohol bath after a total of 5 minutes and drain while rotating the plate at a steep angle.

19. Absorb the excess alcohol by holding the plate almost vertically and wiping the edges with a folded paper towel. Do not touch or scrape the image area. Rotate the plate and wipe all four edges in succession more than once.

20. After the resist seems to have lost most of its sheen of liquid alcohol, quickly dry it with a hair dryer set on cool.

21. Allow an extended drying time for moisture content and relative humidity to attain equilibrium.

TROUBLESHOOTING

Dust Specks and Sunspots A dust speck caught between the resist and the plate will cause a blemish with serious consequences in the etched plate. It affects the resist in an area greater than the dust spot itself and can cause resist failure or leave a crater-type blemish. Sunspots can also be caused by a pit or scratch in the Plexiglas used to sensitize the tissue. The gelatin dries with a little raised bump molded by the scratch. This bump acts like a piece of dust during the following exposures. See "Sunspots" in Chapter 5's Troubleshooting section.

Mottle A mottled appearance visible in the thin shadow areas of a developed resist can be due to vacuum pressure being interrupted, vacuum pressure that is too low, or the tissue being too dry and stiff resulting in contact problems during the screen exposure. This problem is exaggerated if the diffused light source is too close to the vacuum frame. The textured rubber mat on the bottom of the vacuum frame can also impart a mottled (but regular) pattern and should be covered with red or black mat board or stiff card to provide a smooth surface under the tissue during exposure. If there is a texture from the backing, it can usually be seen once the vacuum pump is running.

Air Bubbles Tiny bubbles that are normally dissolved in the water can come out of solution and be trapped between the resist and the plate during the lay-down process, which causes a serious problem. In the plate they will be seen as small unetched pinpoints. They can best be avoided by using 25% alcohol presoak solution. If the problem persists, preboiled or aged distilled water may lessen the likelihood of these bubbles forming.

Figure 6-13 A crease made by the squeegee when the tissue does not lay flat on the copper. This cannot be corrected.

It is also very important to avoid trapping air blisters between the resist and the copper during lay-down. Proper squeegee or roller technique will help prevent this problem.

Lay-Down Problems (creasing, frilling) If the tissue prematurely bonds to the plate in the tray prior to lay-down (i.e., when not in position), it cannot be moved or adjusted without damage. If it bonds with a lump or fold, the squeegee will crease the tissue and ruin it (Figure 6-13). Be sure to lay down the tissue from one end and start the squeegee from that same end. This pushes the water out along the length of the tissue giving a smooth lay-down without blisters or creases. (See Figures 6-3 and 6-4.)

If the edges are frilled they will not lay down smoothly and may lift during development. They may also cause small creases or blisters that will foul-bite during the etch. Be sure that the sensitized tissue is oversized to start with in order to permit subsequent trimming to remove all edge frills. If the tissue is soaked for too long in the 25% alcohol presoak, it can cause edge frilling and also interfere with adhesion (Cartwright 1939, p. 98).

If the tissue does not adhere to the copper, it could be grossly overexposed, exposed to too much heat from the exposure lamp, or too old (fresh or frozen) having fogged due to the dark effect. If the tissue is in good condition, failure to adhere can be caused by presoaking in a 25% alcohol solution that is too cold—less than 10°C (50°F). Also make sure that you use sufficient pressure when squeegeeing the tissue to the copper. Another culprit is a sensitizer solution that has gone bad, through either overuse or, most commonly, being too old.

If the tissue bond typically lets go during early development, some sources suggest placing the freshly laid-down tissue under weight before beginning the development procedure. It is suggested that a blotter and

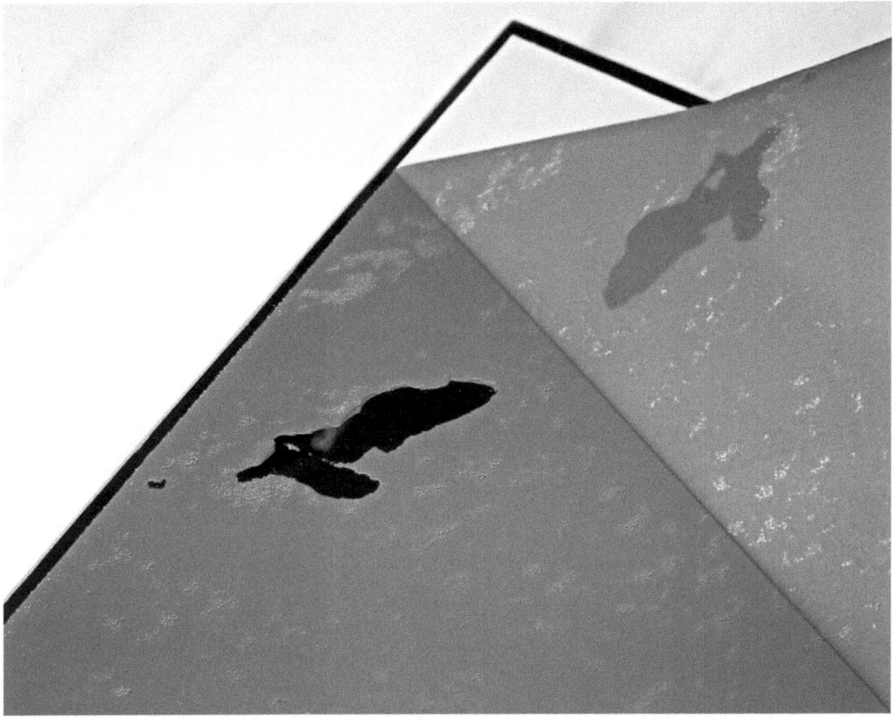

Figure 6-14 Unadhered areas of gelatin pull off the surface of the plate when the paper backing is peeled away during development.

a flat weight like a heavy book be positioned over the freshly laid-down tissue and allowed to rest for 5 to 10 minutes. We have not found this necessary. If the plate is not properly degreased or brightened, the tissue may not adhere properly (Figure 6-14).

Development Problems If the image is not clearing or is hard to wash out, it could indicate overexposure, too much heat during exposure, exhausted sensitizer, or water at too low a temperature. If warming the water to a maximum of 48.5°C (120°F) does not help, try adding sodium bicarbonate (Denison 1895/1974, p. 36). However, it may be best to begin again. If irregular splotchy marks appear in the etched plate's highlight areas, they may have been caused by underdevelopment, especially the unhardened gelatin left behind in the highlights.

Flaws If fingerprints or smudges show up in the gelatin after development, they were likely created by the heat of finger pressure on the front or the back of the tissue before lay-down. Give yourself larger safe edges to avoid touching the image area. Also make sure the presoak solution and the darkroom are not too warm.

Alcohol Problems The 25% presoak will take on a lot of dichromate as it is used and will eventually become quite orange. It is best to replace this solution regularly. You can test the specific gravity and add pure alcohol until the 25% solution attains a specific gravity of 0.970 at 20°C. The final 80% soaking solution will not need to be replaced as often if you add a splash of 100% alcohol after each use. Its specific gravity can be tested to 0.861 at 20°C. Filter both solutions after each use to prevent dust,

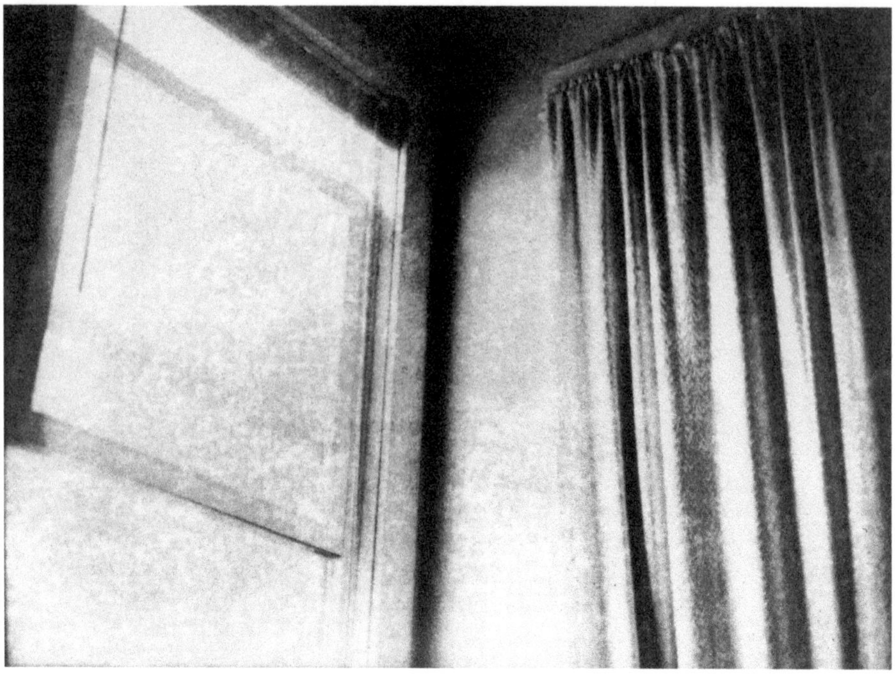

Figure 6-15 Mottle pattern shows up in the highlight areas of this print detail, in this case caused by a contaminated alcohol bath.

hair, and gelatin flecks from contaminating the solution and getting stuck on the gelatin surface.

If the tissue appears to take a long time to dry to a matte surface as you are draining it, the proportion of water to alcohol has probably shifted to where there is too much water. Add more alcohol and test the specific gravity or remix a new 80% alcohol bath. If the trays are left uncovered for long periods of time, the alcohol content will drop due to its high volatility. An optional final brief bath in 100% alcohol is also useful to shorten the drying time of the resist.

If a noticeable mottle appears on the paper backing at lay-down or a fluid-like appearance occurs over the thicker areas of the developed gelatin when drying the resist with a hair dryer, it can be an indication of an old, contaminated, or overly dilute alcohol bath or that the alcohol itself was denatured with the wrong material. This is the problem we found when attempting to use isopropyl alcohol (Figure 6-15).

7 Preparing to Etch

PREPARING THE FERRIC CHLORIDE

A unique aspect of the photogravure process is connected to the properties of ferric chloride, the mordant that penetrates the gelatin to etch the copper beneath. The correct chemical name is iron(III) chloride; however, in this text we will use the traditional term *ferric chloride* to avoid confusion and maintain consistency with other literature. In photogravure, a progression of ferric chloride solutions of varying densities are used to etch the copper. The densities of the solutions are measured on the Baumé scale using a Baumé hydrometer. See Appendix D for the chemical formulae of a ferric chloride solution and its various reactions.

Equipment and Supplies

Look for 40% weight by volume ferric chloride solution, previously called Rotogravure Iron 48 Degree Baumé (acid free). See Appendix G for the desired composition of this solution. Use distilled water to dilute this stock to make the other solutions. You will need six or seven plastic storage bottles, at least 2 liters (2 qts) each being ideal. You will need a set of three to five heavy plastic photography trays and a funnel. Use a glass thermometer and a glass Baumé hydrometer (39° Bé to 51° Bé) with a 250 ml (8 oz) graduated cylinder for the hydrometer. You can make a cylinder with a 31 cm × 3.5 cm (12″ × 1.25″) Plexiglas tube glued to a square of thick Plexiglas as a base. Gradations are unnecessary.

Working with Ferric Chloride

Even though reaction fumes are minimal, work in a well ventilated room. Ferric chloride is a strong irritant. If not rinsed off it will stain everything it touches, even some plastics. Be prepared to end up with a rust-red

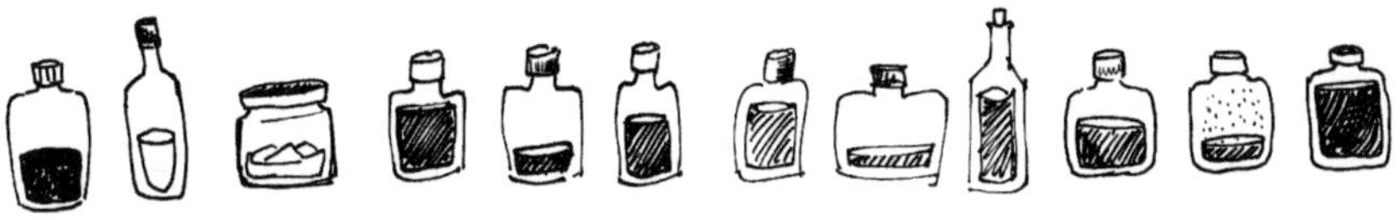

darkroom sink. Always wear long acid-resistant gloves, a rubber apron, and a face shield when working with ferric chloride. Splashing is unavoidable and eye contact is rather painful and dangerous. All plastic equipment such as funnels, beakers, trays, and stirring utensils must not be used for photographic chemicals after being used with ferric chloride. They are contaminated and must be dedicated for use solely with ferric chloride. Glassware is not a problem if well cleaned. Ferric chloride etches all metal including stainless steel, so metal utensils or thermometers must not come into contact with the ferric chloride. For obvious reasons, do not work in a stainless steel sink! Do not dump undiluted ferric chloride down the drain at any time. Ferric chloride spills that are not excessively diluted before being washed down the drain could eventually etch away metal pipes. When cleaning up after etching, be sure to flush the system well.

Use of the Hydrometer

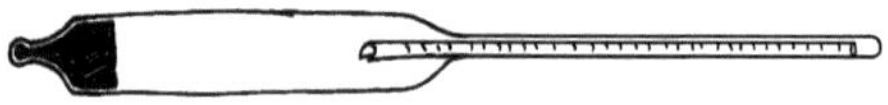

Each mordant must be tested and maintained at a known Baumé. The Baumé of a solution refers to its density or saturation in comparison to water and is another form of specific gravity. See Table 7-1. A hydrometer that covers the range of Baumés from 39° to 51° is necessary. Test the Baumé by pouring the mordant into a *dry* 250 ml graduated cylinder (8 oz capacity). Fill to within a few centimeters (an inch) from the top. Slowly lower the *dry* hydrometer into the solution and release. Be sure that its position is in the middle of the cylinder rather than clinging to the side. Give the hydrometer a spin to dislodge air bubbles. Read the scale at the bottom of the meniscus (Figures 7-1 and 7-2). You can also pour the ferric chloride into the cylinder after placing the hydrometer in it first. It will bob up and down and come to rest at the actual Baumé reading. Test the Baumé levels at a stabilized working temperature somewhere between 20°C (68°F) and 24°C (75°F).

Table 7-1 Relations Between Baumé Degrees and Specific Gravity for Liquids Heavier Than Water at 60°F

Specific Gravity	*° Baumé*
1.465	46°
1.450	45°
1.436	44°
1.422	43°
1.408	42°
1.394	41°
1.381	40°
1.368	39°
1.355	38°
1.343	37°
1.330	36°
1.318	35°
1.306	34°
1.295	33°

CONVERTING SPECIFIC GRAVITY TO BAUMÉ DEGREES

If finding a Baumé hydrometer is difficult, it may be easier to access equipment that reads specific gravity. The following formula (Smeil 1975, p. 88) converts one reading to the other:

$$145 - \frac{145}{\text{Sp. Gr.}} = {}^\circ\ \text{Baumé}$$

For example: A solution reads 1.307 specific gravity, therefore:

$$145 - (145 \div 1.307) = 145 - 110.941 = 34^\circ\ \text{Bé}$$

The chart shown in Table 7-1 might be a useful guide.

Mixing the Ferric Chloride Solutions from Stock 48° Baumé

Start with one bottle 3/4 full of 48° Baumé (Bé) solution and keep it as-is. This is the conditioning bath. Fill a second bottle a little less than 3/4 full with 48° Bé stock and reduce the Baumé by adding a measured amount

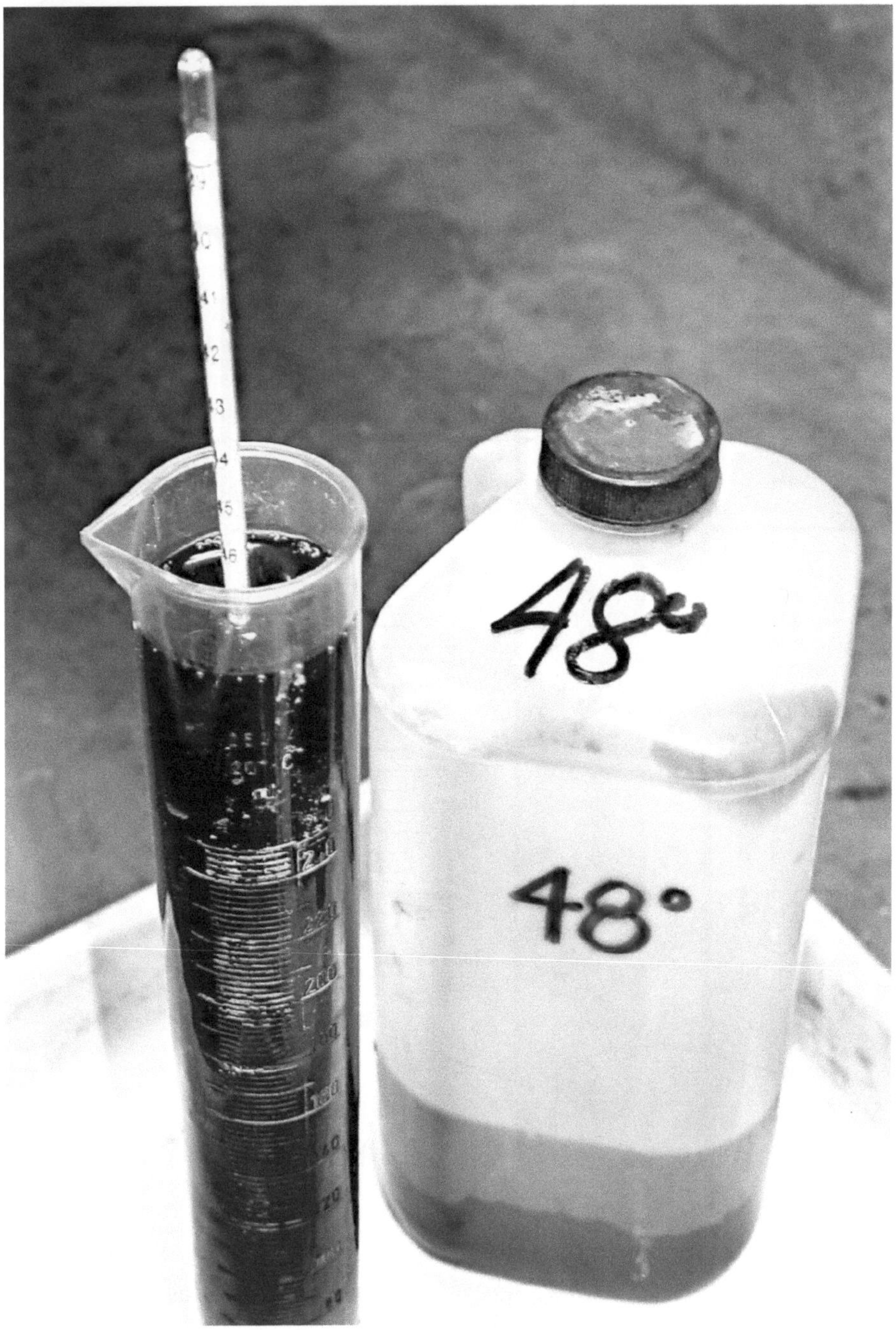

Figure 7-1 The hydrometer and graduated cylinder. Measure the Baumé of the ferric chloride by reading the bottom of the meniscus on the hydrometer.

of distilled water. Start with 75 ml of water to 1500 ml of 48° Bé solution. Mix the solution and pour into the graduated cylinder. Measure the Baumé to see how much it has lowered. Extrapolate from this to determine how much more water is required, if any. Label each bottle clearly.

Repeat until you have at least five bottles of mordant at 48° Bé, 45° Bé, 43° Bé, 41° Bé, and 39° Bé. We have also found that a bottle of 40° Bé and 42° Bé can be useful. Various sources give a usable range of 45° Bé to 30° Bé stating that above 45° Bé will not etch the plate, and below 30° Bé will etch everything at once. Measure the Baumé of all mixes at the

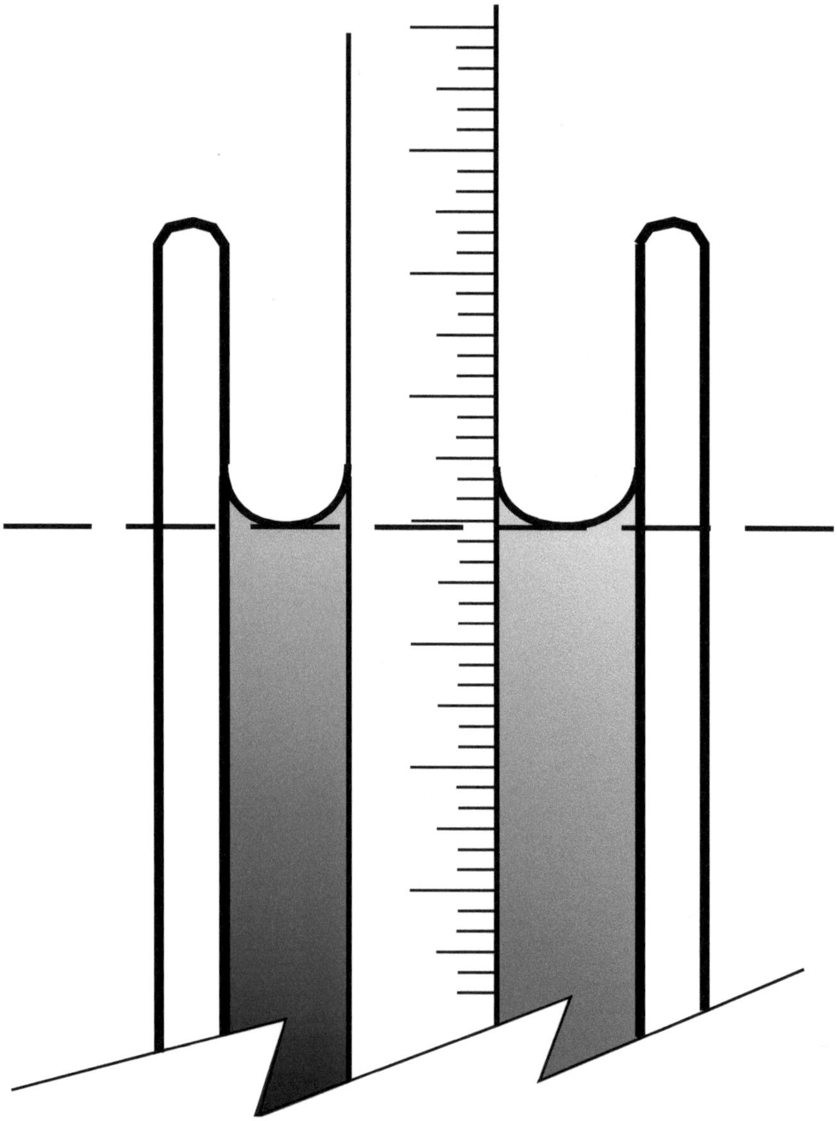

Figure 7-2 Diagram showing how to use the bottom of the meniscus to obtain the correct reading on the Baumé scale.

same temperature. It will be necessary to readjust after a rest period, usually overnight.

Ridding Solutions of "Free Acid"

It is vitally important that the mordant is almost, but not completely, "acid free." It must be free of excess H_3O^+ (aq), a byproduct of dilution with water. The resulting hyperactivity can cause *devils* and overactive etching. One method of removing this acid is through the addition of ferric hydroxide sludge, properly known as Iron(III) Hydroxide. Producing the sludge is a messy and tedious process, and in our experience it is best left to laboratory technicians. See Appendix D for an abbreviated step-by-step description of the procedure and for the chemistry of free acid. You can also take the edge off a fresh bath by dropping thin copper strips into the bottled solution as an alternate to the ferric hydroxide. Even when using top-quality, acid-free ferric chloride it is still necessary to reduce the free acid level of the diluted solution by using either ferric hydroxide or copper strips. Regularly using the solution to etch the copper plates has the same effect as adding strips of copper: it conditions the mordant and controls its hyperactivity. After the baths are well seasoned, further treatment will rarely be required.

Adjusting the Ferric Chloride Working Solutions

After the initial mixing and after normal use, the ferric chloride baths will shift their Baumé readings. As the solutions sit in open trays, some water evaporates and thereby lowers the Baumé. When etching, the movement back and forth from one Baumé bath to another carries with it enough of the other baths to affect each other. Check each Baumé every time you etch.

If the Baumé of a given working solution is lower or higher than required, it should be adjusted. The Baumé reading can be lowered by adding distilled water or, preferably, a quantity of a lower Baumé solution. It can be elevated by adding a quantity of solution already at a higher Baumé. When adding water to ferric chloride, other problems are introduced. The addition of water to the ferric chloride solution encourages the formation of free acid (H_3O^+ (aq)), which in turn has to be mostly neutralized. When the Baumé of a given solution is adjusted by adding a quantity of mordant of greater or lesser Baumé, it should not adversely affect the working solution because no additional foreign material was added to the solution. This is the case only when the added mordant is in good working condition itself.

When making fresh stock or major adjustments to the Baumés, do so 24 hours before use to allow time for any reaction to be completed. Minor adjustments made by inter-mixing various existing Baumés do not normally need a rest period before use. Older, established mordants work in a more consistent, if slightly slower, manner. Therefore, once the working Baumés have been mixed, they can be used for an extended period of time with only minor adjustments. At all times, avoid the introduction of water to the solutions.

When fresh, ferric chloride is a translucent reddish chocolate-cola color. As it is used it becomes darker and more opaque, and when exhausted it has a scummy green cast from an excess of copper(II) chloride (cupric chloride). Monitor the color change and rejuvenate with fresh ferric chloride or replace as needed.

The addition of *anhydrous* ferric chloride powder effectively raises the Baumé without adding any water to the solution. Remember that adding dry ferric chloride creates an exothermic reaction and therefore causes heat. There is a delayed reaction, which requires retesting after a period of rest time. The addition of *hydrous* ferric chloride powder adds water ions to the solution along with the potential of more free acid. It also requires a lot more *hydrous* than anhydrous ferric chloride to raise any appreciable amount of solution one degree of Baumé. We strongly recommend that the 48° Bé is used rather than using dry chemical to raise Baumés. There are serious safety issues associated with the use of dry ferric chloride.

SUMMARY

1. **Start with 8 to 10 liters (2 gallons) of the stock ferric chloride at about 22°C (71.6°F).**
2. **Clean five to seven plastic 1.5- to 2-liter bottles.**
3. **Starting with 48° Bé, fill one bottle 3/4 full and mark "48 Bé".**
4. **Fill a second bottle less than 3/4 full with 48° Bé stock and reduce the Baumé by adding a measured amount of water. Measure the Baumé to see how much it has lowered. Extrapolate from this to determine how much more water is required.**
5. **Repeat Step 4 until you have five bottles of mordant at 48°, 45°, 43°, 41°, and 39° Bé. You might also want to have a 40° Bé and a 42° Bé. Measure the Baumé of all mixes at the same temperature.**
6. **Add small measured quantities of ferric hydroxide sludge to each of the Baumés where a lot of water was added or if there is suspicion of excessive free acid. Adding small fragments of copper will also help to reduce free acid and season the bath.**
7. **After a rest period of 24 hours or more, check the Baumé readings at 22°C (71.6°F) again. Readjust if necessary by inter-mixing.**

STAGING THE PLATE

Before submersion in the ferric chloride, the edges and back of the plate must be protected to prevent etching. Only the image area and the step scale should be left unobstructed so that they come in contact with the ferric chloride. Traditionally, liquid stop-out varnishes or asphaltum were used to stage plates. These materials are messy and require time to dry. We find it easier and cleaner to use adhesive plastics like shelf lining and magic tape.

Equipment and Supplies

You will need 3M Magic tape; self-adhesive plastic shelf lining (Mactac); a burnisher; masking or packing tape; a ruler, triangle, or square; and a fine-tipped Staedtler Lumocolor Waterfast Permanent Black pen. Be sure to use this particular brand and color because some other brands of marker have proven not to be resistant to the ferric chloride. There are also problems with other colors of Lumocolor; even black is not totally resistant.

Procedure

The gelatin resist is *very* moisture sensitive and cannot be touched at all. Even the moisture from your breath can cause severe problems, so avoid breathing on its surface. Use a face mask or shield when working close to its surface. Keep the plate well away from the potential of accidental splashes or gusts of humid air and never touch its surface with your fingers.

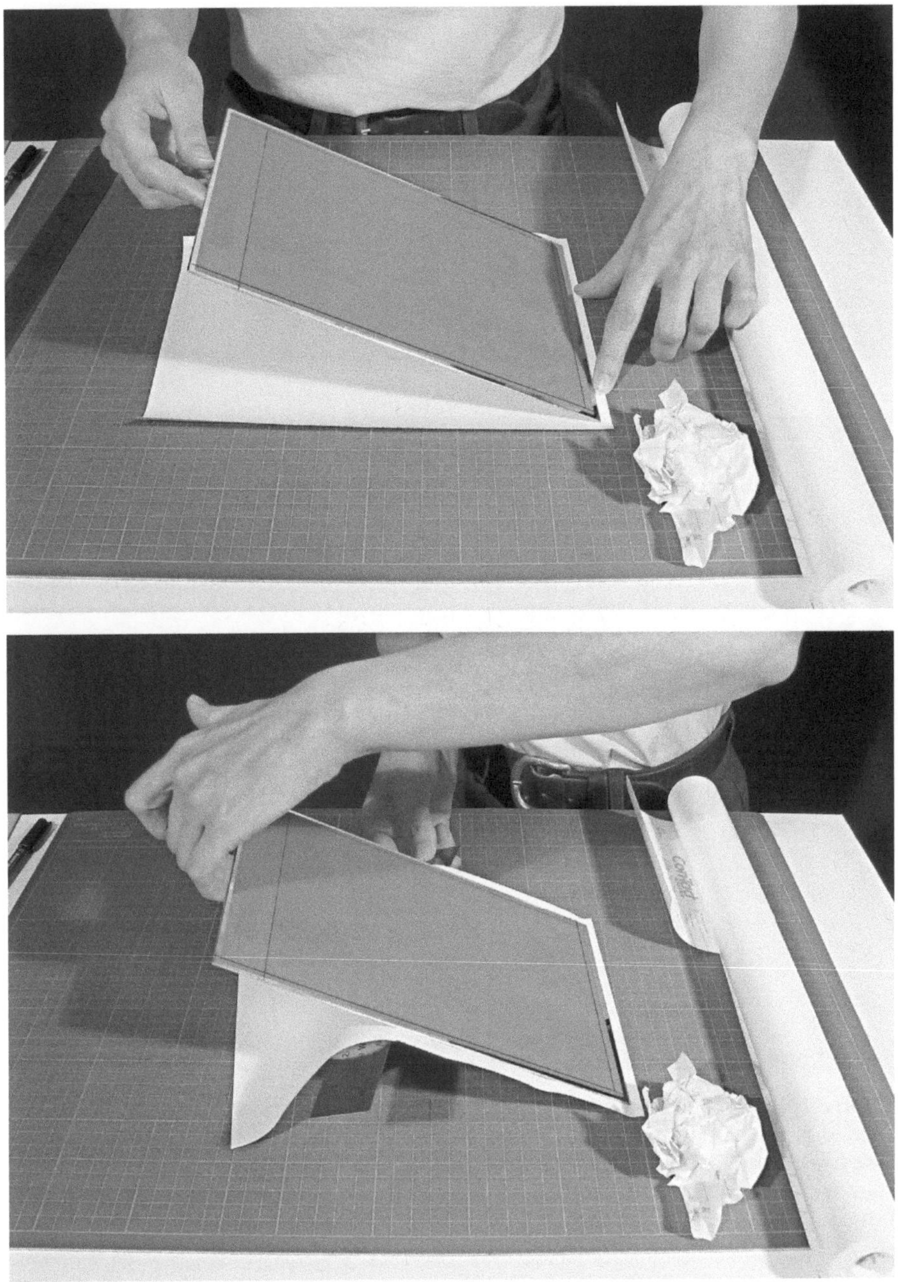

Figure 7-3 Lower one edge of the plate onto the sticky side of a piece of over-sized self-adhesive shelf-linear. Use a ruler to push the plastic up against the back of the plate to prevent the formation of a large air pocket.

The back of the plate can be quickly masked with several layers of packing tape or a sheet of self-adhesive plastic shelf lining. Figure 7-3 shows how to apply a sheet of Mactac to the back of the plate without turning it face down. Remember to be very careful when handling the plate. Do not place it face down in contact with any surface other than clean dry paper.

Next, outline the image and Stouffer Step-Scale with the permanent pen. This provides a visual guide for the strips of magic tape. Cover the plate outside the image and Stouffer Step-Scale with 3M Magic tape.

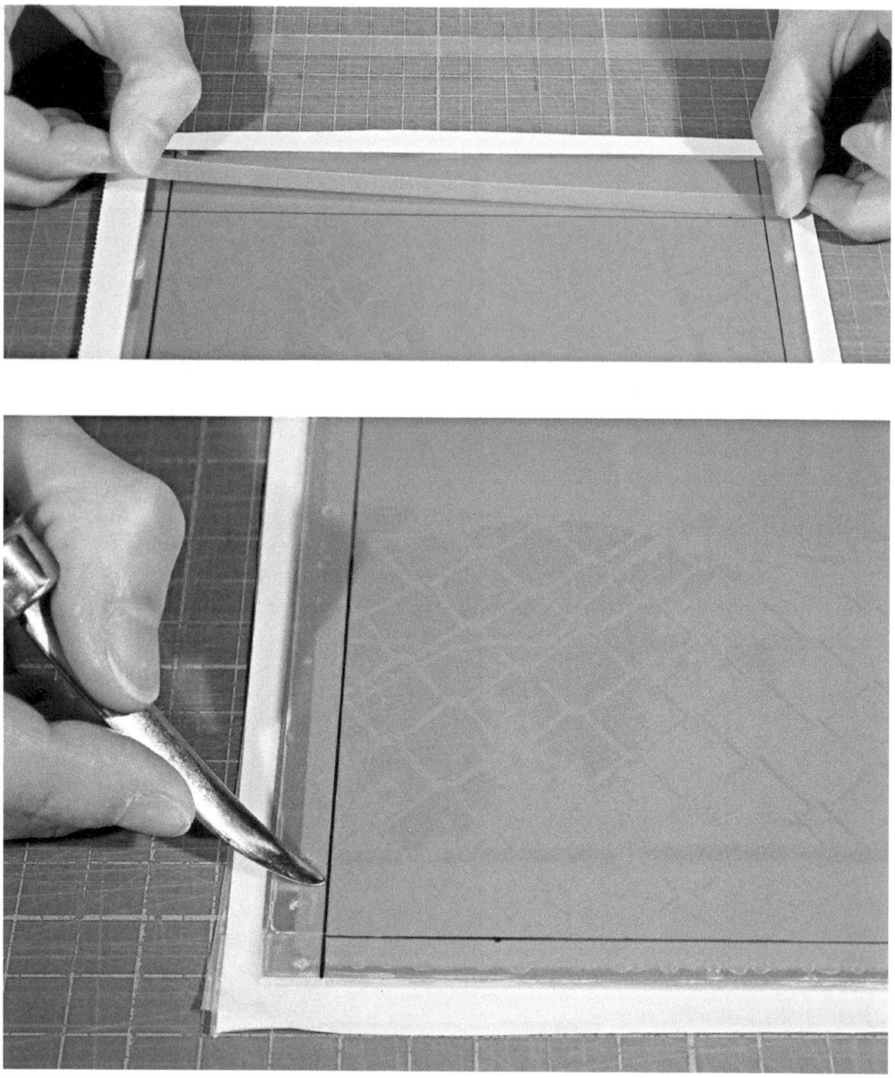

Figure 7-4 Protect the nonimage background with strips of Magic tape placed along the permanent marker pen line drawn along the edge of the image. Use a printmaker's burnisher to seal the edge of the tape. The darkening line makes the outline of the image clearly visible. Be careful not to go beyond the edge and scratch the gelatin.

Be sure to just cover the pen line with the tape. Lightly burnish the tape to ensure a sharp edge and a complete seal (Figure 7-4). Lay more lapped strips of tape down to cover all the borders from the edge of the image to beyond the edge of the plate onto the Mactac backing. When burnishing the very edge of the tape, be careful not to rub the exposed gelatin surface. Use the tip of the burnisher to also seal the little air-space created when one layer of tape crosses over another. This will etch as a line if ferric chloride seeps into it. You can also use a bone folder as a burnisher. It is smooth and light and has a fine tip.

You can use the fine-tipped permanent black Lumocolor pen to spot out small pinholes and flaws. View the plate angled to a bright light to see tiny areas of shiny copper (Figure 7-5). Be careful not to rest your hand on the surface of the gelatin. Use a magnifying visor if that helps. Spot lightly with a tiny dot of ink (Figure 7-6). This will leave a little white

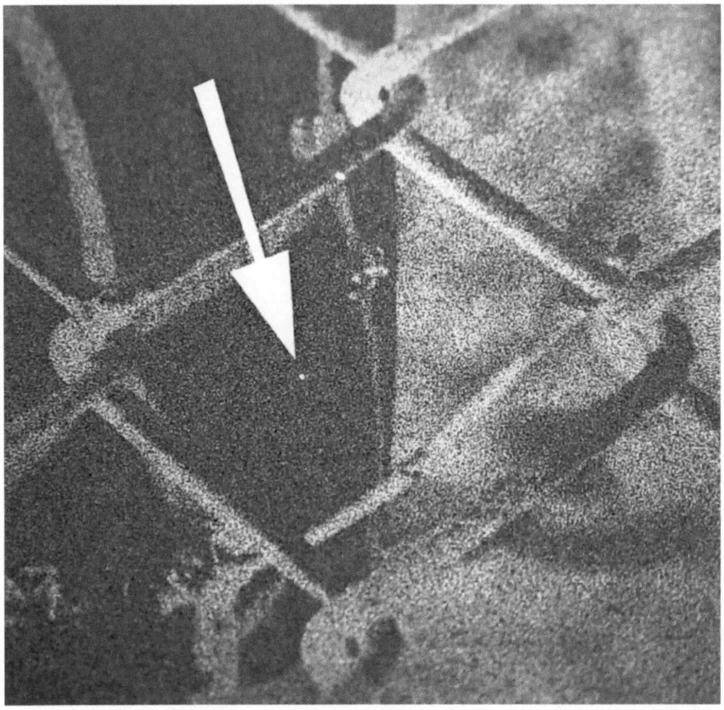

Figure 7-5 A shiny reflection is a clear indication of a pinhole in the resist. This will cause a deep devil if not blocked before the etching sequence.

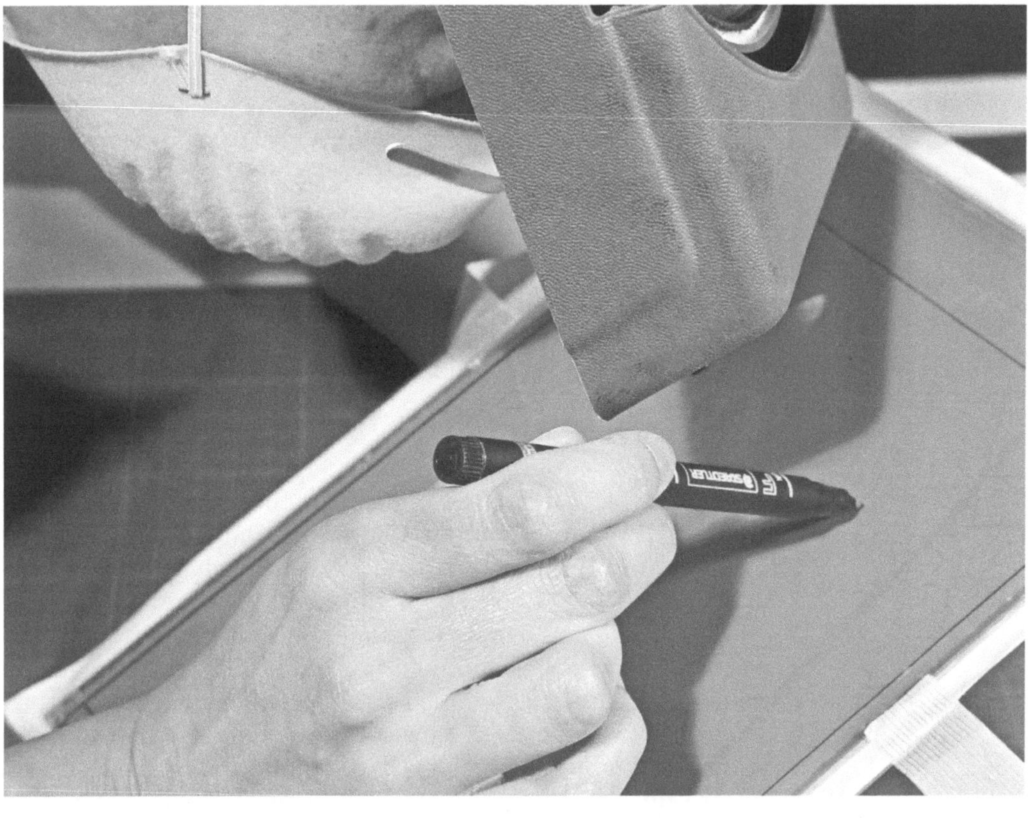

Figure 7-6 A tiny dot of ink from the permanent marker will resist the etchant long enough to prevent a deep pit from being etched. Note the use of a face mask and magnifying glasses, and remember that the hand is *never* in contact with the gelatin resist.

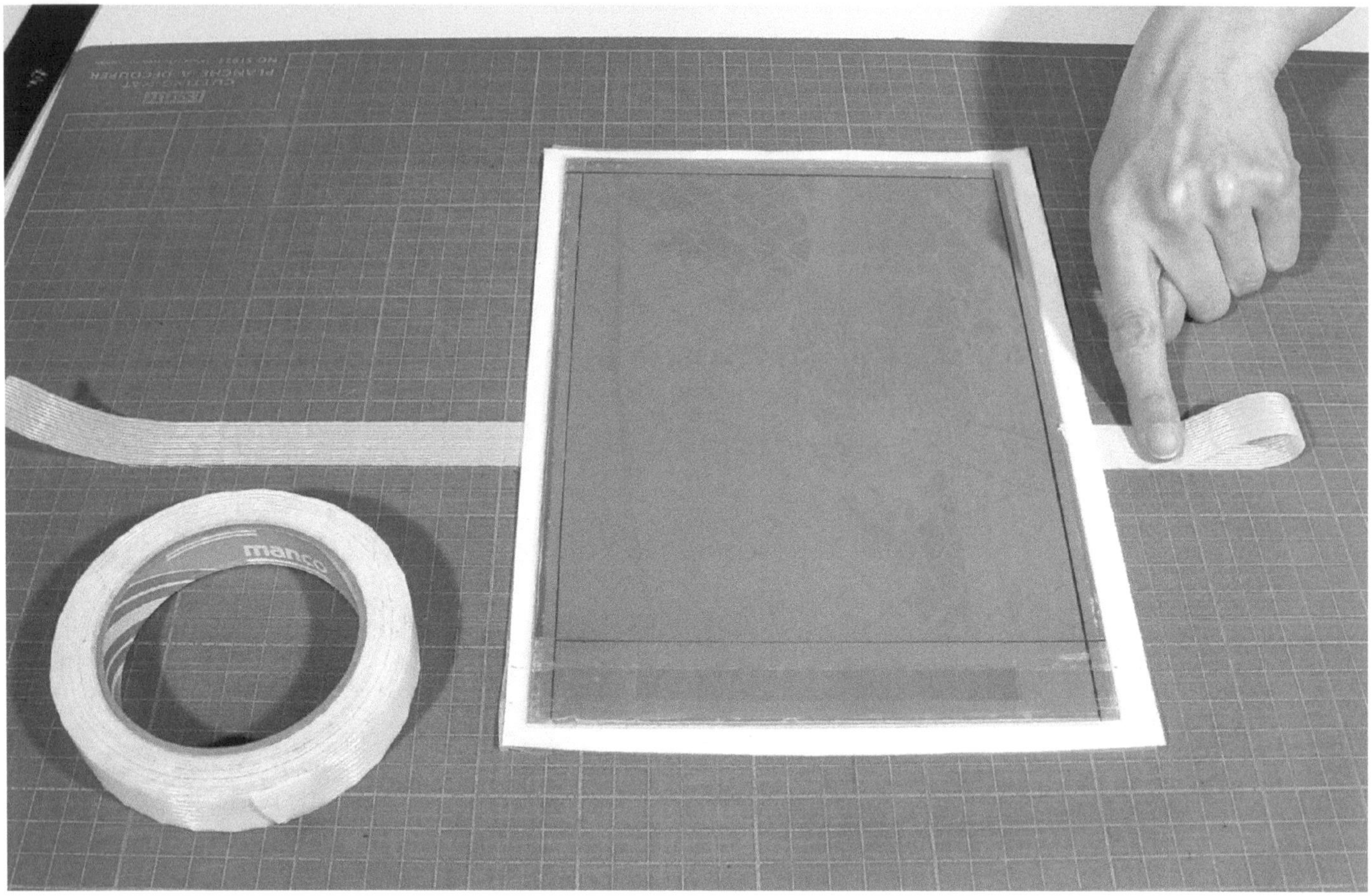

Figure 7-7 Apply a tape handle using strong fiberglass packing tape. Fold it over onto itself and onto the edge of the taped plate to provide a pair of firmly attached, nonsticky handles.

dot when etched and can be easily retouched on the plate. It is much easier to retouch an unetched area of copper than it is to try to fix a deeply etched pit.

Now attach a tape sling to the back of the plate. Cut a 2 cm (3/4″)-wide piece of fiberglass-reinforced packing tape or masking tape about twice the length of the width of the plate or 60 cm (24″), whichever is greater, and lay it face up on a counter. Center the plate onto the tape and drop into place. Fold the tape over itself up to the edge of the image (Figure 7-7). This will leave two tabs on either side of the plate that will make it easy to handle and agitate the plate while in the ferric chloride bath. The location of the tabs should allow you to lift the plate and maintain horizontal balance. These tape handles are necessary because physical contact with the wet resist causes serious damage. If you use masking tape, be sure to apply several layers so that the tabs do not tear off in mid-etch. Use fiberglass packing tape for heavier plates. After the plate is prepared, store it where the relative humidity is stable and there is no danger of dirt or moisture getting onto the gelatin surface.

SUMMARY

1. Cut a piece of Mactac about 2.5 cm (1″) larger than the plate. Peel off the backing and spread out face up on the countertop. Carefully center and drop the plate onto it (face up of course). Alternatively, use a ruler under the Mactac to bring it up in contact with the back of the plate in a smooth progression from one end to the other (Figure 7-3).
2. Define the image and step scale edge with a fine-tipped, permanent black Lumocolor pen. Be careful not to mark the image surface beyond this line.
3. Use long strips of 3M Magic tape to cover the area outside the image. Cover the pen line with the tape. Surround the step scale. Extend the tape over the edge and overlap the Mactac backing by about 1.5 cm (1/2″).
4. Burnish the edge of the tape with a burnisher or a bone folder. Pay special attention to the intersection where tape overlaps tape.
5. Spot out pinholes and tiny flaws with the permanent marker.
6. Attach tape handles across the back.
7. Put the plate aside, far away from any risk of being splashed while you prepare the etch.

8 Etching the Plate

The varying densities of ferric chloride solutions are used to etch the copper plate. There are three reasons to use ferric chloride as the mordant. First, it can be mixed with water to various dilutions for controlled penetration rates through the gelatin resist. Second, it etches the metal beneath without producing gasses or other reactions that would disrupt the gelatin layer. Third, it produces a straight etch, with little lateral etching. The etching process must be controllable and precise so that the depth of the etched wells is inversely proportional to the various thicknesses of the gelatin resist. This results in a tonal equivalent to the film positive. The most common way of etching flat plate photogravures is the multiple-bath method. The plate is advanced through a controlled progression of ferric chloride baths, from less dilute (high Baumé) to more dilute (low Baumé) solutions.

The gelatin resist is still capable of absorbing water, in spite of the fact that its melting point has been raised due to exposure. The rate of absorption is used to control the ferric chloride's penetration and etch. When the copper plate, with its dry hardened gelatin resist, is immersed in a solution of ferric chloride, the water in the mordant causes the gelatin to swell as it is absorbed. There is little water in a high Baumé solution. The diffusion of ferric chloride is slow at this stage. The lower Baumés contain a greater proportion of water so the gelatin absorbs the ferric chloride more quickly—a higher rate of diffusion.

THE PROCESS

The ferric chloride solution migrates slowly through the gelatin resist, starting with the thinnest areas. Moving the plate into increasingly dilute solutions maintains a steady migration rate through the gelatin. This initiates the etch through progressively thicker areas of the resist: first the blacks and shadows, then the midtones, and finally the highlights. Some control of the contrast and detail separation within the plate's tonal scale is possible by adjusting the speed at which the etch progresses. Unlike other forms of intaglio etching, once the plate has been immersed, you must carry the process through to completion.

The ferric chloride solution works by first penetrating into the gelatin resist layer—which is visible as a slight darkening of the resist—and then migrating, or more properly diffusing, to the copper surface. During a second stage, a dark precipitate is produced when the actual etching is taking place and the darkening of the resist in this area is suddenly much more noticeable. Gentle side to side draining agitation is necessary so that the dark precipitate does not block the areas being etched. These two stages are visually distinct, especially in the thinner areas of the resist. When etching the denser highlight areas of the gelatin, the switch from absorption to etch is much more difficult to gauge. Please see Appendix D for a simplified explanation of the highly complex chemical reactions that take place when etching copper with ferric chloride solutions.

To maintain the progress of the etch and separation between the steps of the gray scale, the plate is moved through more dilute baths of ferric chloride (i.e., at lower Baumés). Some sources describe a single-bath etching procedure whereby one tray of ferric chloride solution is progressively diluted throughout the etching process. We find this practice extremely wasteful and hard to control. What do you do with all the low Baumé ferric chloride afterward? Reusable multiple baths are the most cost effective, environmentally friendly, convenient, and controllable means by which to etch.

EQUIPMENT AND SUPPLIES

First and foremost you will need access to a facility with a large, acid-proof, nonmetallic sink (no copper drain pipes or stainless steel drain basket), ventilation, good overhead light, and water. Various darkroom supplies are needed but must be permanently reserved for etching if they are plastic. These supplies include plastic trays and a dedicated funnel, a glass thermometer, a Baumé hydrometer (39° Bé to 51° Bé), a 250 ml graduated cylinder, and a stopwatch or timer. For protection you should wear long acid-resistant gloves, a rubber apron, and a face shield or at least eye goggles. To know the relative humidity, you need a hygrometer. The most important supply, of course, is the ferric chloride prepared to various Baumés. Other minor supplies include cotton balls, washing soda (sodium carbonate), dilute (1:9) muriatic acid (purchased as hydrochloric acid at 20°Bé/31.45% industrial strength), and Brasso. To keep track of everything, chart the etch on a copy of the etching form from Appendix E. Keep a pencil handy and tape the form where ferric chloride drips are not a problem.

PROCEDURE

How to Use the 21-Step Scale During the Etch

The Stouffer 21-Step Scale No. T2115 was exposed into the gelatin tissue at the same time as the positive image and is visible on the resist. It makes the etch easier to control and assess. Find all the comparable densities between the step scale and the positive. Note which steps equate to the highlight detail, the shadow detail, and other important and obvious stages. Write this information on a copy of the etching form shown in Appendix E so you can refer to it during the etch. See "Charting the

Etch," later in the chapter. As the etch progresses from step to step on the scale, you will be able to confirm that the equivalent areas of the image are also etching. The numbered steps make it easier to identify each step and ensure that the etch progresses at a constant and appropriate rate.

Etching Sequence

In describing the following etching sequence, we refer to the resist thickness and characteristics that result from our particular light exposure system, screens, and positives. The characteristics of resists may vary according to the procedures used. We generally work with diffuse light and a screen exposure, so we end up with a more dense resist than one created by a point source or collimated light. In addition, screen exposures contribute to a denser gelatin resist than does the use of an aquatint layer. Therefore, there is no absolute for etching. The following must be taken as a guide to help illustrate the principles of etching rather than as rigid rules or formulae.

The manufacturer recommends that the relative humidity of the gelatin resist should be as close to 60% as possible. This may take some time to readjust if the room is cool or dry. Ideally, store the plate with its gelatin resist in a room with relative humidity of 60% or in a damp-box with an internal relative humidity at 60%. However, we often work in a much lower relative humidity and have not had problems.

The first step in etching is to stabilize the temperature of the various Baumé baths at somewhere between 21 and 24°C (70–75°F) (Figure 8-1). Cold mordants are slow to start their migration through the gelatin, behaving as though they are a higher Baumé. Conversely, warmer solutions

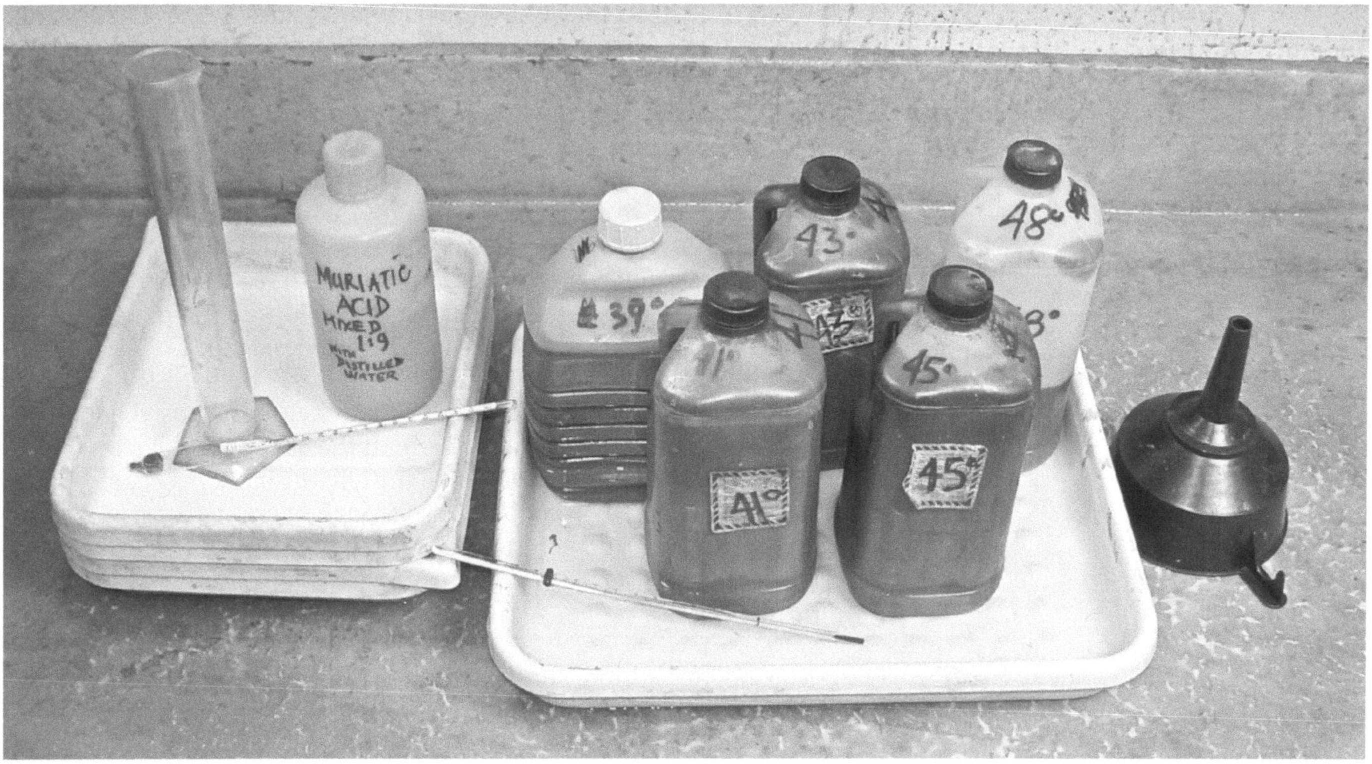

Figure 8-1 Solutions of ferric chloride warming in a tray of water. Ideally the room should be at the same temperature.

penetrate the gelatin faster. Consistent working conditions are very important for repeatable results. If the room and solutions are cool, place the bottles of ferric chloride in trays of warm water to bring them up to working temperature. It is far easier to maintain solution temperatures if the room is also at the same temperature. Once the temperature is stabilized, the Baumés can be checked and adjusted. Have ready solutions of 48°, 45°, 43°, 41°, and 39° Bé (42° Bé and 40° Bé are optional). The etching sequence can now begin. Monitor and maintain the solution temperature throughout the etch.

Explanation of the Role of Temperature and Dilution

Both the temperature and the degree of dilution (Baumé) of the ferric chloride solutions play crucial roles in the way that the ferric chloride migrates through the gelatin. These factors determine when an etch will start and how quickly it will proceed. Somewhat surprisingly, however, if you etch a copper plate that does not have a gelatin resist, moderate changes in temperature and dilution do not result in the same visible differences in the resulting tone. The etch starts immediately and the effects of temperature and strength, while still at work, are less obvious in the printed image.

Starting the Etch

Set out two to four trays. Fill one of trays with the 48° Bé solution and another with the 45° Bé solution. Start with a 48° Bé ferric chloride bath as the conditioning step, even though the gelatin cannot absorb a solution above about 46° Bé. Note: The 48° Bé frequently drops to 47° Bé. This is fine. You only need to get new stock when it goes below 47° Bé. Start the timer and immerse the plate, gelatin resist side up. Soak a dry cotton ball with 48° Bé and gently drag it over the entire surface to dislodge clinging air bubbles. Make one pass in each direction over the plate and then discard the cotton ball. Try to find long-fibered cotton balls. Short-fibered cotton balls tend to fall apart and add debris to the solution. Throughout the entire etch, be very careful that you do not touch the gelatin surface with your gloves and at all costs avoid any water coming in contact with the gelatin. It is also important to agitate the tray by rocking it or by using the tape handles to lift alternate sides of the plate every 15 seconds or so throughout the entire etching process. (See Color Plate 15.) After two minutes in the 48° Bé, move to the 45° Bé bath (Figure 8-2). If the etch begins very quickly, the resist may be underexposed and therefore too thin, or the mordant is too warm. If the etch does not start within 15 minutes, move to a lower Baumé to initiate the etch. A very slow start is indicative of a dense resist, perhaps through overexposure, or it could mean the mordant is too cold or exhausted.

Length of Etch

Once you begin etching, the progression of the etch in areas of black should be advanced slowly so that areas darker than the shadow detail (Steps #12–#13) maintain subtle separation and do not become featureless black. As soon as the shadow detail with a density of 1.75 to 1.9 (near Steps #12–#13) starts to etch, the rest of the etch should take *no more than*

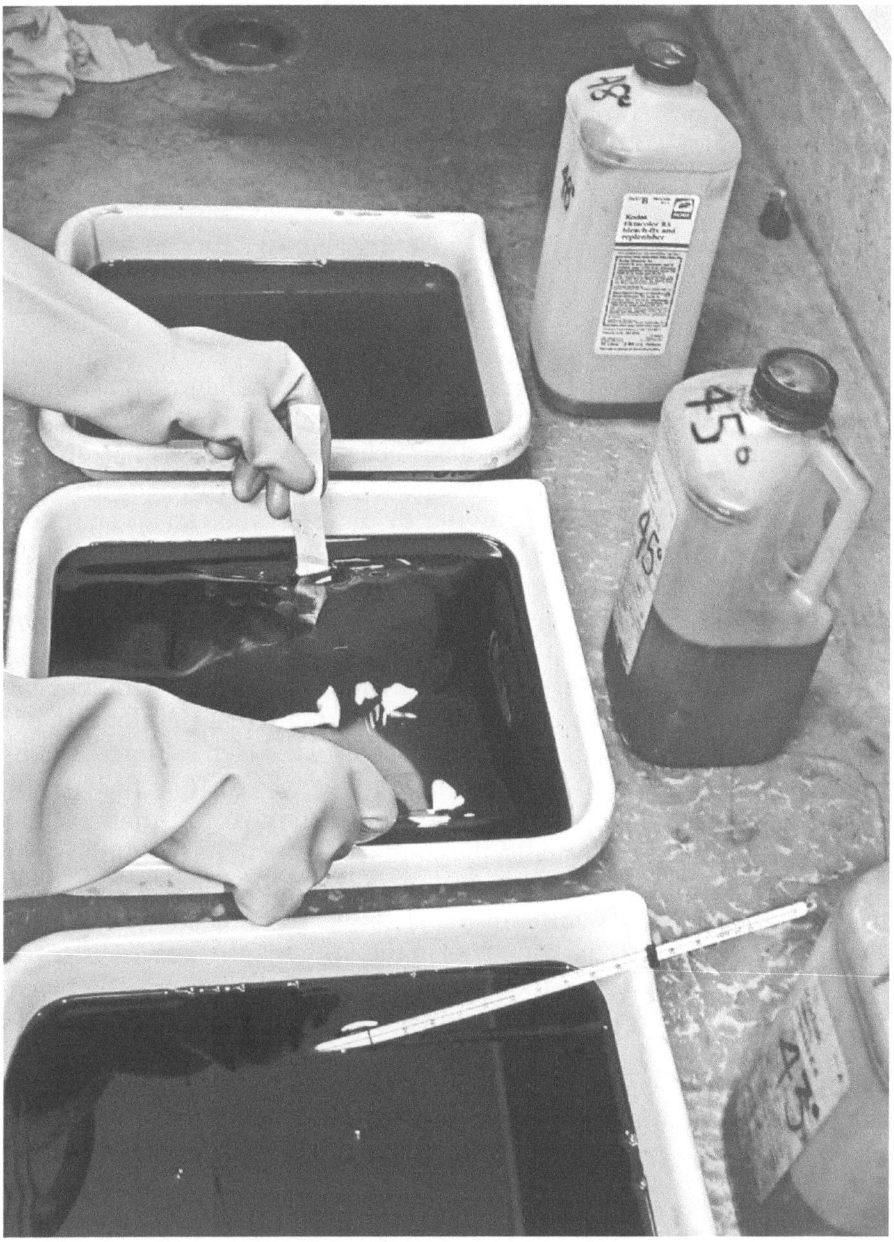

Figure 8-2 Move the plate between trays to shift from one Baumé to the next (or back). Note the glass thermometer.

20 to 25 minutes. A longer etch will result in blocked shadow detail; a shorter etch will result in thin shadow detail or pale blacks. Many sources suggest 20 minutes to etch the image—not including the time to start the etch and establish the black areas denser than 1.80 (Cartwright 1939, p. 112). Sacilotto's suggested times are longer but he, too, suggests 20 minutes when using a very fine screen (Sacilloto 1982, pp. 130, 137). It is also important to know how long the resist can withstand the etch before foul-biting in the absolute black areas. We have found that a total etching time of 30 to 50 minutes from the first indication of gelatin penetration is safe and produces a very rich black. More than 50 minutes of actual etching is very risky and can result in resist failure, foul-biting, or *crevé* when the screen is fine. When referring to older texts on gravure keep in mind that many are directed toward rotogravure, not flat plate gravure.

Bennett (1935, pp. 124–125) clarifies that the rotogravure ink cell is shallower by nature due to the doctor blade's shearing action, whereas flat plate gravure requires a deeper cell due to the aggressive nature of hand wiping, hence the longer etching times. In terms of the actual etching depth, a wide range is given depending on whether it is commercial rotogravure or a very fine dust grain. Regardless, the etch depth is so shallow that it is measured in microns.

Rate of Etch

Control of the rate of etch is achieved by moving the plate back and forth among a sequence of ferric chloride baths of different Baumés. Moving to a lower Baumé advances the progress of the etch and moving back to a higher Baumé arrests the progress of the etch. This does not happen instantaneously but is responsive enough to permit control of the etch. Once you have the shadow detail etching in the 45° Bé bath, you can speed up the etch by moving to the 43° Bé or slow it down by moving back to the 48° Bé. The 43° Bé bath will eventually be required to maintain the progress of the etch as it moves along the step scale. A 41° Bé or 42° Bé bath may be required later to reactivate a slow or stalled progression of the etch. In order to finally etch the highlights, we frequently use 39° Bé, though we rarely go any lower. The main goal is to make sure that the steps do not begin to etch in blocks or groups. Try to control the etch so that *the steps begin to etch one at a time in 2- to 2.5-minute intervals* (Figure 8-3).

The density difference between each of the steps in a 21-Step Scale is an arithmetic function. Visually, the contrast between the light-tone steps appears greater than that of the shadow-tone steps. It may be necessary to exaggerate the value shifts in the shadows in order to maintain the same separation in the dark tones as in the mid or light tones. Thus, you could theoretically control the etch so that there is 2.5 minutes between steps in the shadow detail, 2 minutes in the mid-tone range, and 1.5 minutes between the light tones. Much easier said than done!

Be aware that there is a time delay between the change in Baumé and the rate of etch. It is hard to immediately slow down or speed up the progress when a problem appears. Try to anticipate a needed change and adjust accordingly. If the etch has slowed down, try a brief (30–60 seconds) immersion in a lower Baumé, then go back to the higher one to await the effect. If the effect is still slow, then move into the lower Baumé for a longer period. If you need to slow down a speedy etch progression, you can go back into a higher Baumé for 30 to 60 seconds. Going back two Baumés—for example from 43° Bé to 48° Bé—will effectively stop the progression and is useful in cases of sudden overaccelerated etch. An alternative to going back and forth is to work with more Baumés in smaller increments. Thus, if your etch has slowed in the 43° Bé you can move to the 42° Bé and remain there.

Our approach to etching is to start the etch slowly so that we can maintain separation even in the darkest details, because we use the dark end of the scale a lot in our imagery. The amount of time this takes varies with each plate. Once the shadow details begin to etch, however, we always attempt to complete the remainder of the etch in 20 to 25 minutes. On our Stouffer Step Scale, densities of 1.82 to 0.45 would be approximately equivalent to Steps #12–#13 to #3–#4. In order to make these 10 steps etch within the remaining 20 to 25 minutes, each step must appear at

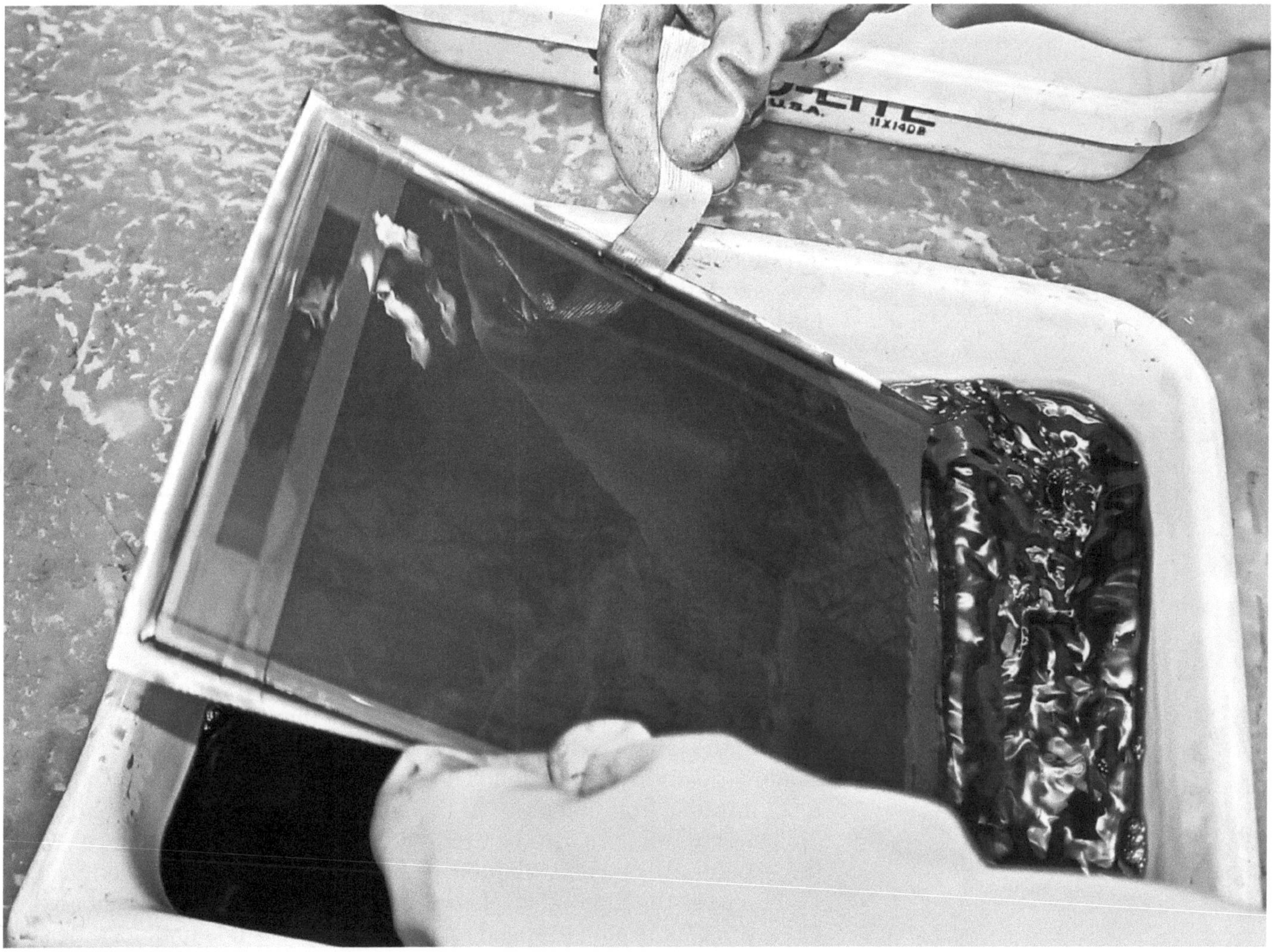

Figure 8-3 Lift the plate from the tray to check the progress of the etch on the step scale and the image.

2- to 2.5-minute intervals so that #3 (brightest detail) is etching when the time is up. (See Color Plate 16.)

Ending the Etch

At the end of the etching time, the highlight detail should have fully etched. This is the point where one must decide when to actually stop the etch. Do you leave it a few seconds longer to establish a detailed—and possibly grayed—highlight, or do you stop it quickly to maintain some unetched copper for spectral highlights? (See Color Plate 17.) The spectral highlights should not etch. Be careful, because this can happen very quickly. Underetching can also result in the loss of important highlight detail. This final judgment is based on the positive, your experience, and your intuition. We have found that to fully establish highlight detail it is necessary to etch beyond the highlight density. For example, if the highlight density reading was 0.45, etch until Step #3 (0.34) is just established (30 seconds). Denison suggests to etch for about 30 seconds "after the whole detail has been obliterated" (1895/1974, p. 83).

If you find that Step #2 etches before the highlights appear to etch on the plate, make the highlights denser in subsequent positives. The densities of the shadow detail, the highlight detail, and the contrast range of the positives should be relatively consistent from one image to the next so that you can learn to use the etch for adjustments. If the positives vary too much, little can be learned as you scramble to use the etch to keep things in control. It is easier to control contrast and image characteristics when making the positive than in the etching stage. However, control of the image is attainable to some degree with the etching procedure. It takes practice to be able to intuitively adjust on the fly as the etch progresses, but this must ultimately be learned. Remember that etching times will vary according to humidity, solution temperature, tissue and solution freshness, exposure densities, and the use of screen exposure vs. aquatint. Keep as many of these variables as consistent as possible in order to simplify the process.

Etching Controls

In an ideal positive, the shadow detail and highlight detail are 8 to 10 steps apart on the 21-Step Scale. When this is impossible to achieve you can try to use etching controls to compensate. In the case where the shadow detail and highlight detail are fewer than 8 steps apart, you can increase the length of time between the appearance of each step after the shadow detail has started to etch. In the reverse situation, when the shadow detail and highlight detail are more than 10 steps apart, you can shorten the length of time between the appearance of each step after the shadow detail begins to etch. The aim is to maintain an etch time of 20 to 25 minutes from the onset of the shadow detail etch to the end.

These controls can also be used within a particular tonal range. Thus, the separation can be extended in the shadow details but reduced in mid tones or the highlights. Having said that, in practice it is not always as easy to get the etch to speed up or slow down as it is in theory. Be sure to have a complete range of Baumés ready before you etch and work with more than one tray at a time, ideally three to four. Don't forget that there will be a delay in the response to a change in Baumé, so anticipate the changes.

Range of Useful Baumés

Our preference has been to work with a series of Baumés that aid in the separation of the shadow details while also giving us the ability to etch the mid tones and highlights within a reasonable time frame. Gelatin will hardly absorb the mordant if its Baumé is above 46° Bé. On the other end of the spectrum, a Baumé in the 30° range will quickly penetrate the entire gelatin resist. We've found that a 47° or 48° Bé conditioning bath is useful. The darkest tones do not begin etching until in the 45° Bé bath. Sometimes a 43° Bé is required to initiate the etch. Areas of undifferentiated black may begin etching in the conditioning bath or the 45° Bé without detriment, provided the total etch time does not cause these areas to overetch and *crevé* or foul-bite. If detailed shadows begin to etch too quickly in the 45° Bé bath, the tissue is either too thin (underexposed) or too moist (not dried or stabilized to the proper relative humidity), or the Baumés are too warm. Conversely, if you need to progress to a low Baumé (41° Bé) to initiate the etch, then the resist is too dense due to overexposure or excessive dryness, or because the Baumés are too cold. The following Baumé-to-step-scale range will vary greatly depending on the type of resist and the conditions. This is a rough guide only. Everyone must find the right combination of working conditions and Baumés.

48° Bé to 45° Bé—etches black and the shadow details—Steps #20 to #12

45° Bé to 43° Bé—etches the dark shadow details—Steps #14 to #10

43° Bé to 41° Bé—etches the mid tones—Steps #11 to #7

41° Bé to 39° Bé—etches the light tones—Steps #8 to #3

39° Bé to 37° Bé—etches the plate tone+—Steps #3 to #2 (if required)

Charting the Etch

During the etching of a plate, it is important to know at what time any given step began to etch. Memory is not enough to keep track of the process. Chart the etch as it happens and use the information to learn what may have gone wrong when the printing results are less than perfect. It is helpful to have an assistant record the data when you are learning the process. Set a timer for one hour (if it counts down) or start a stopwatch (if it counts up) and record the time of each event throughout the etch. Record the time when the copper is moved from one bath to another. Record the time when each step on the 21-Step Scale appears to start etching. Start a countdown so that the steps representing image detail—usually Steps #12 to #3—are all etched within a 20 to 25 minute time limit. A sample form that helps to organize this data is provided in Appendix E. This form was based on one used by Bennett (1927/1973, p.73), but has evolved to fit our working methods.

Rinse and Polish

As soon as you determine that the highlight details are sufficiently etched, it is important to end the etching as quickly as possible. Stop all etching immediately by immersing the plate in a tray of water with a teaspoon of sodium carbonate dissolved in it. A cheap source of sodium carbonate is washing soda (Figure 8-4). Alternatively, even a rinse under running water will cause the ferric chloride–saturated gelatin to blister and slough from the surface of the plate (Figure 8-5). (Warm water is more effective than cold at removing the layer of gelatin.) While immersed, peel off the backing and strip off the tape from the front (Figure 8-6). Be careful your gloves don't have ferric chloride on them. Be sure to rinse them in the tray before handling the surface of the plate. Rinse the plate under running water to prevent further etching and to be sure all the ferric chloride is rinsed away. If a dribble of ferric chloride, even dilute, runs across the plate, it can leave a printable mark in light areas. Immediately dry the plate with paper towels and move it to another room to protect it from the splashing of cleanup.

After the plate has been washed clear it will look quite disappointing because the image will lack definition and everything will look oxidized.

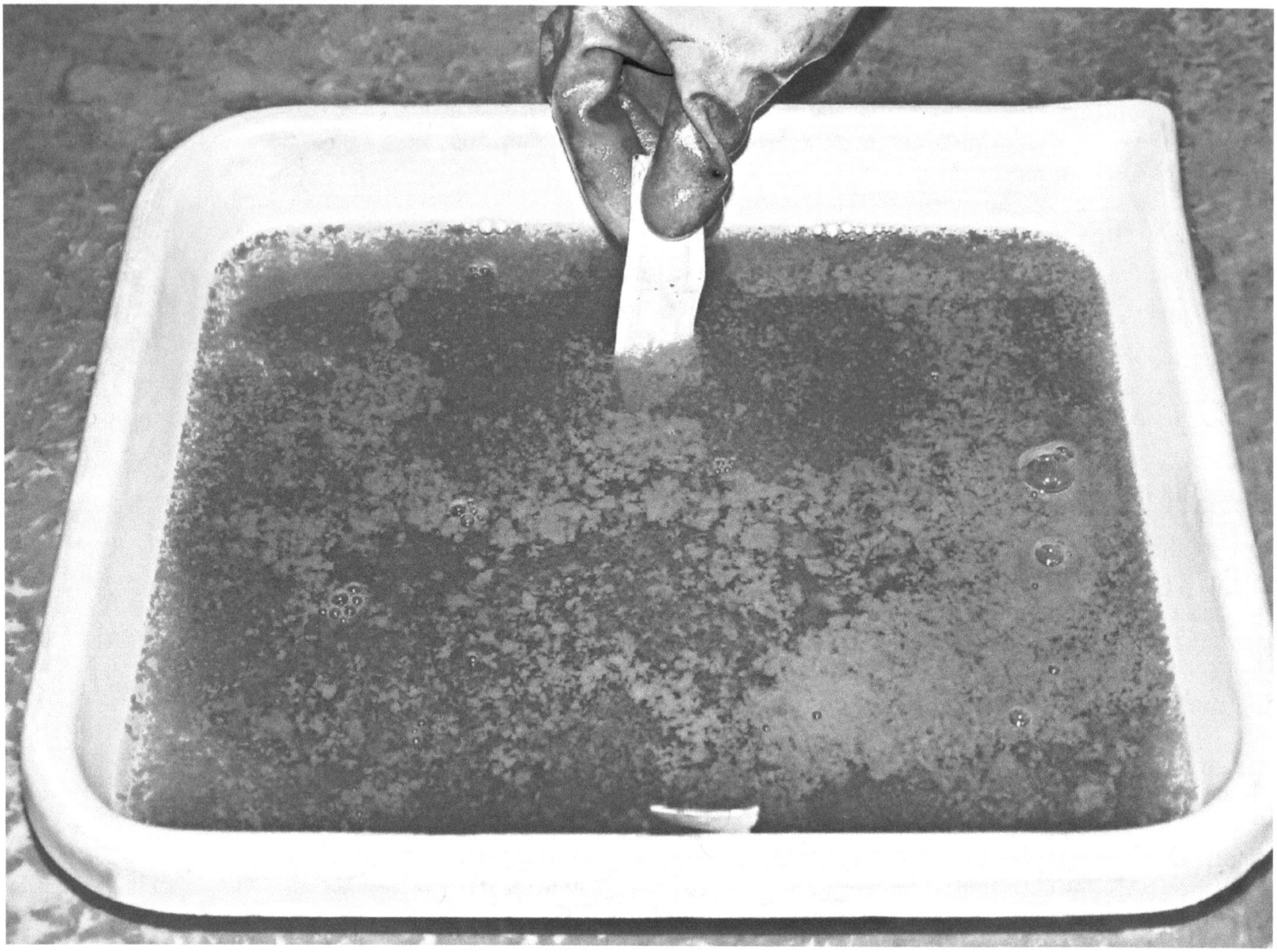

Figure 8-4 Sudden immersion into a tray containing a solution of sodium carbonate (washing soda) causes an immediate chemical reaction that effectively halts the etch.

There will also be some stubborn areas where the gelatin resist will not wash off, even under warm water. This happens in particular in areas protected by tape during the etching process. Dampen a paper towel with a bit of dilute (1:9) muriatic acid and rub the old gelatin resist. The resistant resist will easily rub off. If you quickly immerse the plate into a dilute muriatic acid bath (1:9), the copper will brighten and give a much better indication of how the etch went. (Color Plate 18.) Do not leave it in the muriatic acid bath for more than a few seconds because it can cloud the highlights and bright areas. Rinse the plate well in water to remove all vestiges of the acid. If a dust grain aquatint was applied it should be removed at this point with the appropriate solvents. Finally, a quick and gentle polish with Brasso can make an incredible difference. Be careful not to overpolish because this will lighten etched tones. Use naphtha to remove the black deposit that remains after Brasso. At this point, if the plate has already been beveled, it is ready to proof with ink.

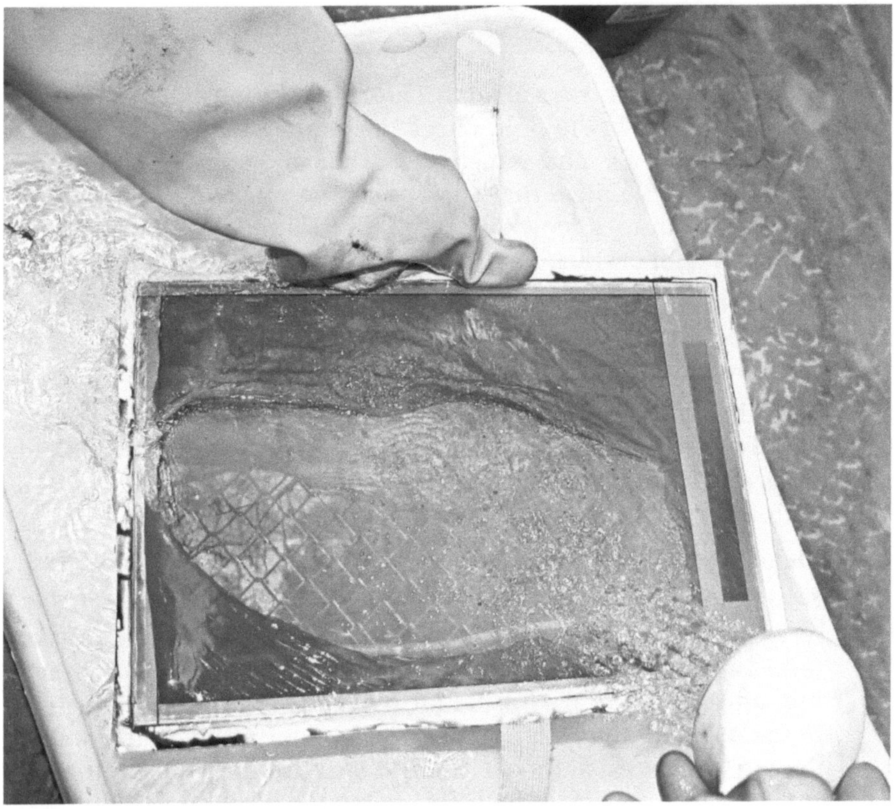

Figure 8-5 Water causes the gelatin to immediately fail and slough off the plate. Work in a tray full of clean running water.

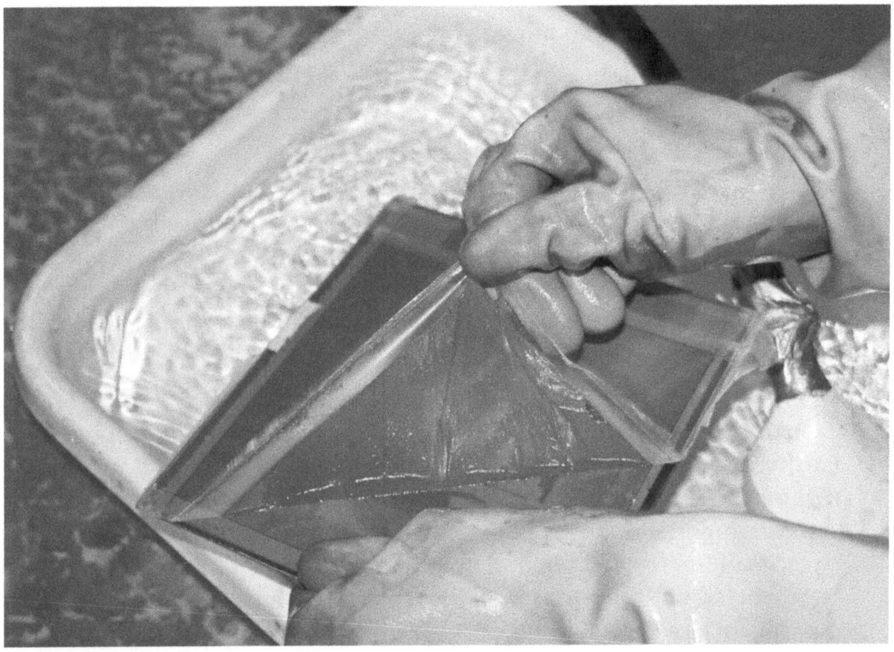

Figure 8-6 Peel off the backing and stripping tape either under the running water (as in Figure 8-5) or in a tray of sodium carbonate solution. Sometimes the gelatin will come off at the same time.

SUMMARY

1. Set up two to four trays of ferric chloride solutions, depending on available space in the sink. Check the Baumés (48° Bé, 45° Bé, 43° Bé, 41° Bé, 39° Bé, and sometimes 42° Bé and 40° Bé) and adjust if necessary. Stabilize the working temperature of the solutions and the room to somewhere between 21 and 24°C (70 to 75°F). Keep the temperature consistent throughout the entire etch.
2. Set up a blank copy of the etching chart (Appendix E) and a pencil in a convenient spot where ferric chloride drips will not be a problem.
3. Start the timer with the 48° Bé bath. Immerse the plate and immediately go over the surface very gently with a cotton ball soaked in ferric chloride to dislodge air bubbles. Leave the plate in this bath for 2 minutes, agitating every 15 seconds.
4. Move the plate to the 45° Bé bath and continue regular agitation.
5. Keep the plate in the 45° Bé until all steps from #21 through #15 have begun to etch slowly. The first signs of etching should not take longer than 15 to 20 minutes. If it appears to be slow starting, move along to the 43° Bé for a couple of minutes, returning to the 45° Bé if the etch begins rather quickly. Depending on the density of the resist, the 48° Bé may serve only as the conditioning bath. Use 45° Bé and 43° Bé to etch the blacks and shadows.
6. As the etching slows down, begin to use the 43° Bé bath. Again, move back and forth between the 45° Bé and the 43° Bé if necessary to control the speed of the appearance of the next step (#12, then #11, etc.). Once again, you may need the 41° Bé at this stage to restart a stalled etch. The alternate procedure is to use a 42° Bé next, but don't go back. Move on to the 41° Bé as needed.
7. Use successively lower Baumés, down to 39° Bé, to start the etching in the denser areas of resist. Go back and forth between Baumés to initiate and then hold the etching.
8. The aim is to have an even progression of etching from step to step on the Stouffer Scale, ideally 2 to 2.5 minutes between steps. Once Step #12 has begun to etch (an average density for shadow detail), the remainder of the etch should be completed in approximately 20 to 25 more minutes. The entire etch should take approximately 30 to 50 minutes. Longer total times will run the risk of open biting the darkest areas, especially in fresher mordant.
9. Once the highlight detail (usually Step #4, #3, or sometimes #2) fully etches, be prepared to remove the plate from the mordant quickly. Immediately rinse well under running water or, preferably, immerse in a tray of sodium carbonate solution. Remove all traces of ferric chloride and do not let any ferric chloride drip onto the surface of the plate. The gelatin resist should rinse off at this point or can be helped along with a quick rinse in 1:9 muriatic acid (purchased as hydrochloric acid at 20° Bé/31.45% industrial strength) to remove any stubborn resist and to brighten the plate.
10. Lightly polish the plate with Brasso. Remember that firm polishing will lighten some tones, so be gentle at first. Now, you can proof it!

TROUBLESHOOTING

Mordant Problems

Baumé Readings that Change Baumé readings that drop from one session to another may be caused by accidentally adding water from wet utensils or splashes. Evaporation can cause Baumé readings to rise when mordants are left out for long periods of time or are in an area with a strong air flow or high temperature. Check the Baumé levels of each bottle regularly.

Dirty Solutions Over time, the solutions may become dirty with sediments, gelatin particles, dust, and hair. These particles will interfere with a clear view of the etch as it progresses through the steps of the Stouffer Step Scale. We use a vacuum system (with a heavy beaker and ceramic filter funnel using coffee filters) to this debris from filter the solutions. The ferric chloride solutions are simply too thick to be filtered by gravity alone. It may be possible to run them through a few layers of cheesecloth or a cotton ball in a funnel if you do not have access to a vacuum filter setup.

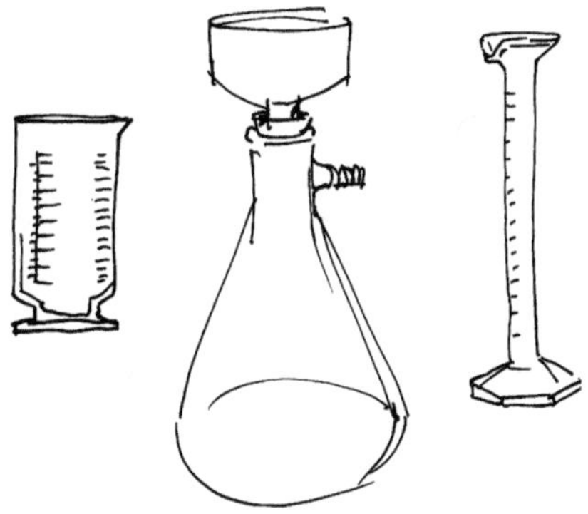

Exhausted Solutions Over an extended period of use, the solutions may change to a scummy greenish brown, losing the transparent chocolate/cola color they had when fresh. If they are overused, they will be saturated with iron(II) hydroxide and copper(II) chloride and should be changed or rejuvenated. This can be done by adding fresh stock ferric chloride, or by adding salt or (preferably) hydrochloric acid, both of which add the important chloride ion. When the mordant is weak or exhausted, a chalky precipitate may appear on the surface of the resist. This is copper(I) chloride or cuprous chloride ("white etching," Cartwright 1939, p. 120). The best cure is to refresh the bath. If minor, these precipitates will have little or no effect on the image.

Resist Problems

Excess Free Acid Too much free acid in the ferric chloride can cause rapid and uneven etching, foul-biting, devils, and failure of the resist. If the gelatin resist is well exposed, properly developed, and dried and yet the shadow areas begin to etch almost immediately upon immersion in the etch, the problem may be due to free acid. Free acid also increases the possibility of resist failure by foul-biting in the shadow areas, which results in a loss of copper surface. The only sure way to immediately reduce the free acid is to add a small amount of ferric hydroxide sludge to the bath. (See Appendix D.) When the etching solutions are well used, or if copper strips are added, the edge of an overactive mordant will be diminished.

Lateral Etching If blisters or raised loose areas of swollen resist form during the etch, it is an indication of a severe foul-bite occurring under the resist. The copper lands have been cut off by lateral etching and the resist is no longer attached to the plate (Figure 8-7). You can be sure the image is ruined at this point. Check for a too active mordant or one that is high in free acid. Alternatively, flaws may have occurred when the

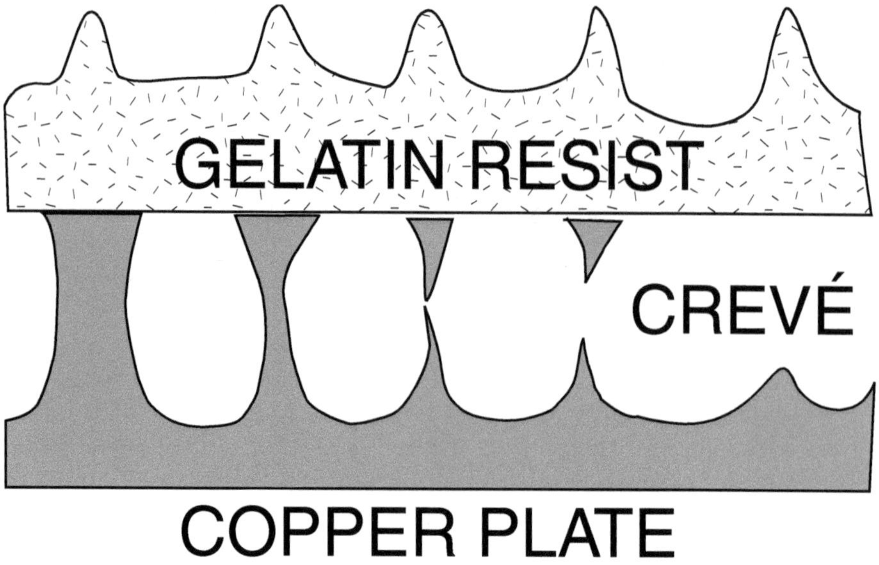

Figure 8-7 Lateral etching causes the tops of high points to be cut off short and prevents ink-holding wells from being properly formed. This open bit area will print lighter in tone than it should.

resist was adhered to the plate. A greasy or tarnished spot on the plate could result in premature resist failure. If an aquatint was applied, it may have been undercut or let go, again possibly due to insufficient fusing, degreasing, or high free acid content in the mordant (Figure 8-8).

Pinholes Small pinholes can result in deeply etched pits that can spread into devils. Spot them out with a fine-tipped permanent marker on the dry resist prior to etching. It is easier to correct the resulting white dot than to deal with a black pit or devil. See the following section, "Plate Flaws," for more on devils.

Step-Scale and Image Discrepancies If a small area on the positive with a specific density reading and the corresponding step on the 21-Step Scale do not etch at the same time, there are two likely reasons. First, the area on the positive are made up of various densities and the densitometer has given an average reading. Second, the color of the step scale and the positive do not match and this has affected the density of the gelatin resist during exposure.

Technique Problems

Overly Quick Etch When Stouffer Scale steps begin to etch too quickly and in groups there will be no definition between different tones. This means you moved the plate to a low Baumé too soon, or left it in a low Baumé too long, and speeded up the penetration so much that it jumped steps. If the tone steps appear too slowly, with as much as 4 minutes between them, then you should move the plate to a lower Baumé to

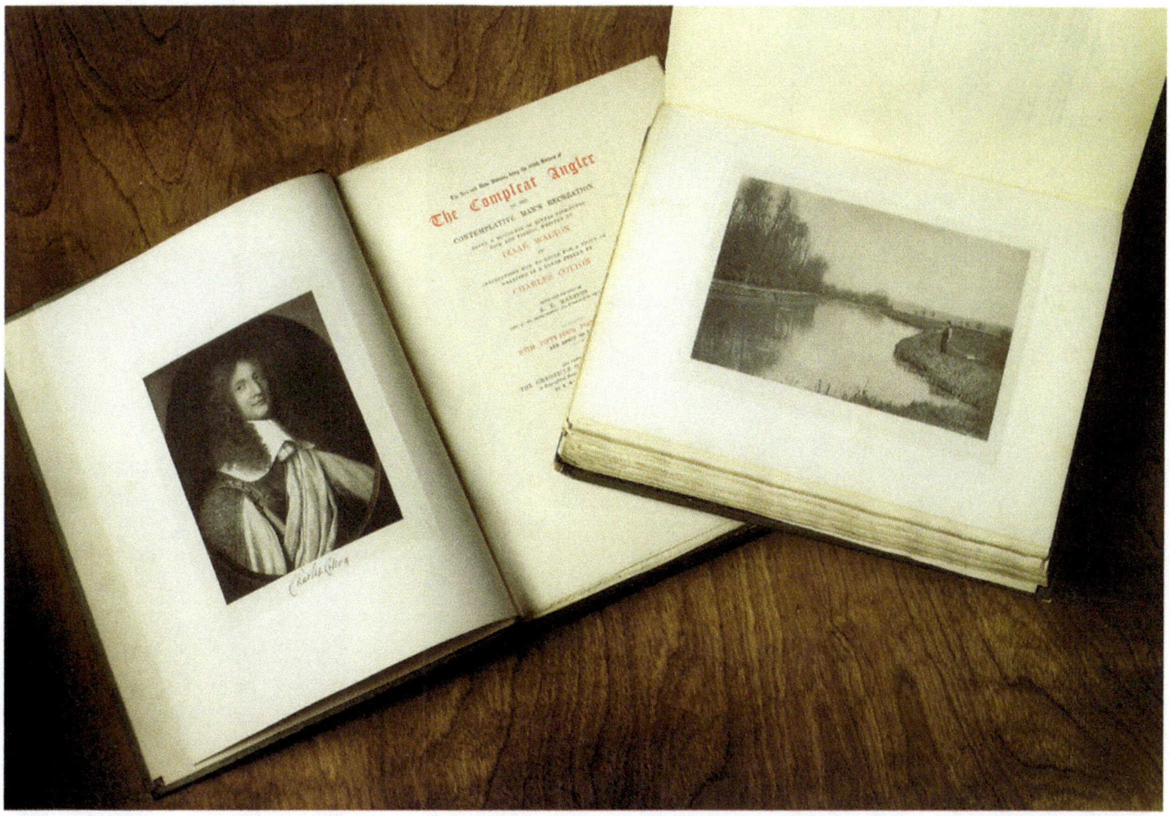

Plate 1 The 100th edition of Izaak Walton's *The Compleat Angler*, Vols. I and II, 1888. Photogravure plate (*right*) by Peter Henry Emerson. See also Figure 1-3. (Photo by Mark Katzman, Ferguson and Katzman. From private collection.)

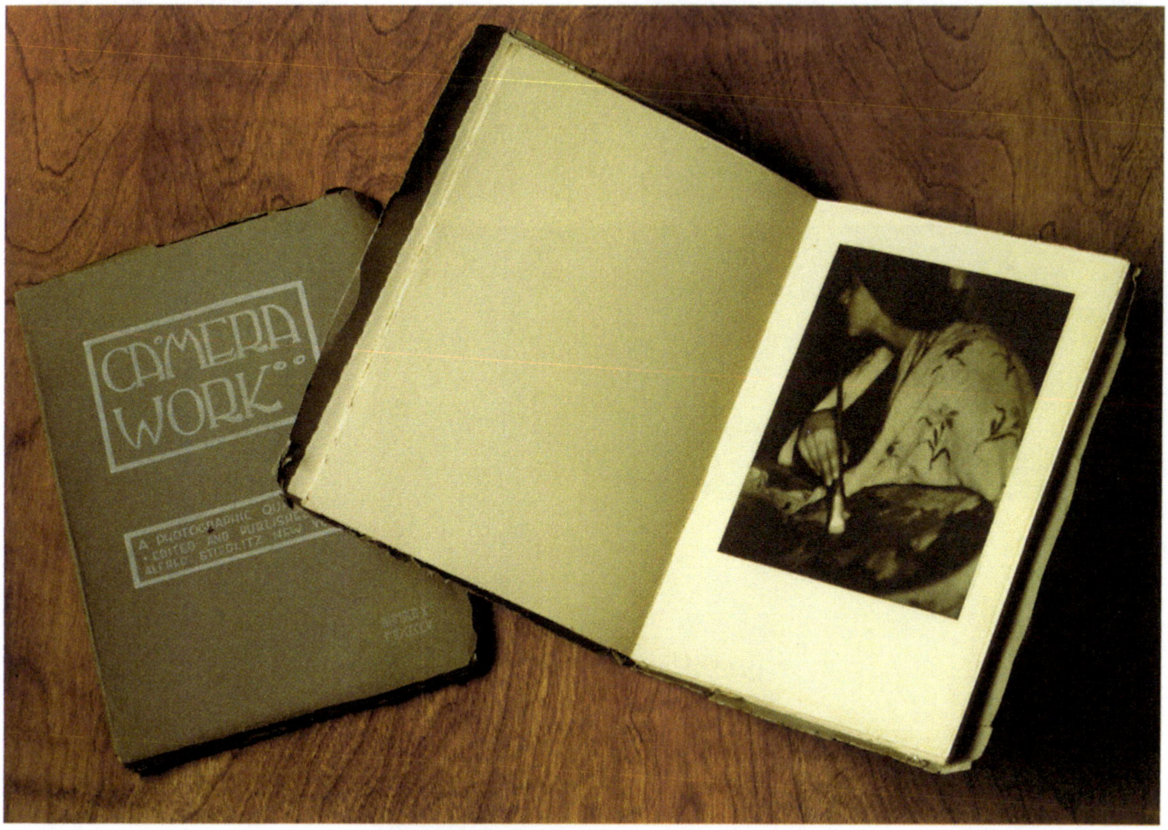

Plate 2 *Camera Work 29*, open to gravure image by George H. Seeley, *The Artist*, 1910. (Photo by Mark Katzman, Ferguson and Katzman. From private collection.)

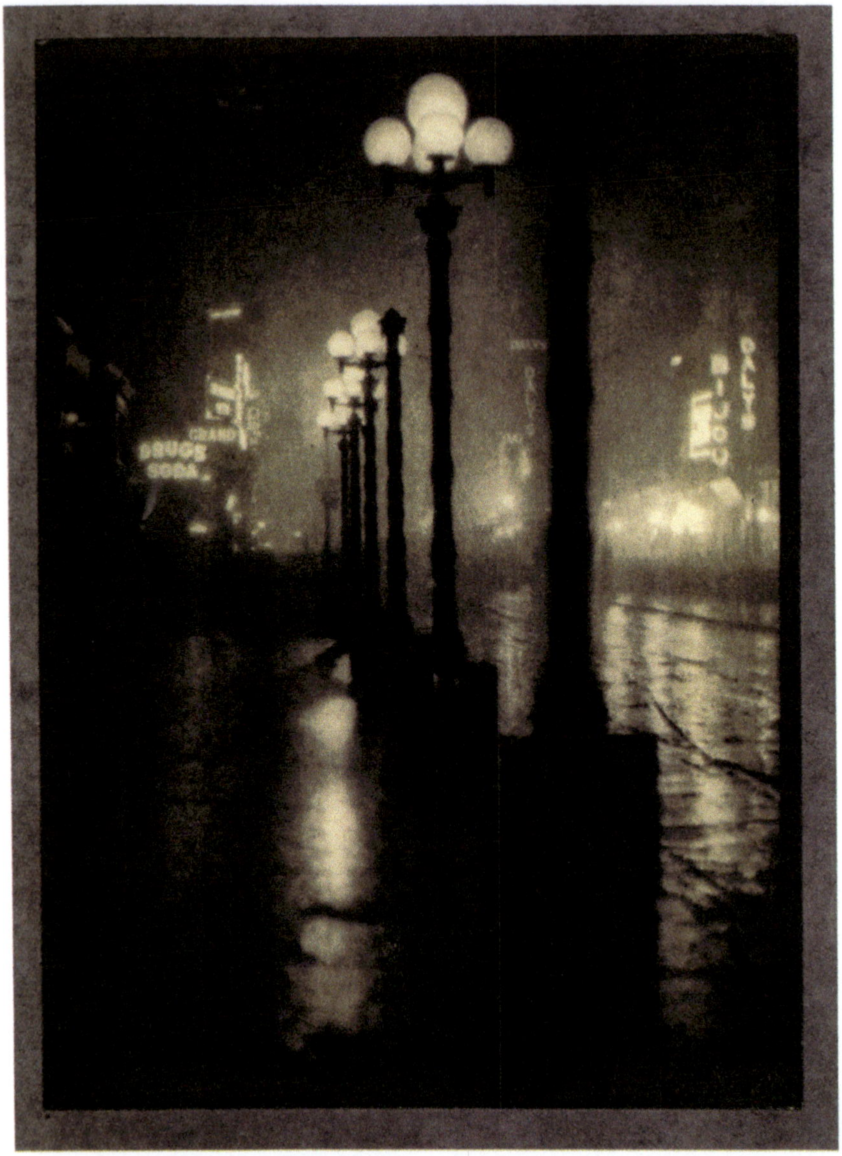

Plate 3 Alvin Langdon Coburn: from *New York*, 1910 (21.6 × 15.3 cm). (Photo by Mark Katzman, Ferguson and Katzman. From private collection.)

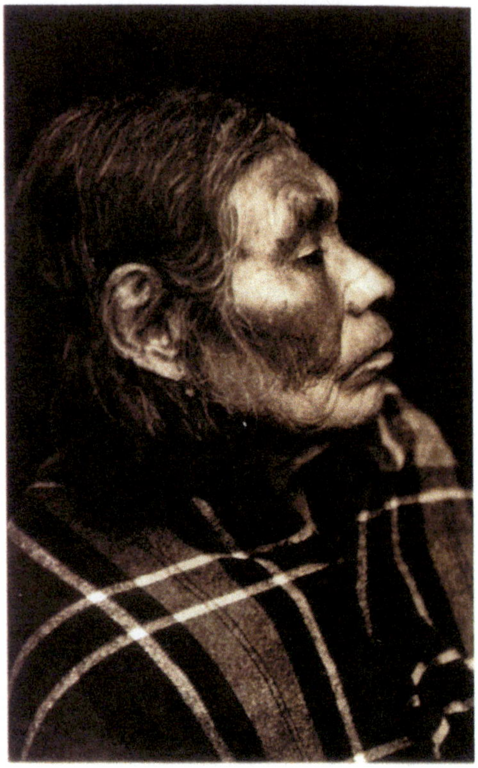

Plate 5 Edward Sheriff Curtis: *Chinook Female Profile*, from *The North American Indian* (18.4 × 12 cm). (Copyright © 1910 by E.S. Curtis.)

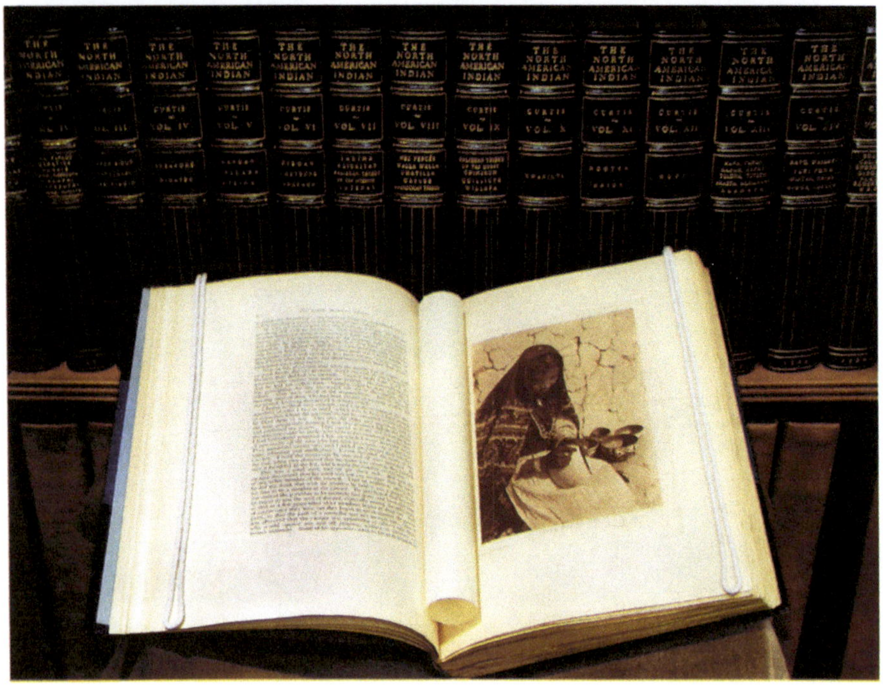

Plate 4 Edward Sheriff Curtis: *The North American Indian*. (Photo permission of Special Collections and Archives, Wesleyan University, Middletown, CT.)

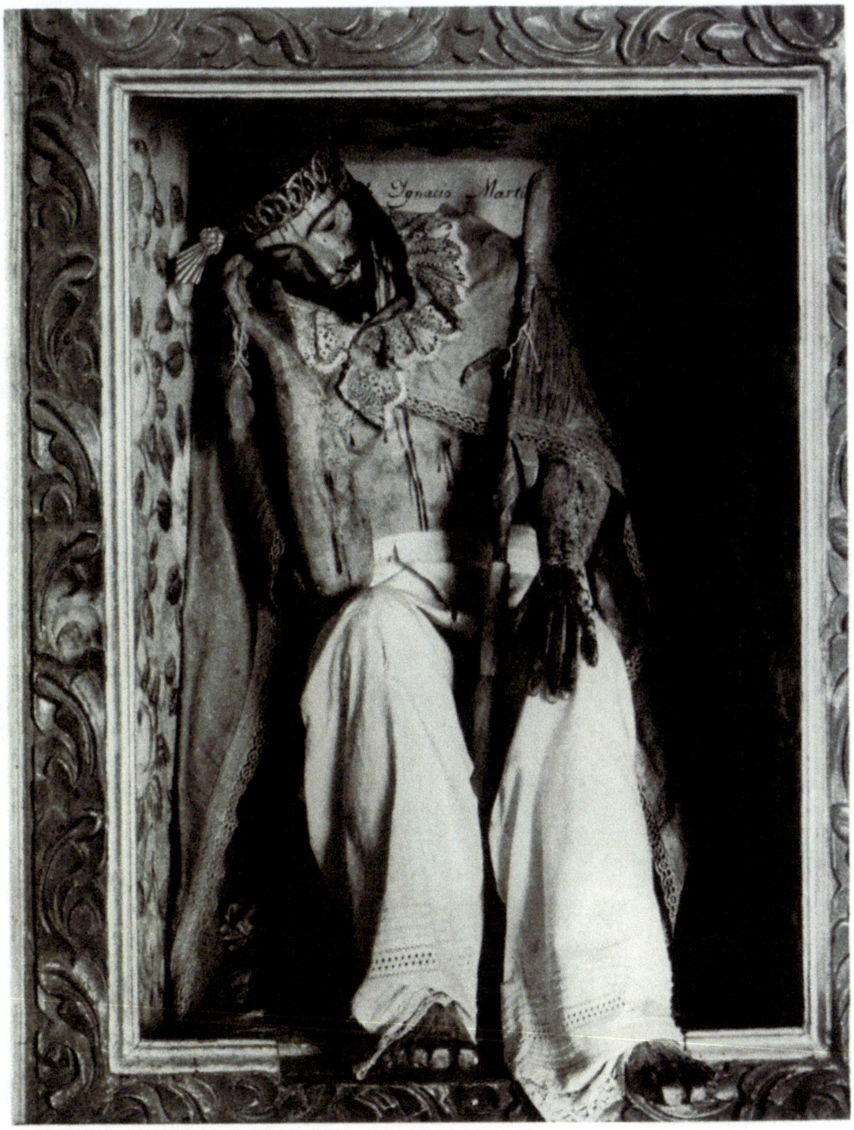

Plate 6 Paul Strand: *Cristo, Tlacochoaya, Mexico*, 1933. (Copyright ©Aperture Foundation Inc., Paul Strand Archive. From private collection.)

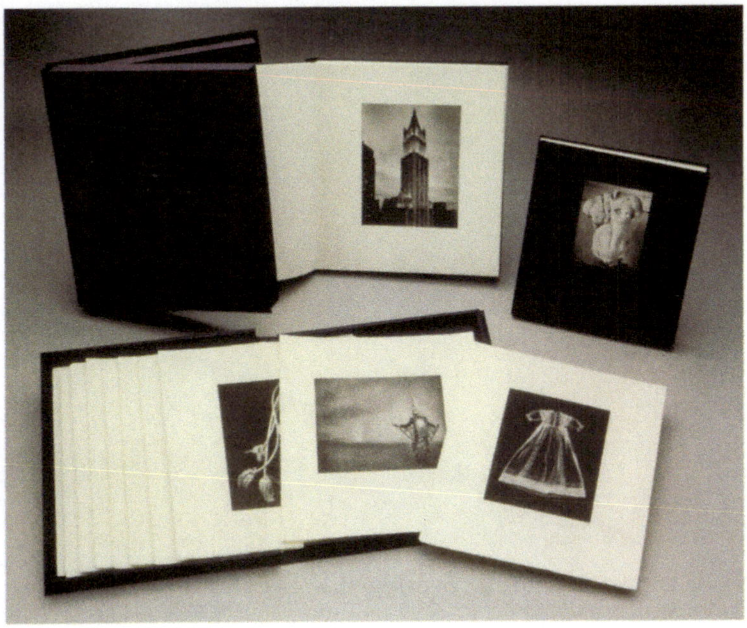

Plate 7 *21st: Journal of Contemporary Photography*, Volume Two. (Copyright © *21st: Journal of Contemporary Photography*. Photo courtesy of Steven Albahari.)

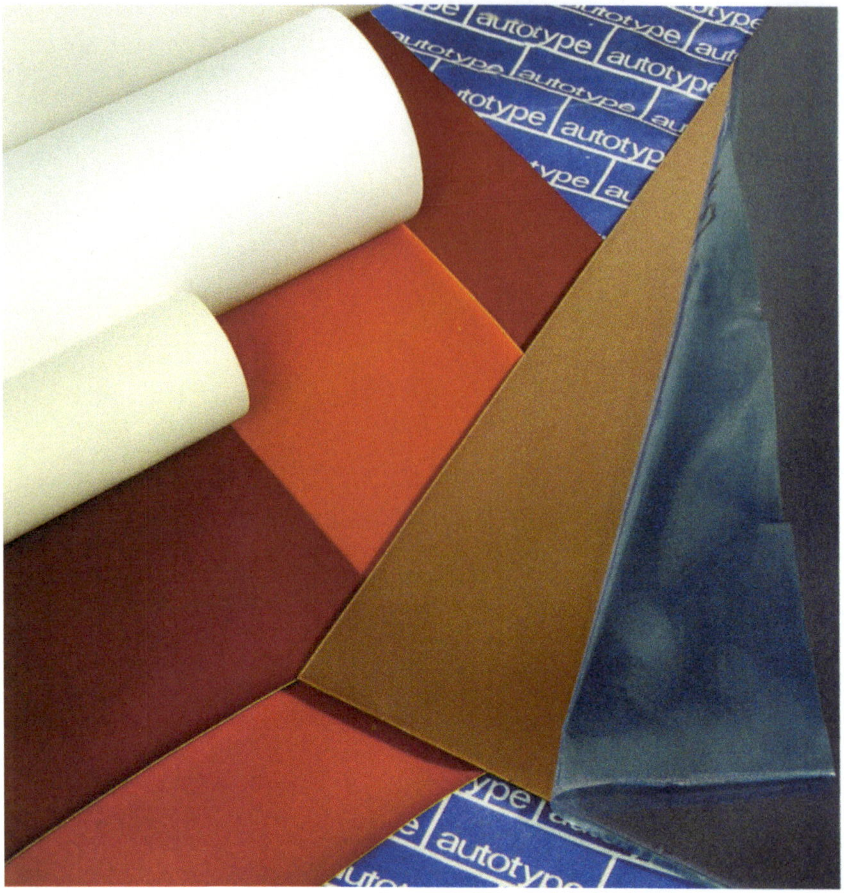

Plate 8 Autotype Gelatin Pigment Papers G25 and G35 (gelatin tissue), hand-made gelatin (carbon) tissue, and mirror-finished copper with the protective plastic layer partially peeled back.

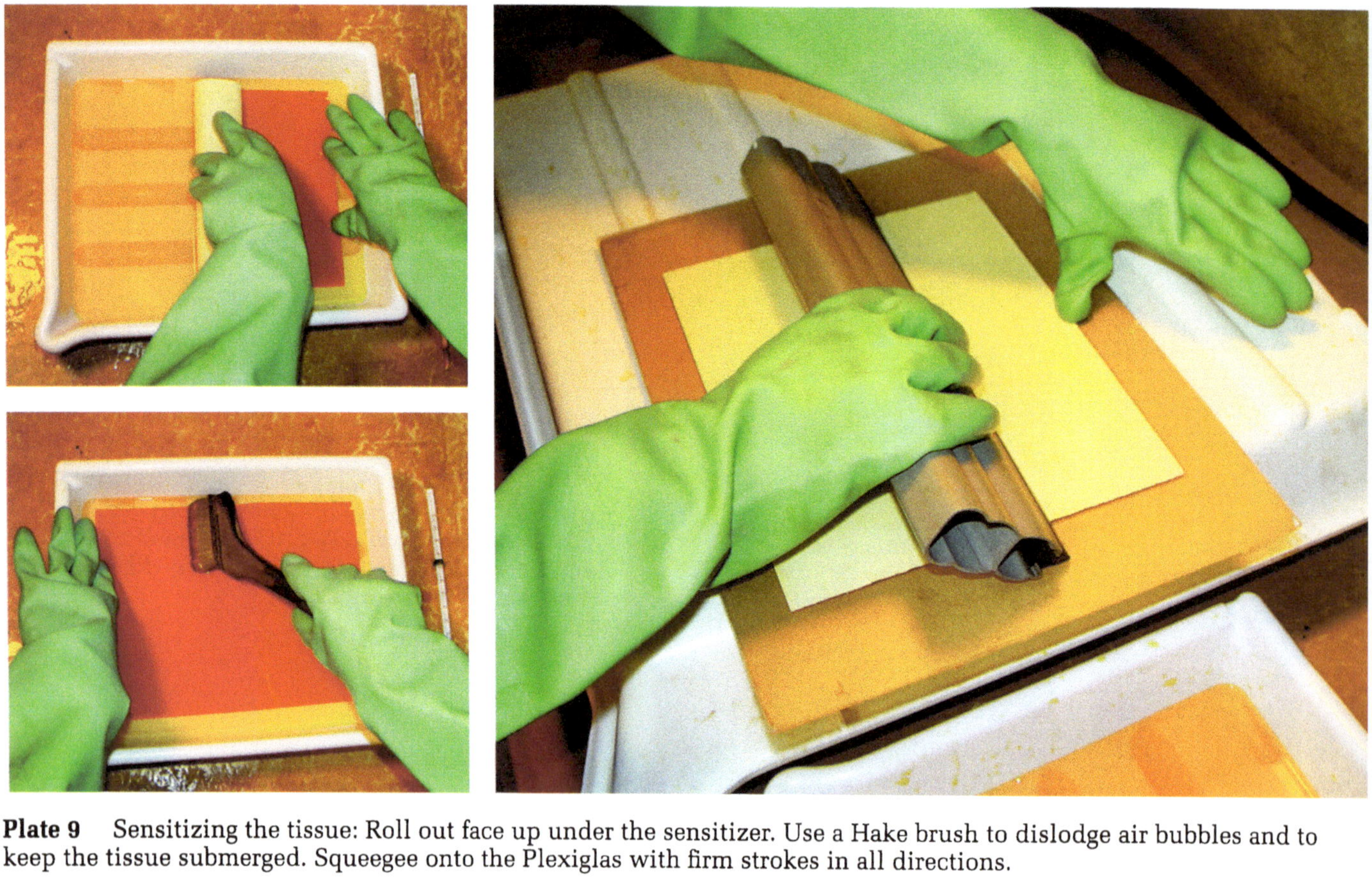

Plate 9 Sensitizing the tissue: Roll out face up under the sensitizer. Use a Hake brush to dislodge air bubbles and to keep the tissue submerged. Squeegee onto the Plexiglas with firm strokes in all directions.

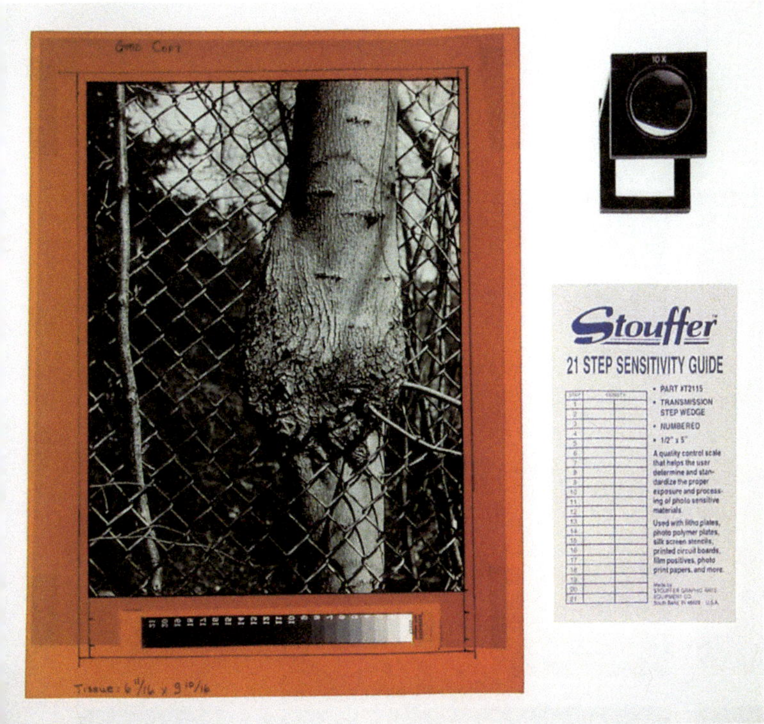

Plate 10 Strip the positive and step scale into goldenrod paper and mark it for positioning the gelatin tissue.

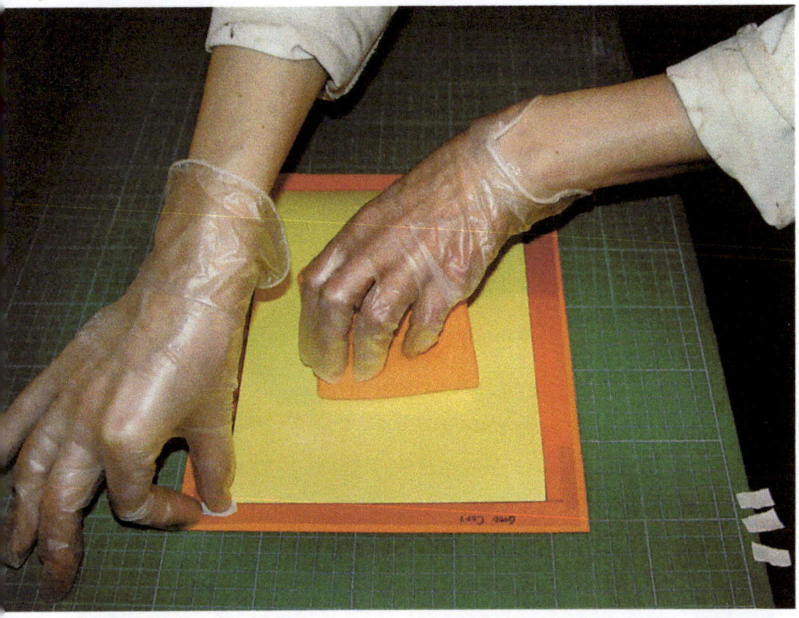

Plate 11 Hold the tissue in place with a pad of folded cloth and tape it in position onto the positive array. Note the use of gloves due to the dichromate in the paper backing.

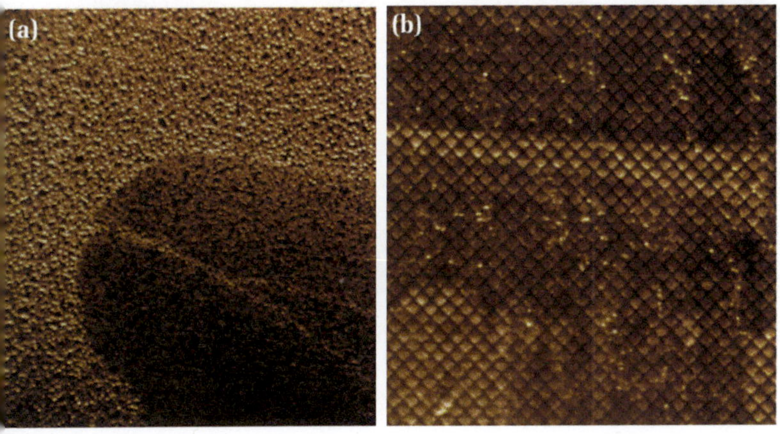

Plate 12 Micrographs of a plate's surface showing the texture created by random screen (**a**) versus a commercial rotogravure screen (**b**).

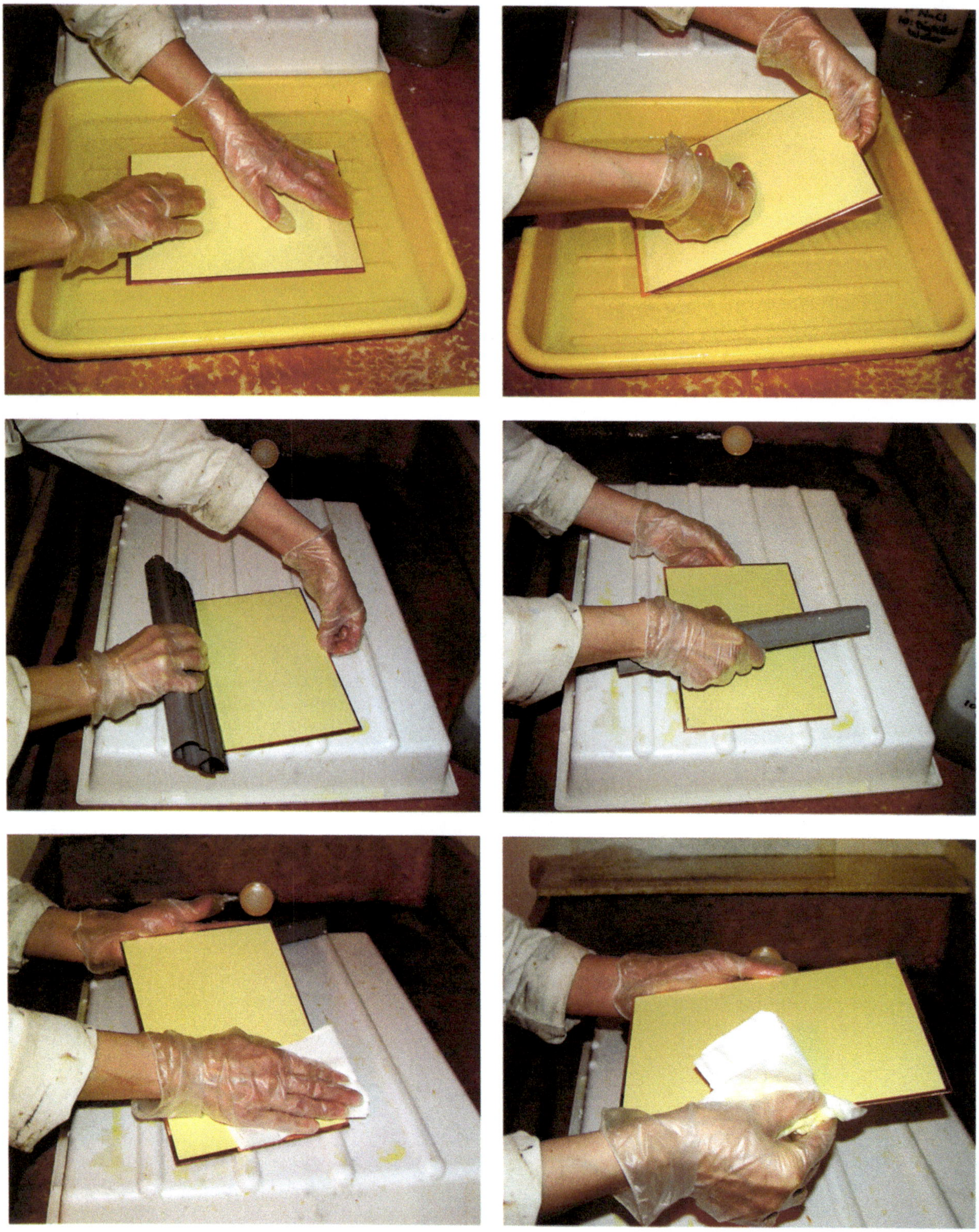

Plate 13 Adhering the tissue to the copper plate: Carefully position the tissue, tilt the plate, then squeegee in four directions. Use a paper towel to wipe the back and edges. Gloves are still necessary.

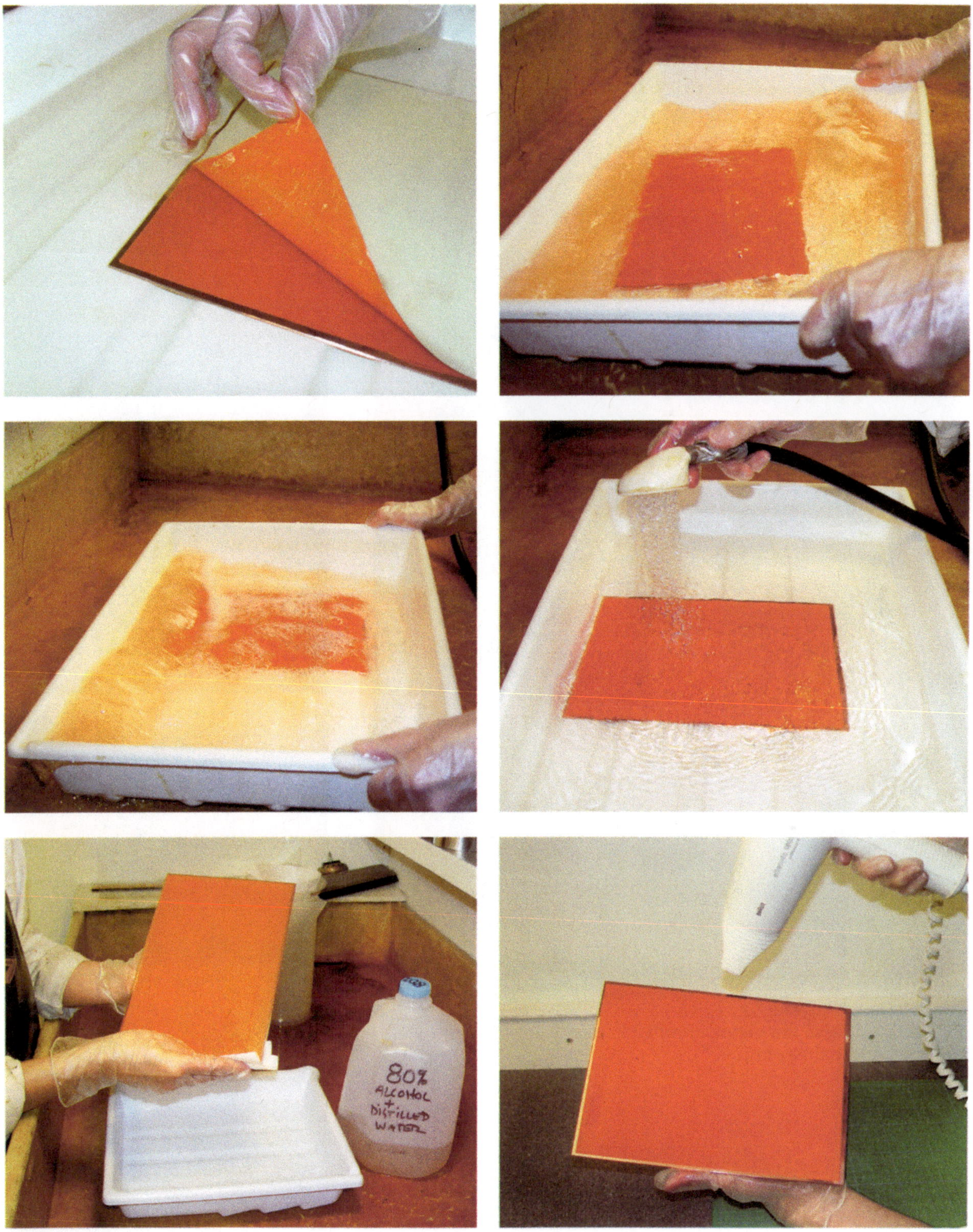

Plate 14 Developing the gelatin resist: Gently peel off the backing paper after 3 to 4 minutes, wash vigorously, soak up the alcohol from the edges, and finish with a hair dryer set to cool.

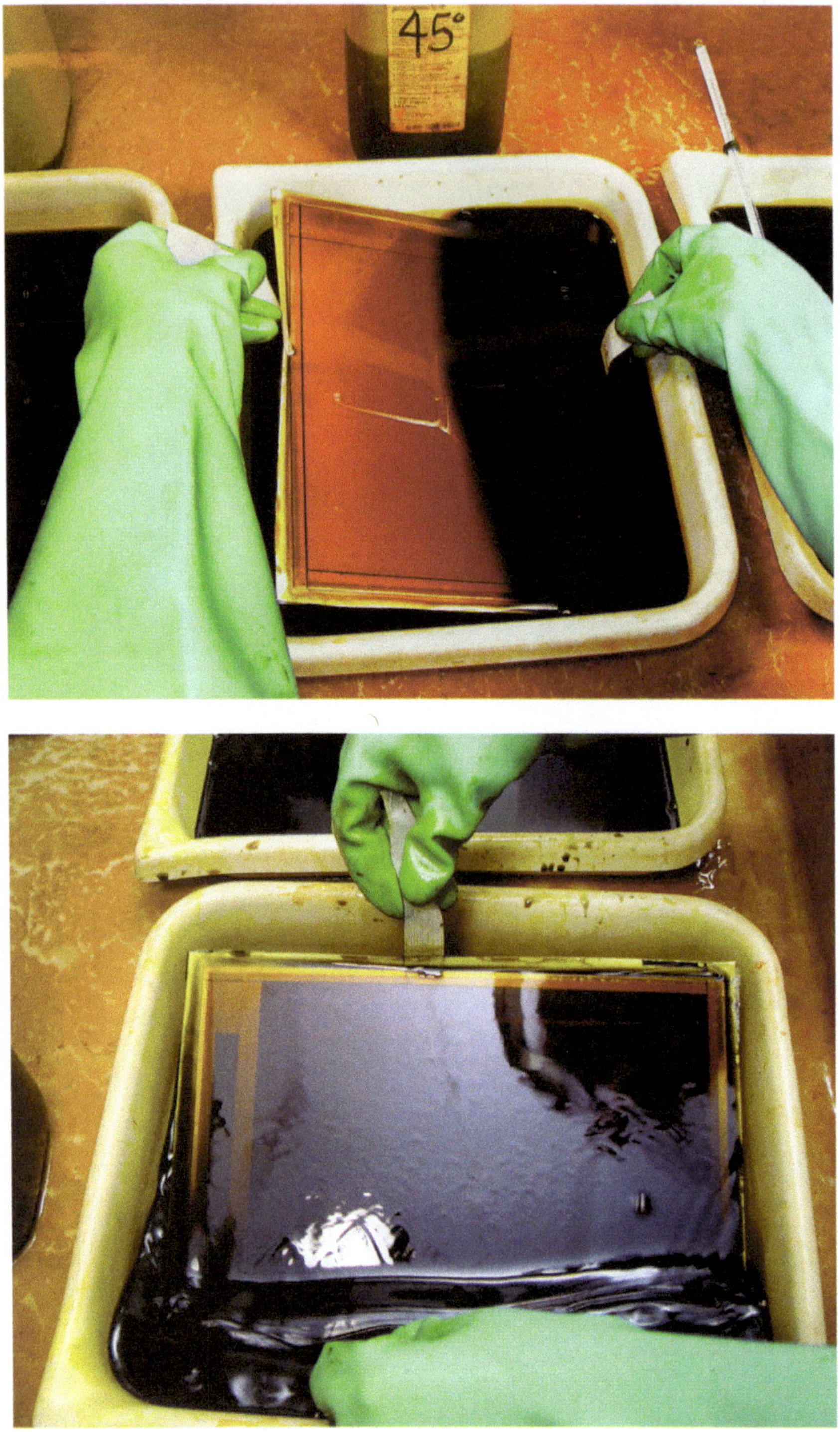

Plate 15 Etching: Raise the plate periodically to check the progress of the etch and to agitate the solution.

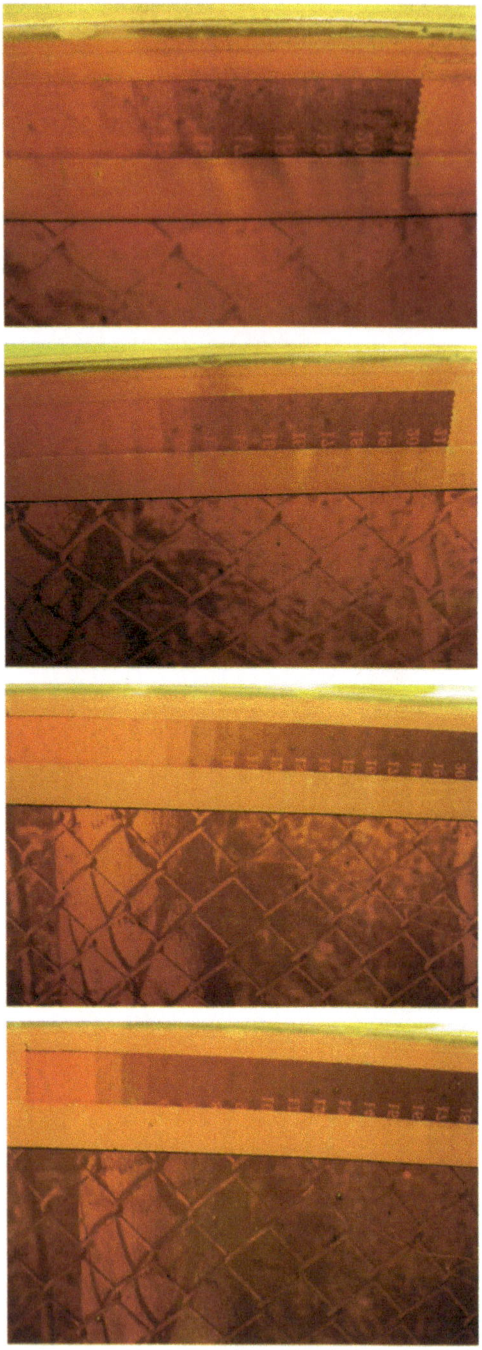

Plate 16 The etch progresses up the step scale one step at a time. Note how the image changes as the steps advance. See also Color Plate 17.

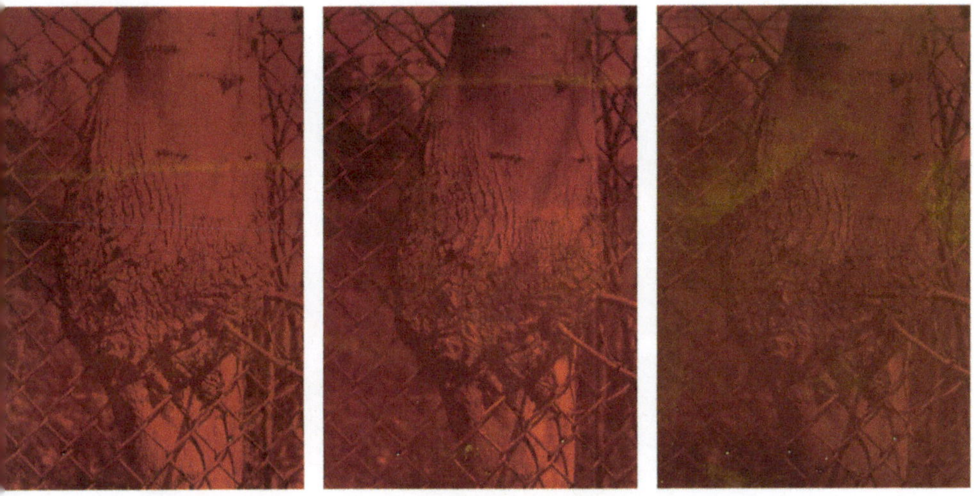

Plate 17 Surface of gelatin progressively etching. Note how the image loses definition and only the brightest highlights have not yet darkened in the last image.

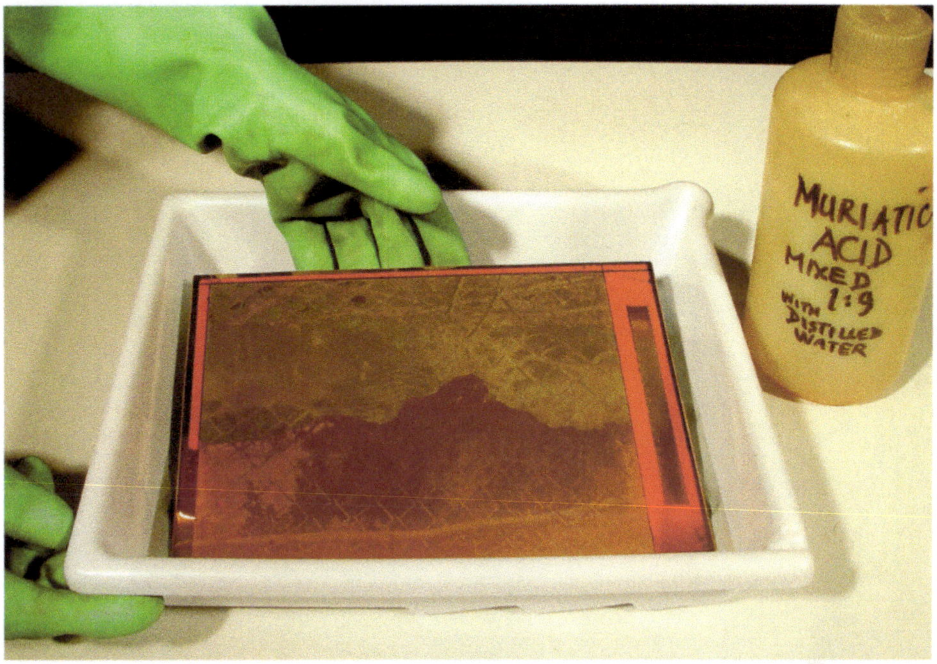

late 18 The tarnished plate rightens dramatically as it enters a :9 solution of muriatic acid. This vill also allow the remaining elatin resist to be easily removed.

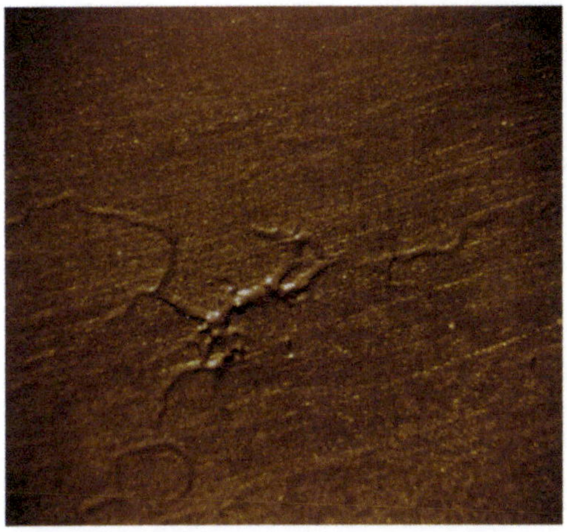

Plate 19 Micrograph of plate surface showing a devil.

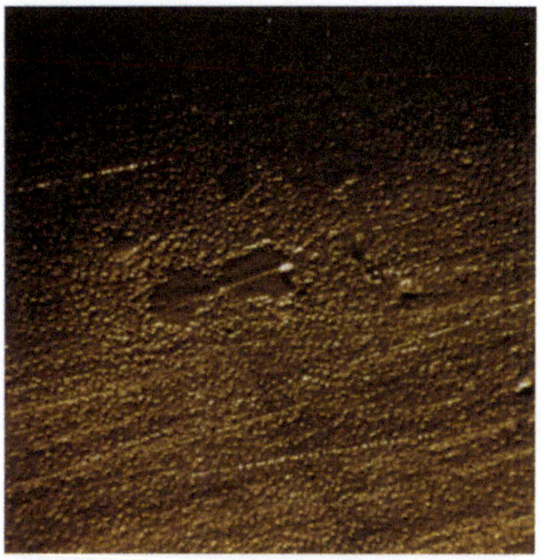

Plate 20 Micrograph of plate surface showing stopped-out areas that require stippling to match background tone.

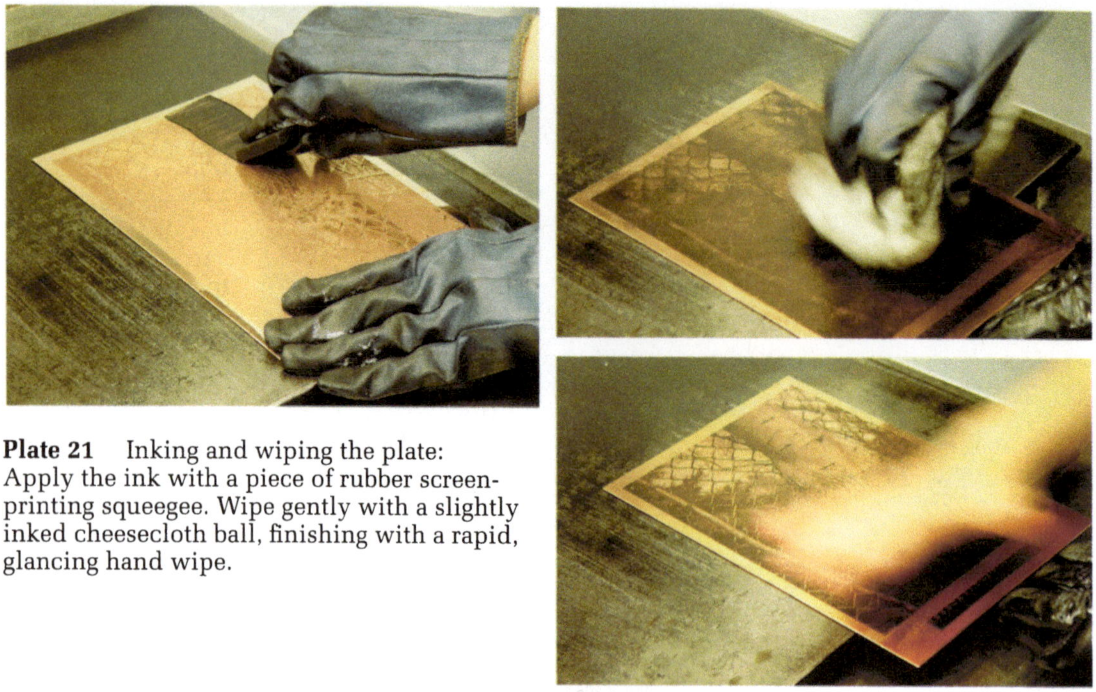

Plate 21 Inking and wiping the plate: Apply the ink with a piece of rubber screen-printing squeegee. Wipe gently with a slightly inked cheesecloth ball, finishing with a rapid, glancing hand wipe.

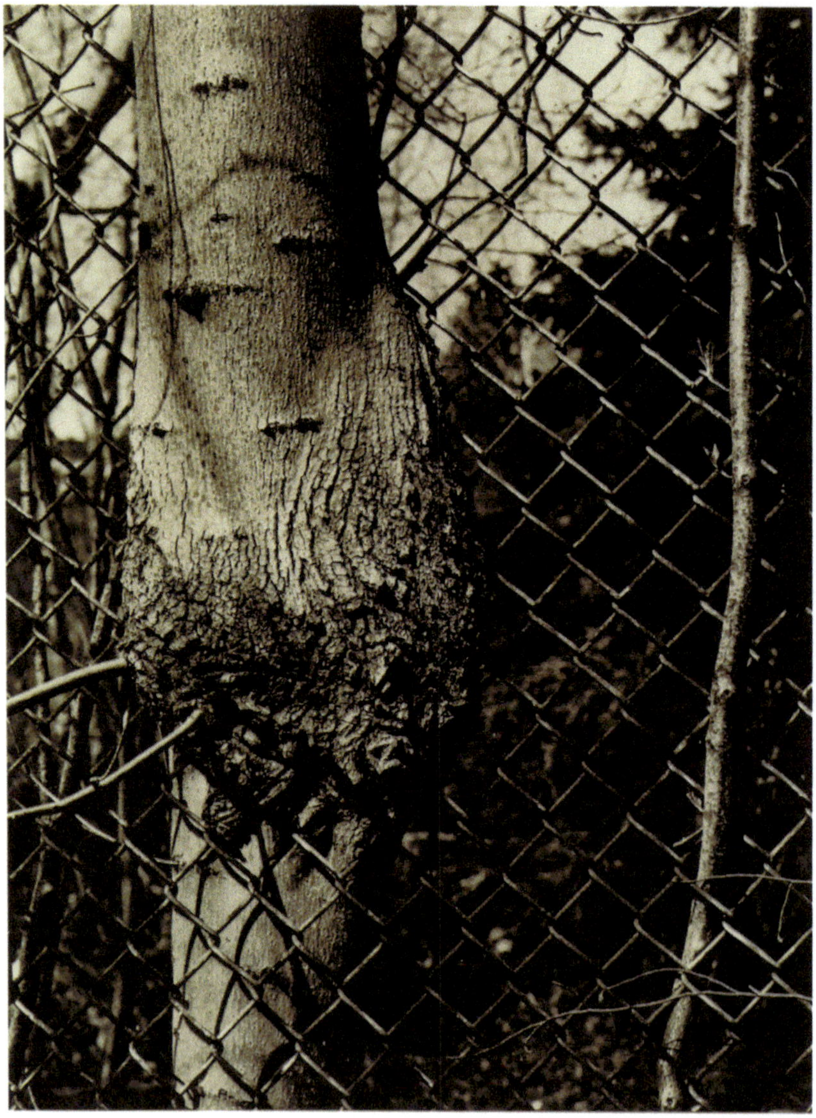

Plate 22 David Morrish: *Penetration*, 2002 (21.3 × 16 cm). Printed on thin Gampi *chine collé*. (Copyright © David Morrish.)

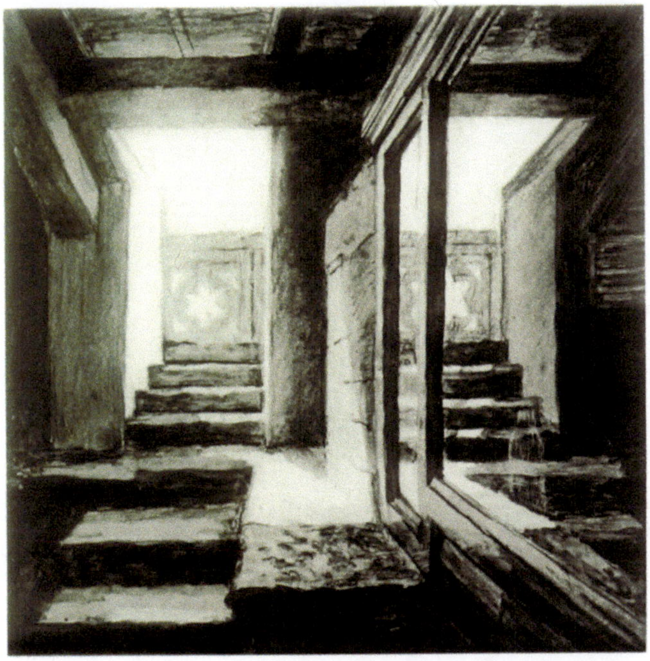

Plate 23 Marlene MacCallum: *Firenze*, 2002 (13.5 × 13.5 cm). Direct gravure from Mylar positive drawing. (Copyright © Marlene MacCallum.)

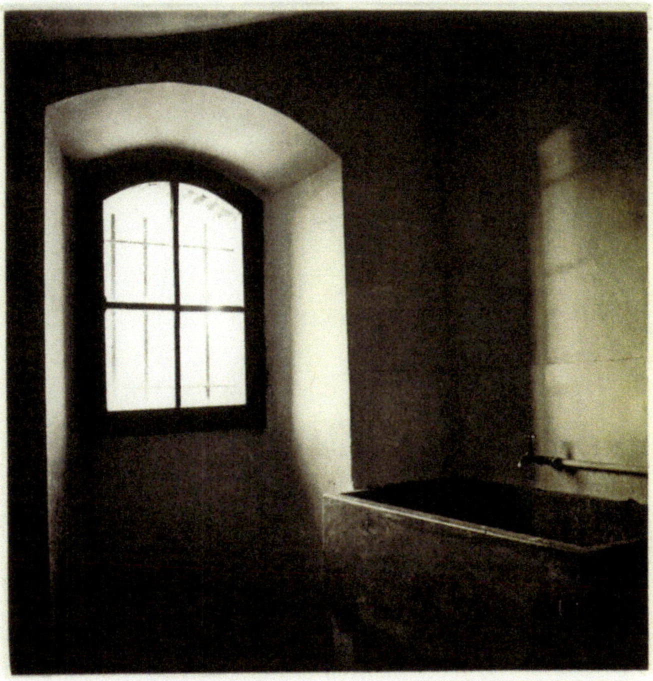

Plate 24 Marlene MacCallum: *Siena*, 2002 (14 × 14 cm). *Chine collé* of Twinrocker Yale on Lana Gravure paper. (Copyright © Marlene MacCallum.)

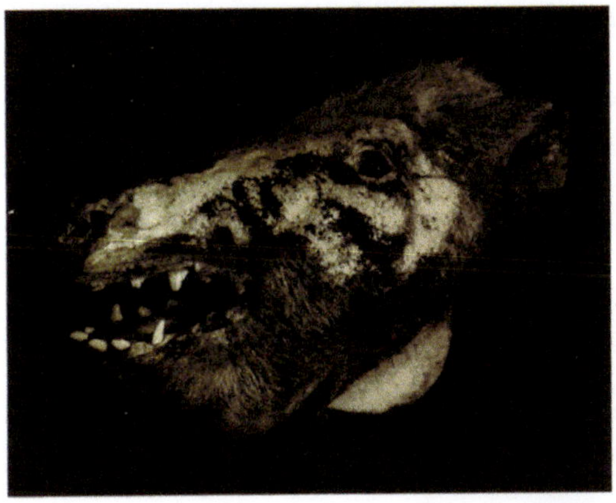

Plate 25 David Morrish: *Swine I*, 1995 (16 × 20.5 cm). Unaltered dust-grain photogravure. (Copyright © David Morrish.)

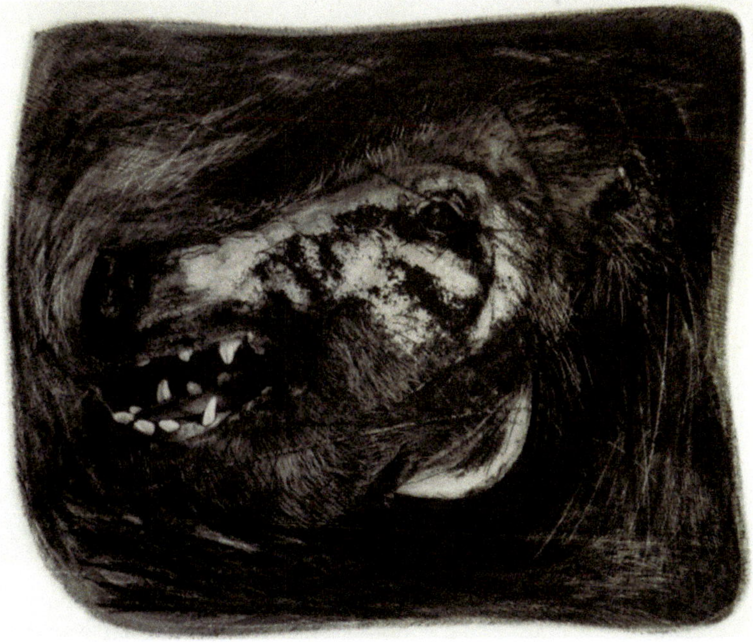

Plate 26 David Morrish: *Swine II*, 1995 (18 × 22 cm). Plate from Color Plate 25 altered by radical dry-point and even some mechanical grinding. (Copyright © David Morrish.)

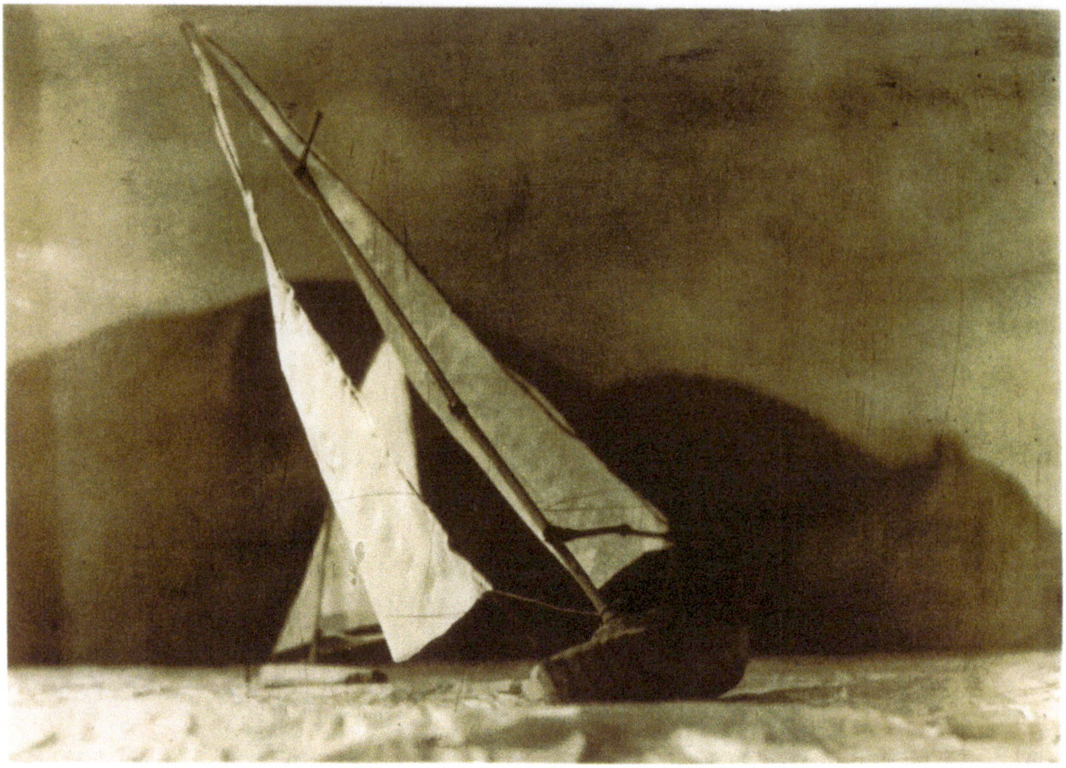

Plate 27 Lothar Osterberg: *Sailing the Golfo Paradiso*, 1998 (46 × 66 cm). (Copyright © Lothar Osterberg, reproduced with permission of the artist. Photo courtesy of L. Osterberg.)

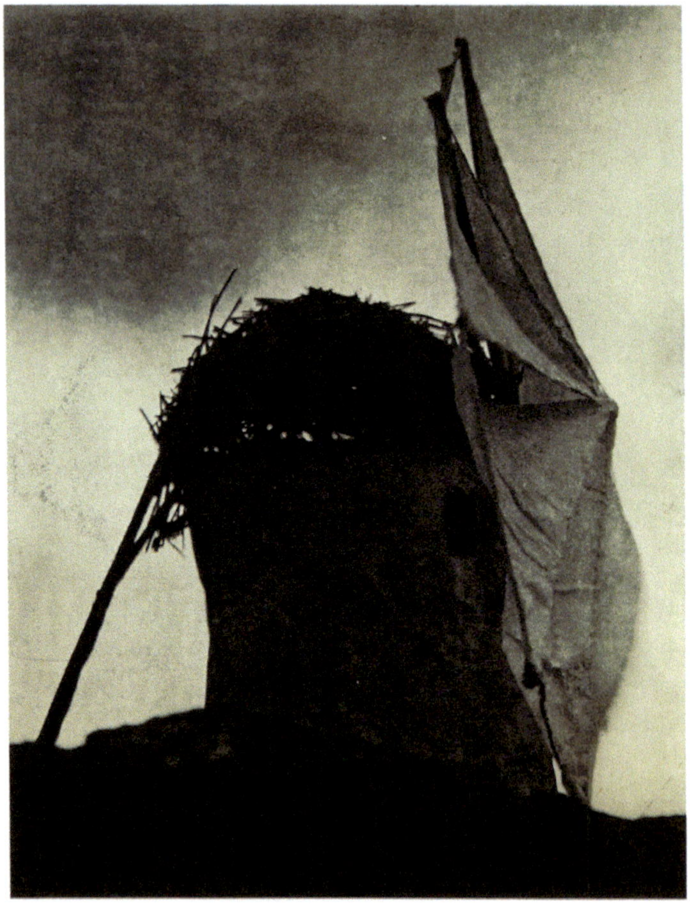

Plate 28 Lothar Osterberg: *Spanish Windmill in Lacoste, France*, 2001 (36 × 28 cm). (Copyright © Lothar Osterberg, reproduced with permission of the artist.)

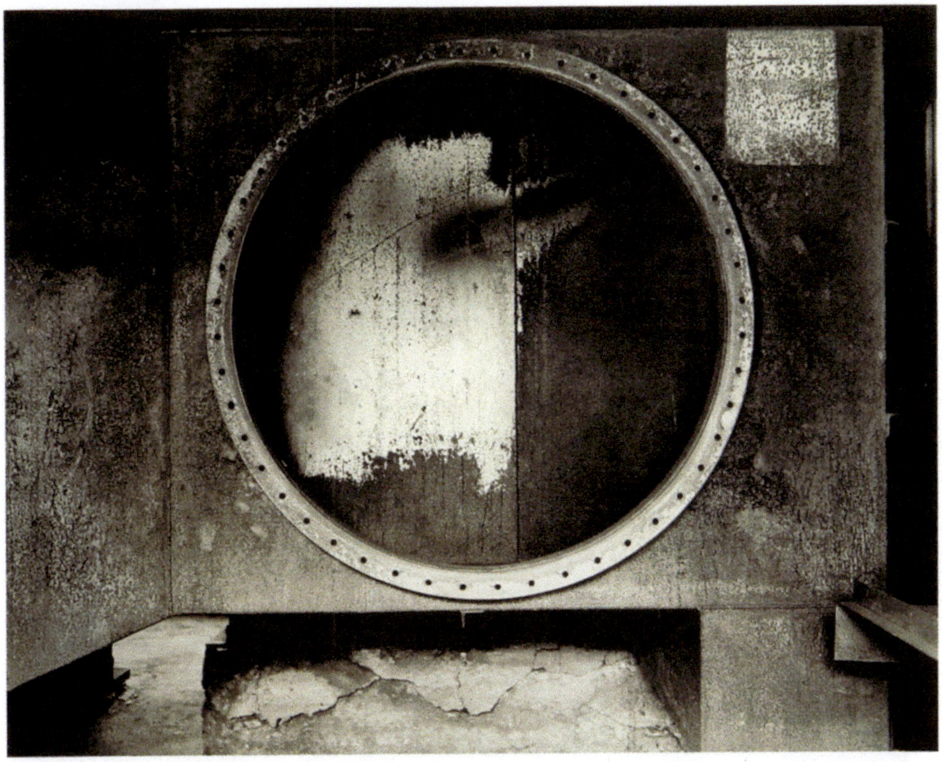

Plate 29 Steven Dixon: *Mine Site No. 5*, 2000 (25.5 × 33 cm). (Copyright © Steven Dixon, reproduced with permission of the artist.)

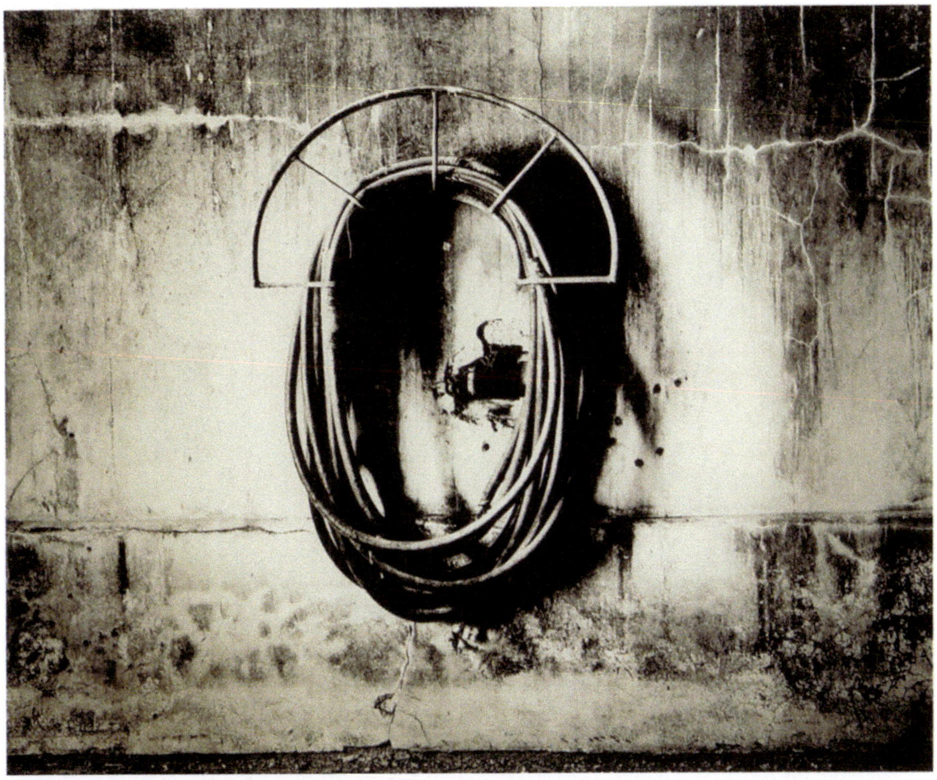

Plate 30 Steven Dixon: *Mine Site No. 17*, 2001 (40 × 50.3 cm). (Copyright © Steven Dixon, reproduced with permission of the artist.)

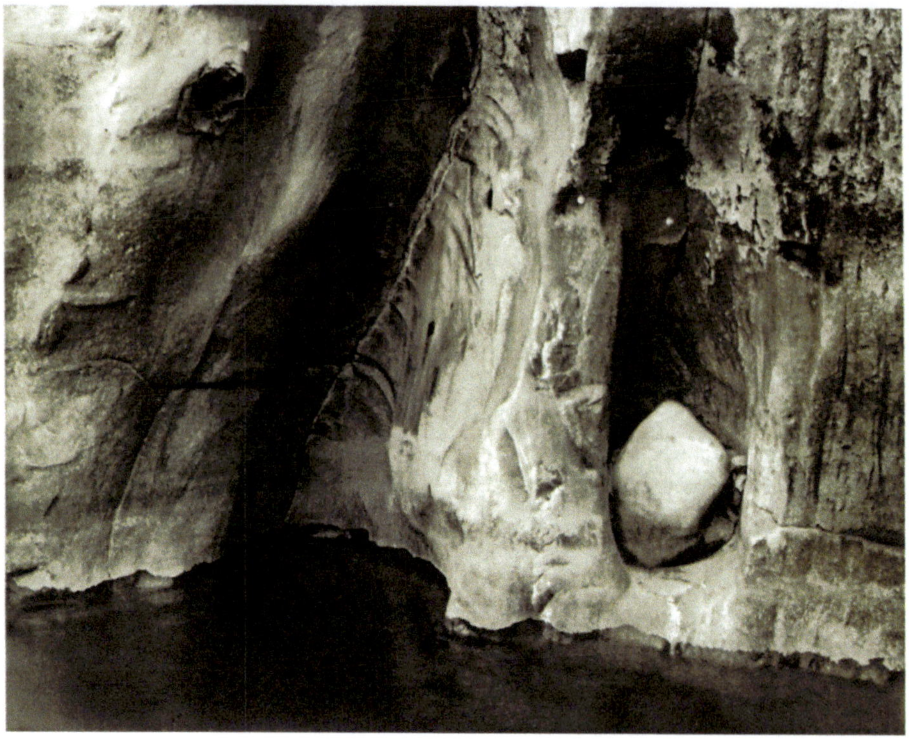

Plate 31 Jon Goodman: *Shinumo Creek*, 1999 (25.4 × 31.8 cm). (Copyright © Jon Goodman, reproduced with permission of the artist.)

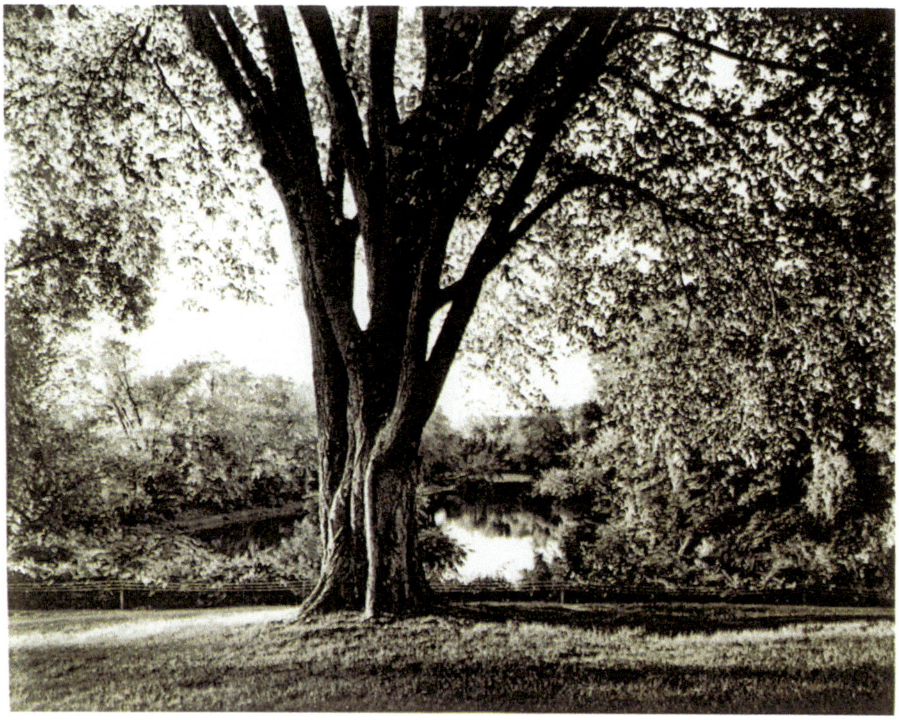

Plate 32 Jon Goodman: *The Elm Tree*, 2000 (22.4 × 29.0 cm). (Copyright © Jon Goodman, reproduced with permission of the artist.)

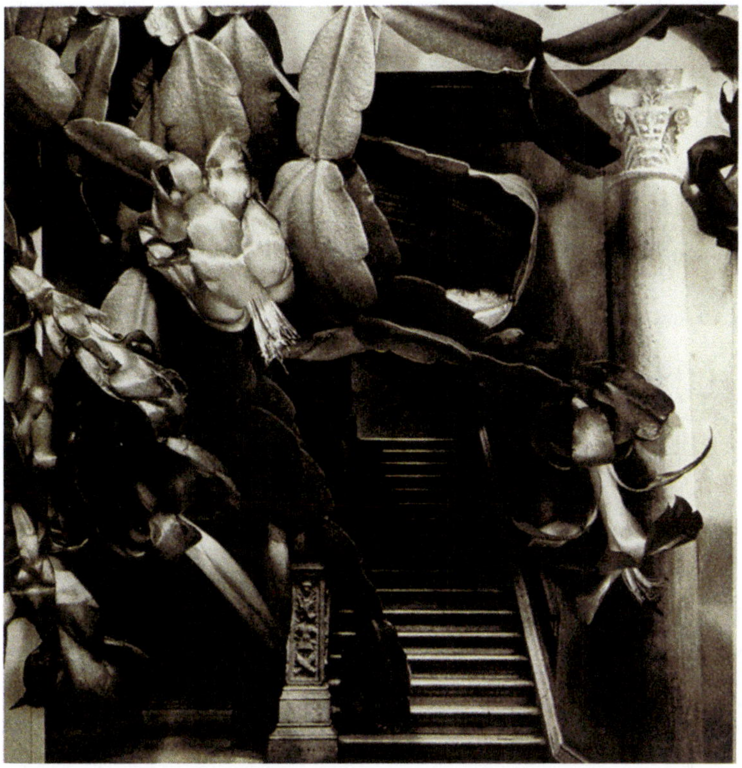

Plate 33 Marlene MacCallum: *Strange Chambers: Stairway*, 2001 (25 × 25 cm). (Copyright © Marlene MacCallum.)

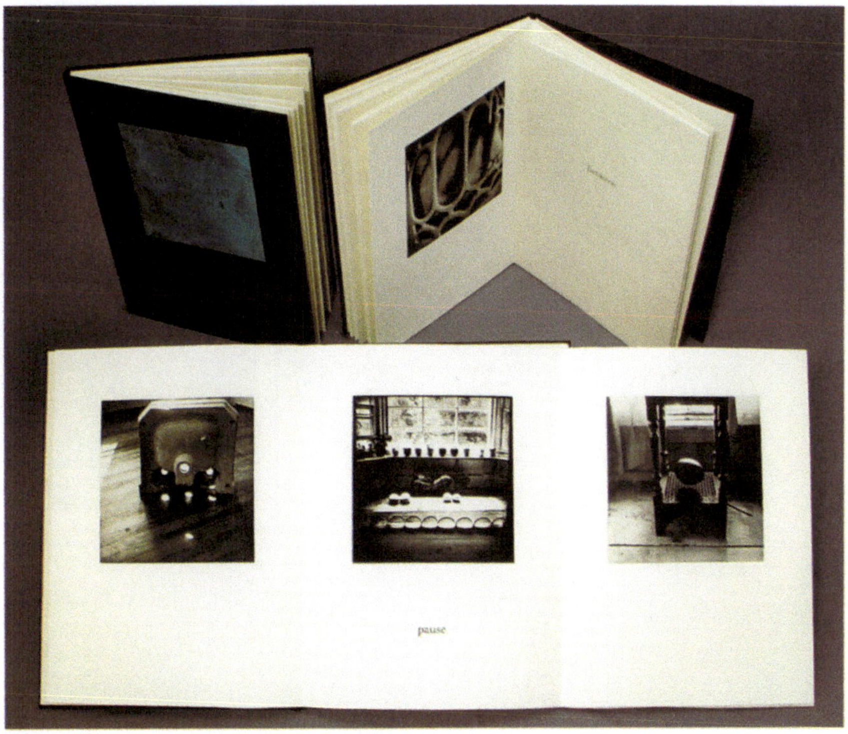

Plate 34 Marlene MacCallum: *Domestic Arcana*, 1999 (book closed: 37.5 × 28.5 × 2 cm). (Copyright © Marlene MacCallum.)

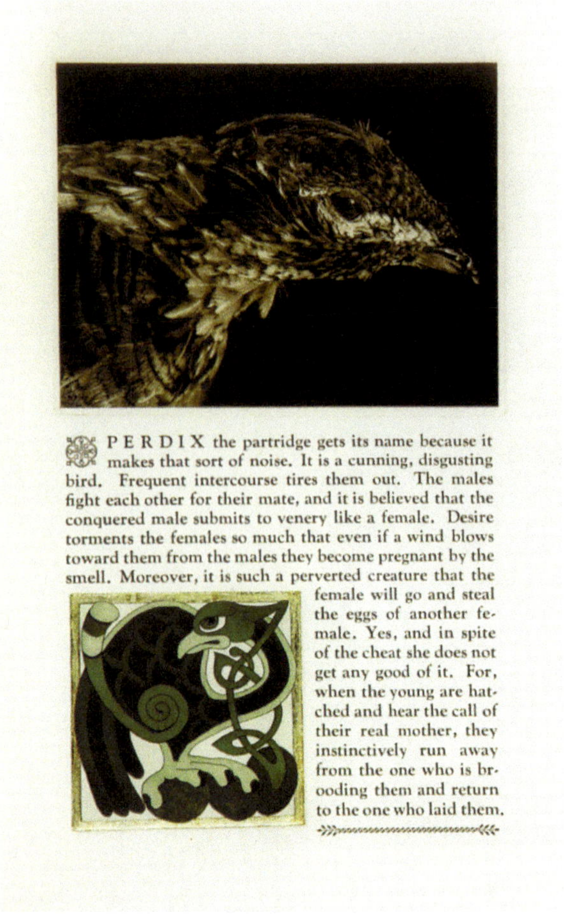

PERDIX the partridge gets its name because it makes that sort of noise. It is a cunning, disgusting bird. Frequent intercourse tires them out. The males fight each other for their mate, and it is believed that the conquered male submits to venery like a female. Desire torments the females so much that even if a wind blows toward them from the males they become pregnant by the smell. Moreover, it is such a perverted creature that the female will go and steal the eggs of another female. Yes, and in spite of the cheat she does not get any good of it. For, when the young are hatched and hear the call of their real mother, they instinctively run away from the one who is brooding them and return to the one who laid them.

Plate 35 David Morrish: *Perdix*, from *Bestiarum Excerptum*, 1999 (29.5 × 17.5 cm). Photogravure plate, intaglio plate, letterpress type, and 22K gold leaf. (Copyright © David Morrish.)

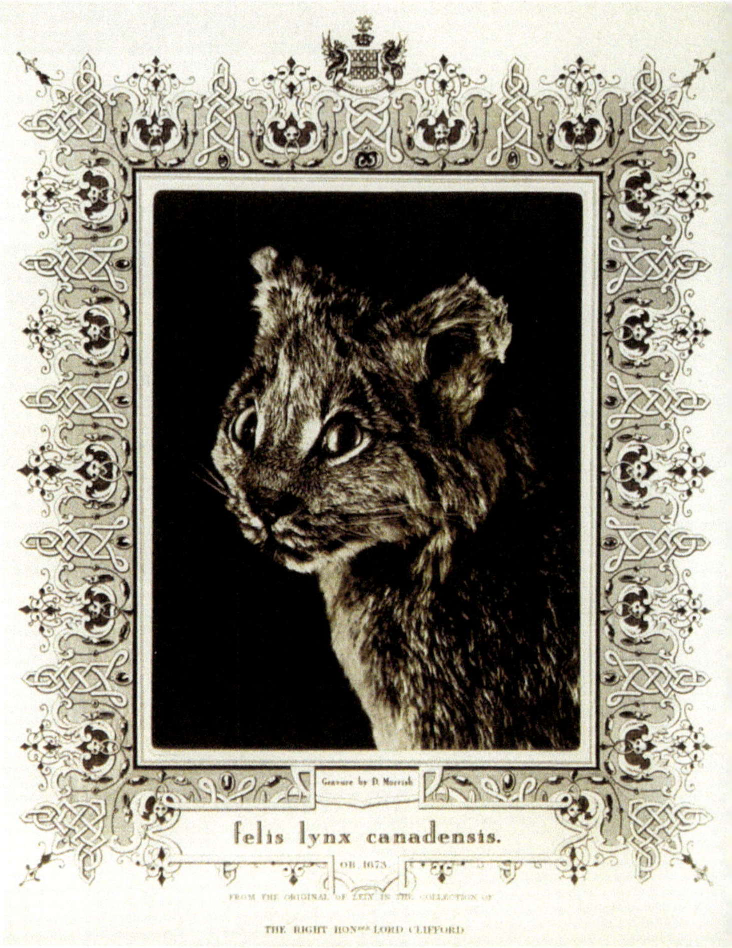

Plate 36 David Morrish: *Canadian Lynx*, from *The Gallery*, 2000 (35 × 30 cm). Two nested photogravure plates, each inked differently. (Copyright © David Morrish.)

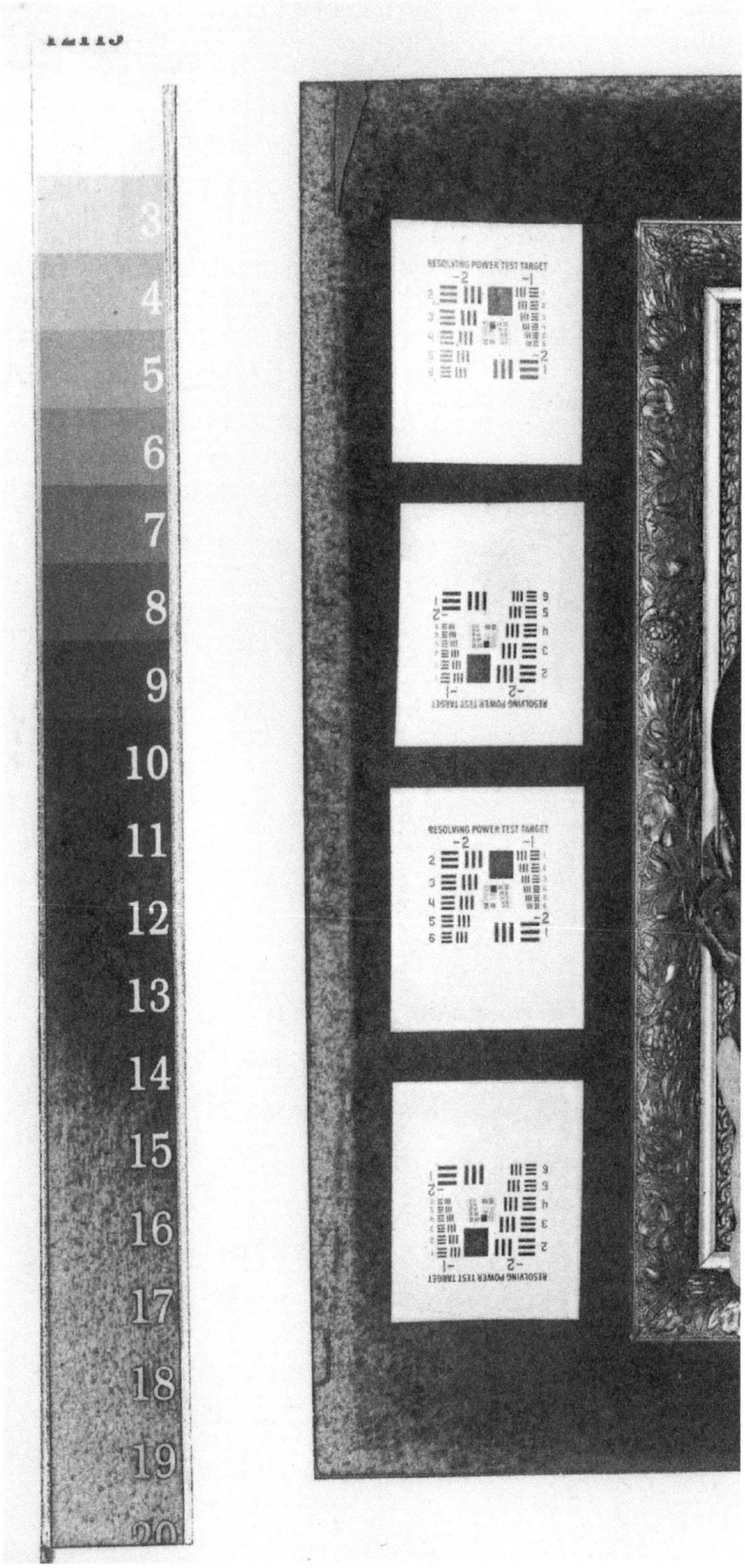

Figure 8-8 Example of a dust-grained plate where the dark end of the scale and the image's shadow areas have foul-bitten.

speed it along. If it speeds up too much, return to the higher Baumé for a short time.

Overly Slow Etch If the progression of the etch seems to slow down, the temperature of the Baumés may have dropped. Monitor and maintain a consistent working temperature.

Water Contamination During an etch you might see the sudden appearance of a dark, heavily etching area unrelated to the underlying image. This can happen if you get water on your glove, which then drips onto the surface of the plate (Figure 8-9). If you wet your gloves in the sink, you must dry them before reaching for a plate resting in a mordant tray.

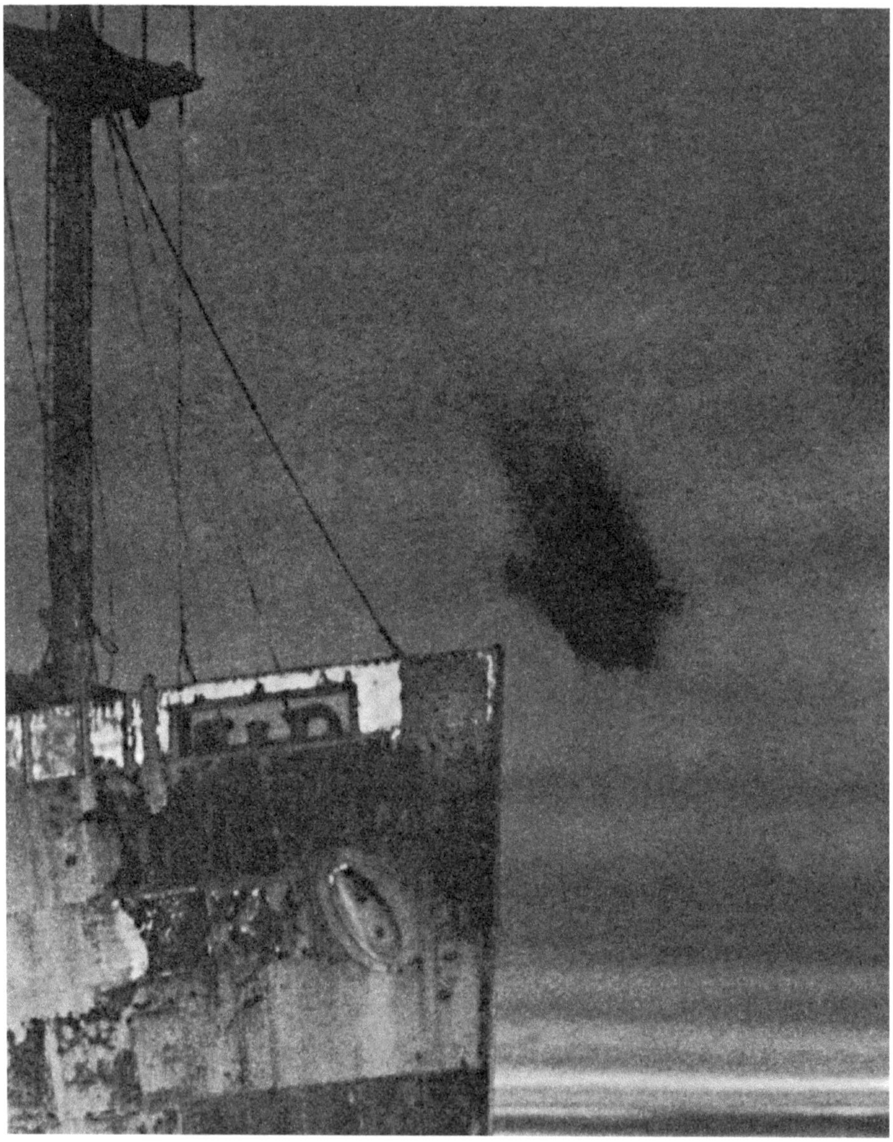

Figure 8-9 This dark spot appeared after a drip of water from a glove caused the spot to commence etching too early. This resulted in an overetched black area in the middle of a dark gray-toned, sky which is very difficult to correct. (Detail from a larger image.)

Highlight Streaks Streaks or splotches sometimes appear in the highlight areas after the plate is cleaned and dried. They are usually caused by ferric chloride running over the surface of the plate while you are washing off the resist or removing the tape and backing. Be sure to use a lot of running water, or better yet, submerge the plate in a solution of sodium carbonate to neutralize the ferric chloride. When the backing is peeled off, there is sometimes a seam of ferric chloride that has soaked under the tape and backing. Wash this away from the image area, not over it.

Plate Flaws

Mottle For areas of mottle, refer to the troubleshooting sections of Chapter 5 or Chapter 6.

Devils The deep spidery or tree-like flaws that sometimes appear in the dark areas of a plate are appropriately called *devils* (Figure 8-10 and Color Plate 19). They are sometimes an indication of out-of-control etching through pinholes in the resist (Figure 8-11). The sideways travel that forms the legs of the spider (or branches of a tree) is also an indication of an overactive mordant, probably with a high free acid content.

Splotches Dark areas in the gelatin resist or drying marks in areas not related to the original exposure can result in rapid local etching and then corresponding darker marks in the print. These may be seen as splotches in random areas of the plate or a darker streak like a pale birthmark in the highlight areas. It is usually caused when the resist has not dried evenly and there are still areas with a high moisture content (Figure 8-12). Splotches due to excess moisture can be caused by breathing on the plate when staging or spotting it in preparation for the etch. Use a face shield or dust mask when working very close to the plate's surface. Unevenness or

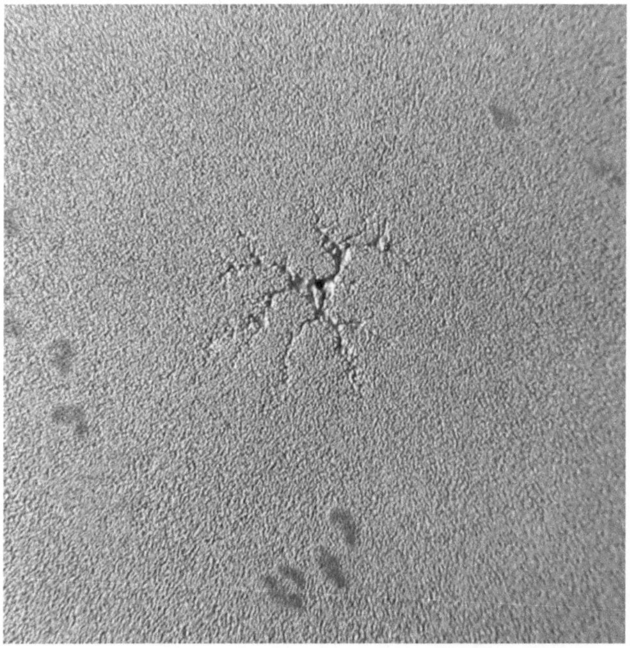

Figure 8-10 A good example of a devil, in all its sinister glory. See also Color Plate 19.

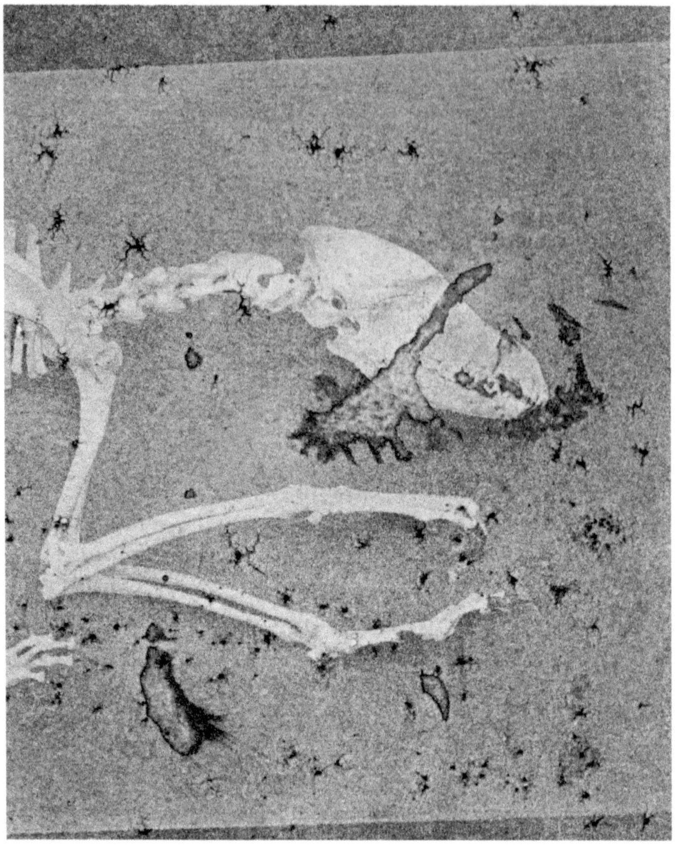

Figure 8-11 Devils and pits in the lighter areas of a print. They usually occur in the blacks, so are far less visible. Many can be prevented by stopping out pinholes with a Permanent Black Lumocolor pen. (See Figure 7-6 in Chapter 7.) (Detail from a larger image.)

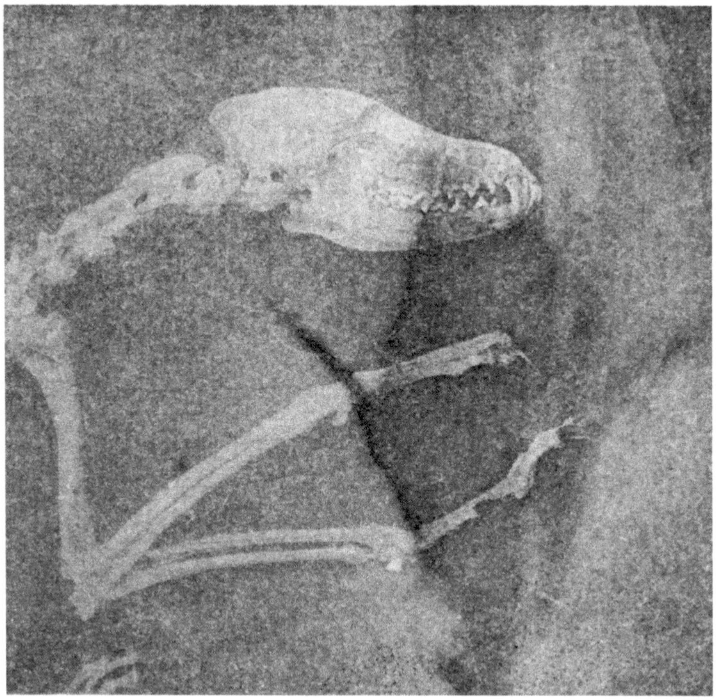

Figure 8-12 Ferric chloride penetrated quickly through a portion of the resist that contained more moisture and caused a dark splotch on this print. (Detail from a larger image.)

mottle can also appear if the tissue was sensitized before it has been fully acclimatized and has an uneven moisture content. It could also be an indication of underdevelopment if it appears in light tones.

Foul-Bitten Borders Deep open bite (foul-bite) lines can occur at the corners of the image on the border of the plate. When the tape staging is not burnished down all the way to where it overlaps, the mordant can creep up the seam and create a line. This is usually not a problem if you plan to trim the plate to be borderless. If you want clean borders, however, you will have to scrape and burnish the line. Be very careful to protect the delicate image area nearby.

9 The Printing Process

The stage in which you are finally presented with the fruits of your labors is the printing stage. This is the most satisfying part of the process because the final results are immediate and concrete. A lot of effort has led to this point and the temptation to rush is strong. Try to approach the printing stage with as much care as the earlier stages of your journey.

Photogravure printing is basically the same as intaglio printing and uses many of the same materials and equipment. The difference lies in technique and the quality of the materials and equipment necessary for a good print. The surface of the plate is very fine and the etch is shallow. To print all the subtle details and tonal variations of a gravure plate, you need high quality sensitive paper (dampened), high quality ink made of finely ground pigment, good woven wool blankets, and an intaglio press capable of exerting high pressure. The material and equipment requirements are significant enough to the success of the print that if they are not taken into account nor fully understood, the plate's potential will never be fully realized.

PAPERS

The choice of an appropriate paper for a photogravure is a critical part of the process, technically and aesthetically. Quality rag papers are always recommended because they have strong fibers that can hold up under heavy printing pressures and are archival. An unsuitable paper can be one of the factors that cause mottling, pale blacks, abrupt tonal gradations, or the loss of subtle highlight details.

It is a good idea to get a thorough paper catalog from one of the many suppliers that publish them. We have tested several papers and the information we have collected is provided in Appendix G. This is only a guide based on our facilities, equipment, and working practices. These factors can vary greatly so each printer should experiment with a variety of papers. Of even greater significance is the individual artist's preferences. Moreover, each plate and image will have its own idiosyncrasies and may require a different paper. Also note that paper mills do occasionally change owners and this may result in changes to a paper's characteristics or composition.

Here are some basic considerations to keep in mind when selecting a paper:

- The paper must be pH neutral to be archival. Preferably, it should be 100% cotton or rag (Western papers) or other inert plant fiber (Asian papers). Some are buffered (elevated pH) to ensure nonacidity over prolonged exposure to airborne contaminants.
- In addition to the fiber composition, another factor to consider is the sizing and coatings. Heavily sized or buffered papers may present problems. Photogravures are almost always printed on presoaked or dampened papers to soften the fibers of the paper so that it can conform to the minute depressions in the plate and pull out the ink. Determine whether the paper is sized (externally or internally) or unsized (waterleaf). This is important because it affects the soaking or dampening procedure.
- The weight of paper varies greatly. It is possible to use the entire spectrum, depending on the context of the print. The heavyweight versions (250 to 300 grams per square meter or g/m^2) give a print that has a greater sense of being an object. Text weight (100–175 g/m^2) is appropriate for tipping in or as part of a book work. Thin, tissue-like papers (30–80 g/m^2) give a very sensitive printing and are generally used for *chine collé* (see Chapter 10, pp. 159–162), or in conjunction with another heavier backing paper. The original issues of *Camera Work* used a thin tissue-type paper that was then bound with a heavier dark paper for background color.
- The finish and color of the paper is also important. We have found that smooth finishes work best. A hot press finish is smooth whereas a cold press finish is more textured. Another important factor in selecting paper will be the desired tone. There is a huge range from bright white to cream, to subtle tone, to highly colored. The choice of ink color combined with paper tone can have a dramatic effect on the image.
- The cost of fine paper is an unavoidable reality. Generally speaking, the best papers are often very expensive, so many of the papers appropriate to printing photogravure are costly. An ideal paper is one that will give a print that has rich blacks and smooth tones, and that preserves the resolution of fine details in both the highlights and the shadows. Proofing should be done on good paper to provide an accurate print in order to avoid the danger of making an incorrect judgment or unnecessary changes to the plate.

INKS AND ADDITIVES

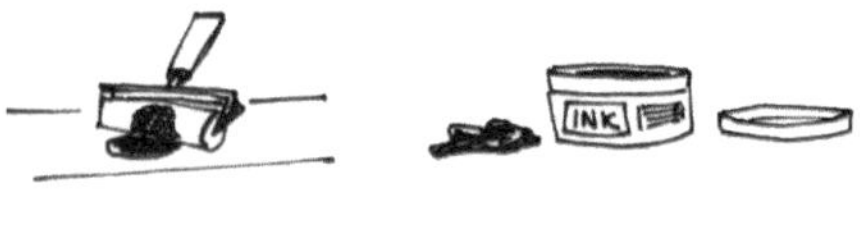

Most intaglio inks will work for printing the photogravure plate. The finely ground pigments of expensive inks will print better and are less abrasive. See Appendix F for a description of several types of inks and their wiping characteristics. The inks are mixed to a looser consistency for photogravure plates than for other intaglio plates. Stiff ink gives a harsher print with more contrast. Ink that has burnt plate oil added to it gives a softer, more veiled print with smooth tonalities and increased plate tone. Overly loose or oily inks can be stiffened with the addition of magnesium carbonate. Transparent base can be added to the ink mix to open the shadow detail. More intense inks will increase the contrast and make the blacks and dark tones more dense. Easy Wipe Compound or Miracle Gel Reducer can be used to make the surface ink release more easily and helps prevent the need for aggressive wiping. They should never be used in excess, however, because over time they can leave greasy stains in the print.

We suggest that you have a supply of three or four basic high quality inks. We have found that the Gamblin line of inks has excellent characteristics for photogravure. Presuming that black will be the most common ink used for proofing and editioning, have a selection of cool and warm blacks and intense and translucent blacks. You should also have a transparent base extender. Color inks can, of course, be used, either by themselves or to tint a translucent black. Multiple colors can be used either *à la poupée* or for multiple plate printings. Some pigments will, however, oxidize on contact with copper. If you want to work with a lot of color, get a good book on color intaglio printing.

The proofing process is the stage where the correct ink type and consistency are worked out. We recommend that you mix a small quantity of ink and then modify it to change the translucency or to adjust the viscosity (amount of oil). Be sure to keep accurate notes that list the types of ink, the additives, and the precise proportions of the various components used so that the results can be repeated. We always use the same brand of bone black ink for our first proof. This slightly transparent ink maintains open shadow detail and the constancy lets us compare and contrast the etch with previous plates.

SOLVENTS

You will need solvents to clean up the plate, the tools, and your work area when using ink and asphaltum. Common paint thinners or mineral spirits are usually used for the bulk of the cleaning. Always wear gloves and work with adequate ventilation when using solvents. Also be aware of the safe disposal of oily rags and newsprint. We use old newspapers as a base on which to clean plates so that solvents and inks are not spread all over the countertops. The newsprint absorbs spills and is a good surface on which to wipe excess ink from tools. Layers are removed as they become saturated or inky. Place in a garbage bag and remove it to the outdoors awaiting disposal. Under no circumstances should you leave a bag or can full of rags or papers containing oils, wet inks, or solvents anywhere indoors. It is a fire hazard and a source of fumes. A very safe alternative to volatile, flammable solvents can be as simple as vegetable or baby oil, but these can be slow to act and leave a very oily residue. The new soy-based (ester) solvents do not produce fumes and can be used to clean the ink from a plate or for general cleanup but they often leave an oily residue. These products, however still require skin protection and rags must be disposed of in safety containers. Other useful household products include non-toxic bio-degradable degreasers which clean very well. Naphtha is a nongreasy, highly volatile solvent that removes all residues. It is very useful to achieve a cleaner plate and squeaky clean tools. Acetone can be used to clean out stubborn or dried ink from a plate. Alcohol is useful for removing the rosin from an aquatinted plate. Never use dangerous solvents like lacquer thinner, turpentine, benzene, gasoline, or methyl hydrate.

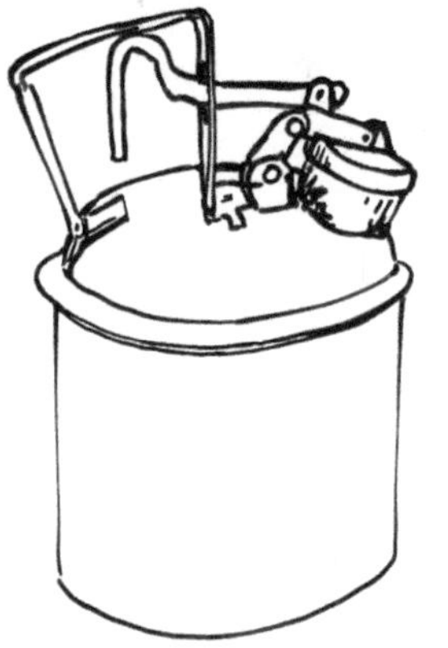

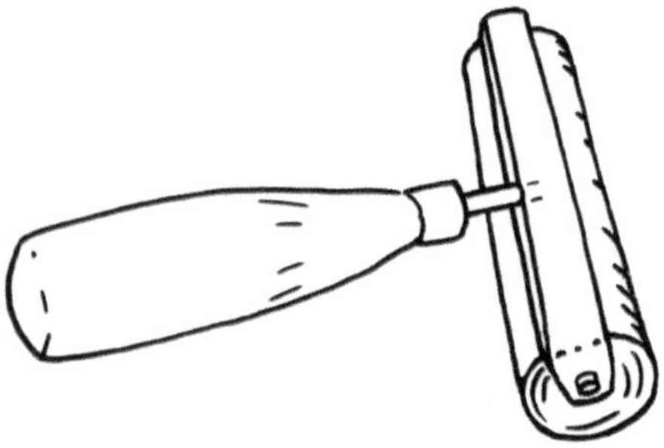

INKING AND WIPING SUPPLIES

We use two methods to initially ink the plate. The most gentle and safe method we have found is to use a soft rubber brayer. Alternatively, a small square of rubber screen-printing squeegee blade is excellent to spread ink on a plate. Do not use a square of mat board as a squeegee because it may contain embedded particles of grit.

Tarlatan is traditionally used to wipe intaglio plates but it is heavily starched. Hot-soak the tarlatan to remove the starch and soften the fabric in order to prevent scratches on the delicate surface of a photogravure plate. We prefer to use cheesecloth because it is soft and will not harm the plate.

Several pieces of cheesecloth or very soft tarlatan are needed, each about one square meter (yard) and rolled into a fist-sized wad with a flattened smooth bottom surface. It is best to have one previously used or inky cheesecloth, one moderately used cheesecloth, and one fairly clean cheesecloth. Keep the cheesecloth or tarlatans soft by storing them in a plastic bag. Do not wipe a gravure plate with a hard or crusty cheesecloth or tarlatan because it will scratch the delicate surface.

Other materials needed in the inking process are soft cotton rags for cleaning the edges and magnesium carbonate or baby powder for a clean, dry hand wipe at the end.

THE INTAGLIO PRESS, PRESS BLANKETS, AND BLOTTERS

A high quality intaglio press is the most important item of equipment you will need to print photogravures. The press must be capable of exerting very high and even pressure. Its rollers and press bed should be free of dips or distortions. The press is obviously a very expensive item and a major commitment for a home studio. A practical alternative is to become a member of a cooperative printshop in order to access its equipment and facility.

Use only high quality woven wool press blankets for printing photogravure. Ideally the blankets should be new, finely woven, and clean. Old worn blankets or blankets that are permanently hardened are unsuitable for photogravure. We use two very thin (<2 mm [1/16″]) blankets against the paper, and one 3 mm (1/8″) blanket above that (the pusher). Some printers may use even less. It is possible to use a sizing catcher blanket against the paper but this will need to be changed and cleaned fairly often. We reserve our set of blankets for photogravure printing only. If you are printing in a cooperative printshop, we recommend that you purchase your own blankets and bring them with you each time.

Good paper blotters are needed to remove excess water from the printing paper just prior to printing. Use cotton blotters that are pH neutral (acid free). Have enough for blotting and a second unwrinkled set for drying the prints.

MAKING THE FIRST PROOF

Paper Preparation

Paper comes in various standard sizes, often with two deckle edges and two torn edges. To determine the paper size, establish appropriate borders around an image with some thought given to minimizing waste. Resize or divide the paper to the finished dimension. A torn edge rather

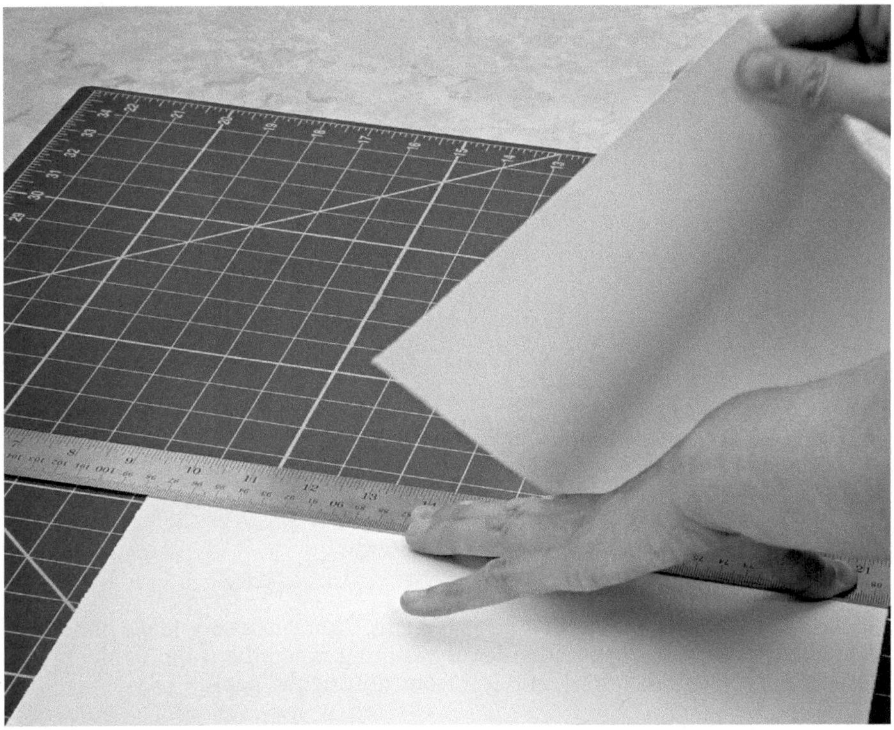

Figure 9-1 Tear a sheet of paper using a long, sharp-edged ruler. Be very careful of slippage.

than a cut one is common because it mimics the deckle edge and alludes to handmade paper (Figure 9-1).

Paper often has an internal grain running in one direction. This is apparent by the differences in the way it folds and tears. It also effects the amount of stretch when being printed and the direction of pronounced shrinkage when drying. Try to have the grain direction of all sheets used for an edition oriented in the same way. This is even more critical when multiple plates are being printed in registration on one sheet.

The sizing in many printing papers impedes ink transfer because the dry paper fibers are too stiff. It is important to soften the paper fibers by soaking the paper so that the fibers are flexible and can pick up all the ink from each minute depression in the plate. The usual practice is to submerge pieces of paper in a large sink of clean room-temperature or cold water (Figure 9-2). Soaking can be brief or can take most of the day depending on the paper type and its coating or size characteristics. As a preferable alternative to sink-soaking, the paper can be damp-bagged overnight. To do this, use the soak-and-wrap method:

1. Tear several sheets to size.
2. Soak the paper for 20 minutes to 1 hour in a clean sink with fresh water.
3. Have a large square of clean garbage bag plastic ready.
4. Remove the paper from the water and let it drip; do not blot.
5. Stack the sheets on the plastic.
6. Wrap the stack of paper in the plastic and tape the seams shut. Wrap it in a second layer or insert it into a bag.
7. Let it sit flat overnight.

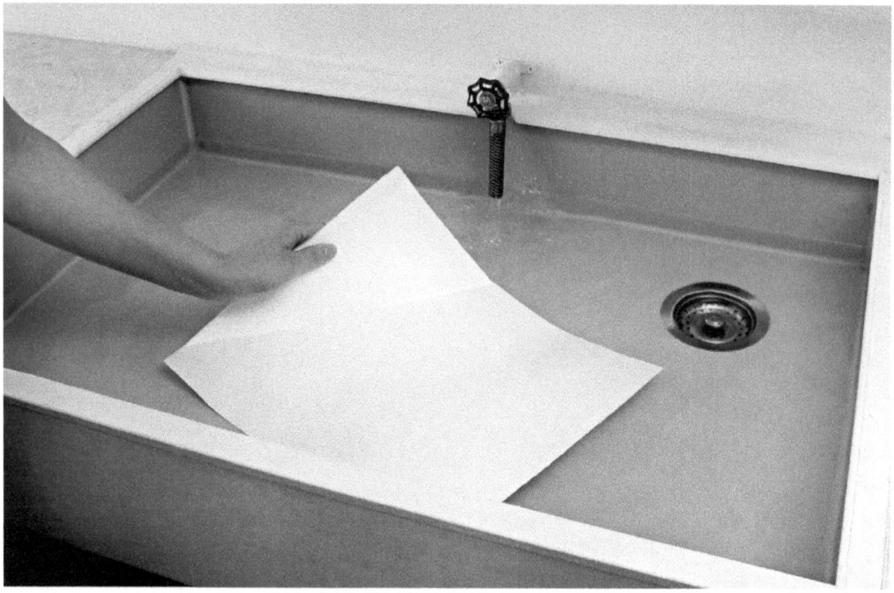

Figure 9-2 Soak the paper in a large flat sink. You can easily make one using plastic laminated plywood and silicone caulking. Keep the water fresh and clean to prevent stains, mold, and grit from ruining the paper.

The resulting paper will be thoroughly softened and ready to use. This method saves time and space because a large quantity of paper can be prepared easily. Note that foxing or mold may result if the damp paper is left wrapped in plastic for too long or if clean procedures and proper precautions are not observed. If printing is going to be suspended for a period of time (more than a few days), spread the soaked paper out on blotters to air-dry. It will keep indefinitely once dry. It can be easily resoftened with a short soak because most of the sizing has been removed during the first soak.

Papers with little sizing require a short soak. Simply immerse this paper in a sink of cool or tepid water for 30 seconds to 2 minutes. Blot before printing. Waterleaf papers have little or no sizing and simply need a quick dip or just a misting with an atomizer filled with water. Mist both sides of the paper while the paper is lying in a dry sink or on a clean counter. Blot just before printing. The paper chart in Appendix G gives soaking recommendations.

Ink Mixing

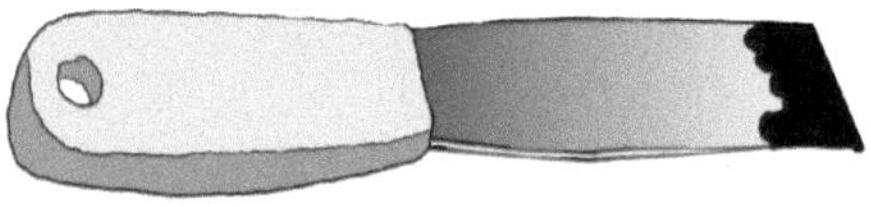

To start, be sure to wear solvent-resistant gloves or barrier cream when working with inks and solvents. Use a piece of 7 mm (1/4″) thick glass as an ink mixing surface. Place it over a sheet of white paper. Be sure the edges are ground or taped. Set out daubs of ink and additives along the top edge of the glass (Figure 9-3). Use the rest of the glass to mix and roll the ink. Modifiers such as Easy Wipe Compound or plate oil can be added in small quantities only if needed and mixed in well. The Easy Wipe helps prevent the over-wiping that comes from scrubbing the plate to remove resistant or sticky ink. Plate oil also makes the ink easier to wipe initially but more difficult to remove all of the plate tone. Work or mix the ink using a palette knife or 3 cm (1 inch) putty knife to loosen it up. It is important to work the ink before applying it to the plate (Figure 9-4).

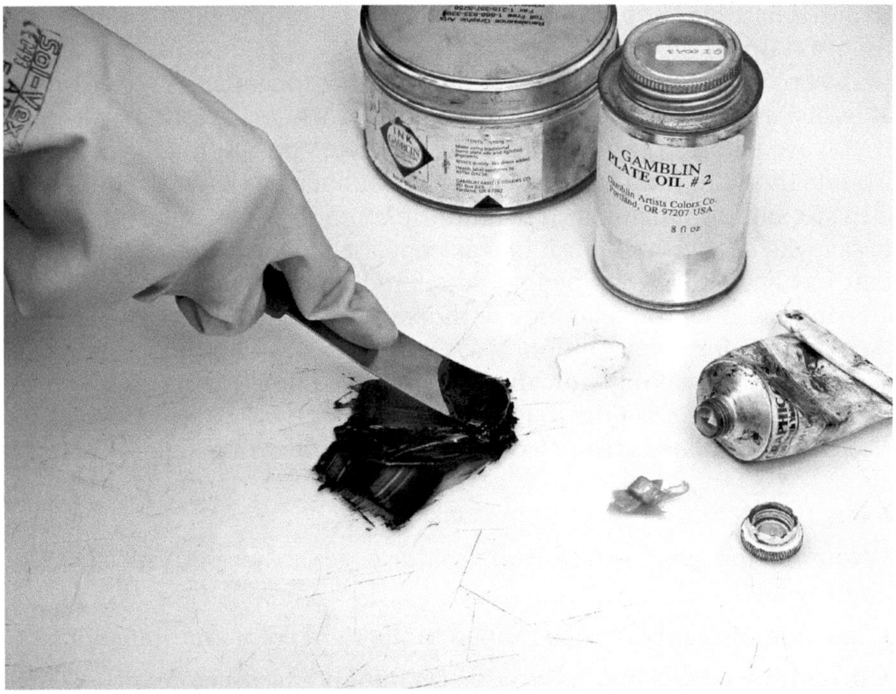

Figure 9-3 Ink additives and ink on a glass slab.

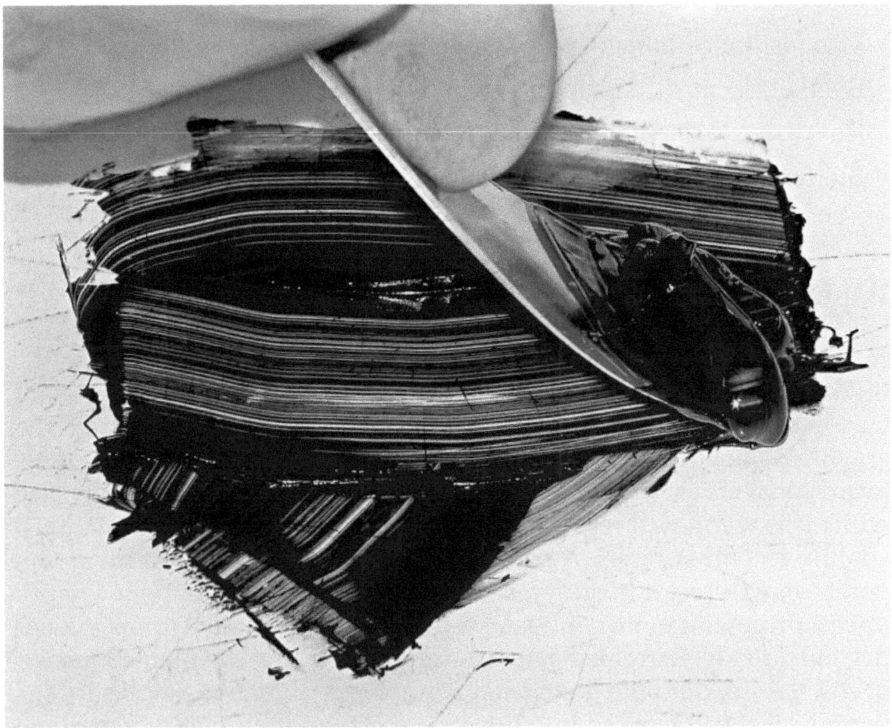

Figure 9-4 Work the ink back and forth to loosen it, to separate out dried lumps, and to mix ingredients.

Remove particles or bits of ink skin because they can cause problems if they end up on the plate. If you are using more than one ink, work them with separate putty knives before mixing them together. You will need at least 5 ml (1 tsp) of ink to coat a 20.5 cm × 25.5 cm (8″ × 10″) plate once if you are using a piece of rubber squeegee to apply it to a warm plate. Double this amount if you apply it with a rubber brayer. You would usually mix more at one time so that multiple proofs can be inked. It is easy to save the ink by wrapping it in wax paper or a foil-coated butter-wrapper that has been cleaned. Shadow detail and contrast can be adjusted by altering the ink's transparency with extenders or transparent base. See Appendix F for a description of various inks and their characteristics. Books on printmaking are valuable sources of detailed information on intaglio printing. (See the Reference Bibliography.)

When your plate is ready to print, have the following materials ready:

- Rag paper (torn to size and presoaked)
- Your photogravure plate (check that the edges are properly beveled and burnished)
- Ink on a glass mixing slab (mixed to the right color and intensity)
- Ink additives at hand (Easy Wipe Compound, burnt plate oil)
- Paint thinner, naphtha, newsprint pad, and rags
- Solvent-resistant gloves, print apron, and barrier cream
- Square of soft rubber squeegee or a rubber brayer
- Cheesecloth or softened tarlatan (three pieces)
- Magnesium carbonate, whiting, or baby powder
- Blankets in position and press set to appropriate pressure
- Hot plate set to "blood warm" (see description under "Wiping")
- Blotters for the wet paper, a wooden rolling pin, and a drafting brush

Wiping

A printmaker's hot plate can be as simple as an electric hot plate with a thick plate of steel (1/4 to 5/16 inches thick and at least a foot square) on top of the element(s) to diffuse the heat. Warm the etched copper gravure plate with the hot plate set to a very low temperature described by printmakers as "blood warm"—or warm to the touch. Wiping the plate while it is warm will result in ink that is easier to remove and will give slightly smoother tones. A cold wipe, or wiping the plate at room temperature, will result in stiffer, more resistant ink and a tendency to produce a print with more contrast. Do not overheat the plate or the ink will dry too quickly. Extreme heat will bake the ink into the plate. Squeegee or roll the ink onto the plate from a number of directions to ensure that the ink gets well worked into the etched pits and hollows on the plate's surface (Figure 9-5). Start with the inkiest cheesecloth or tarlatan and work the ink into the plate with a gentle downward twisting motion (Figure 9-6). This pushes ink into the recesses while lifting some of the excess ink off the raised areas. Next, wipe the surface of the plate using a sweeping semi-circular stroke. Start the cheesecloth just off the edge

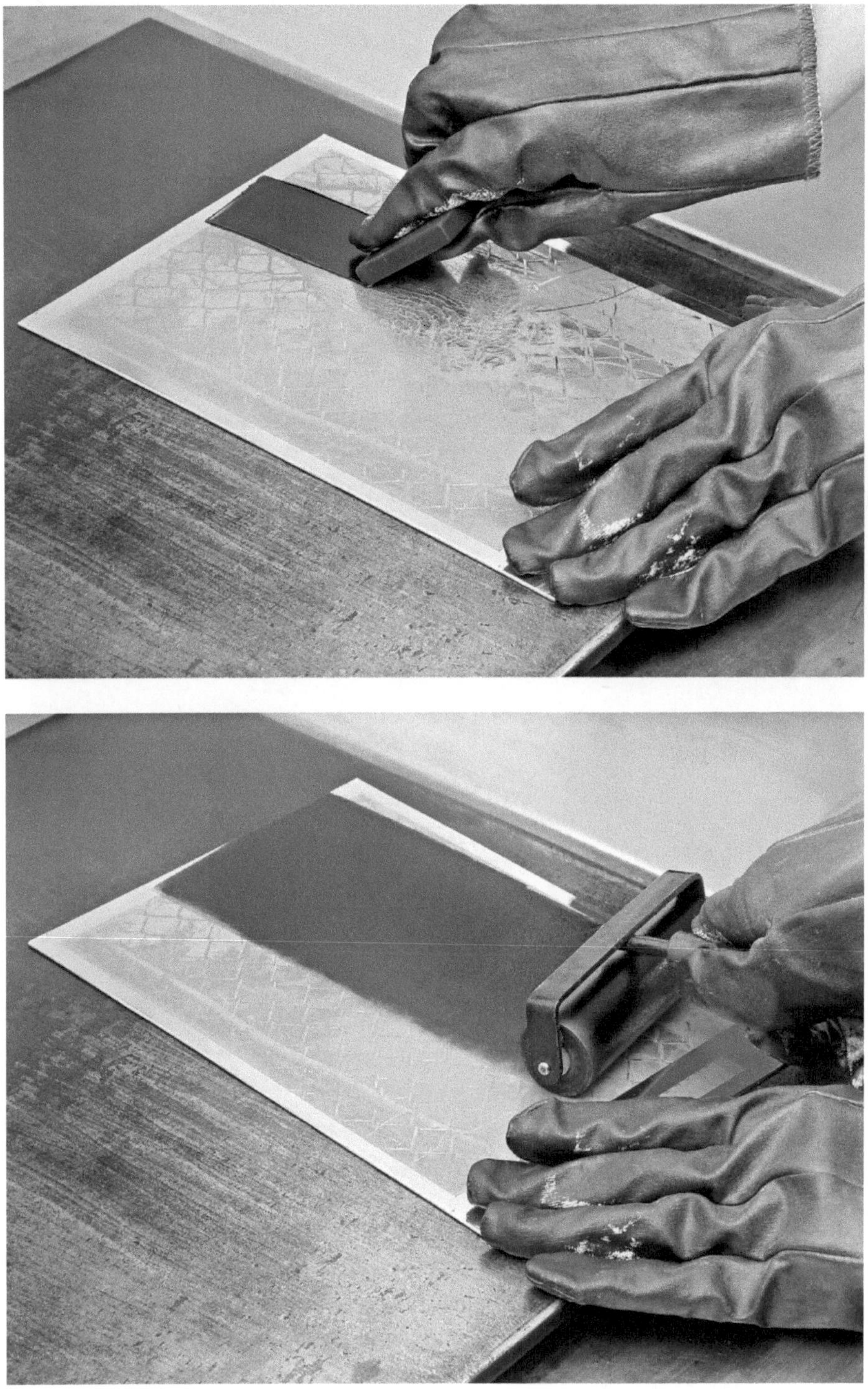

Figure 9-5 A soft piece of squeegee rubber can be used to spread and work the ink into the image area of the plate. Alternatively, a soft rubber brayer will spread a thinner layer over the plate. Be sure that enough ink is spread and worked into the image so that it is totally hidden under a dense layer of ink.

of the plate and release the pressure toward the center before lightly lifting it off the surface with a smooth follow-through stroke (Figure 9-7). Do not scrub. The plate will quickly lose excess ink and the image will begin to emerge. After most of the excess ink has been removed, switch

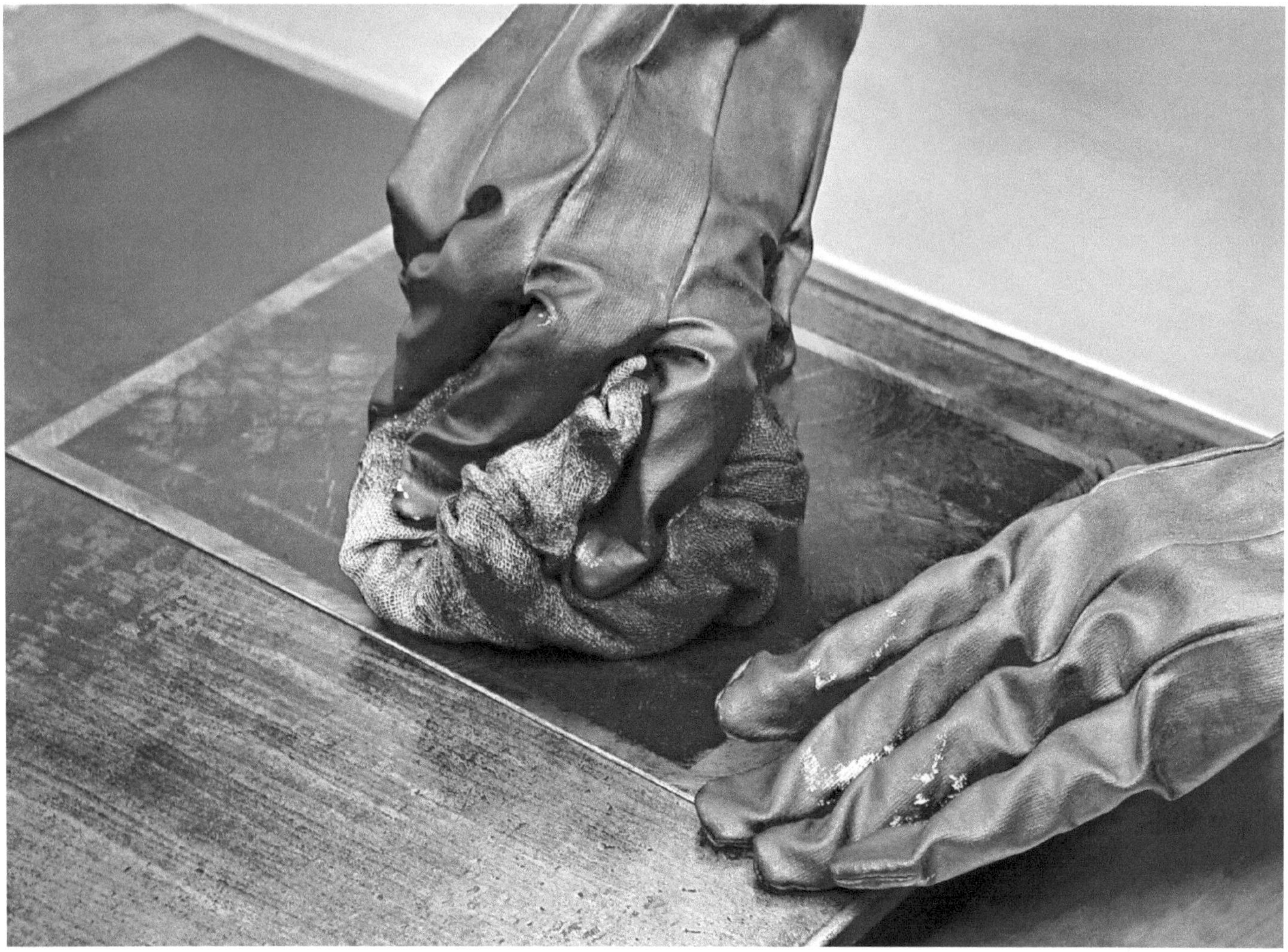

Figure 9-6 Use the balled up cheesecloth to twist the ink into the face of the plate with downward pressure.

to the next cheesecloth and continue wiping in the same way but with even less pressure, all the while turning the plate so that each stroke wipes a different segment. Also, keep reforming the cheesecloth pad so that the inky surface is turned into the middle and a fresh surface is smoothed across the wiping side of the pad (Figure 9-8). Remember that wiping too slowly with heavy pressure will remove ink from the pits and hollows, causing an over-wiped plate.

Use the cleanest cheesecloth or tarlatan to quickly and lightly give a final wipe. At this point, there should be very little surface ink remaining on the polished unetched areas. All that is left to do after this is a hand wipe to bring out the highlights and smooth the tones. If you haven't already done so, apply barrier cream to your hands and wrists to prevent absorption of the ink into your system when hand wiping. Dry your hands well. Use the fleshy part of the side of your palm in a quick stroking swipe towards your chest. The wiping stroke should be light and quick, drawing your palm towards you across the plate (Figure 9-9). Do not abruptly press your hand down onto the plate nor lift suddenly from its surface.

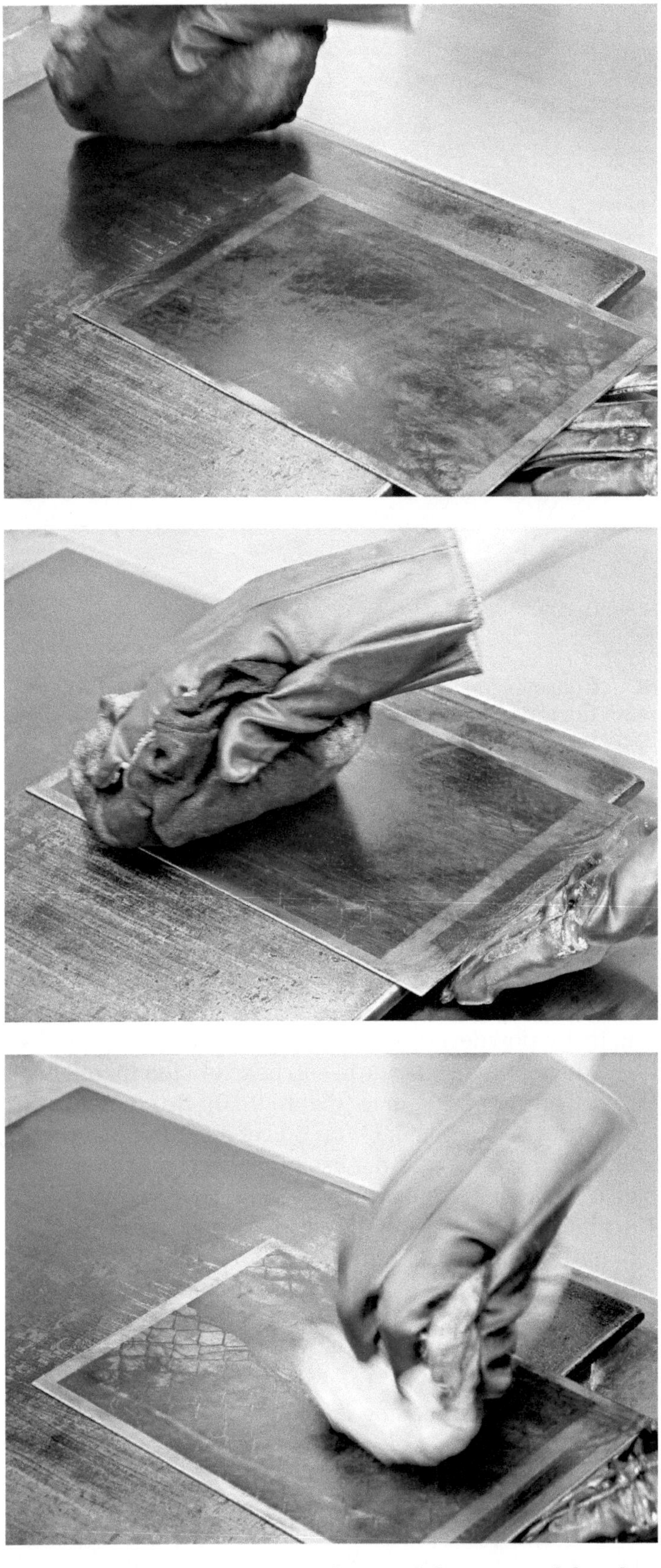

Figure 9-7 A sweeping motion is used toward the center of the plate to wipe the ink. The cheesecloth is lifted while still in motion.

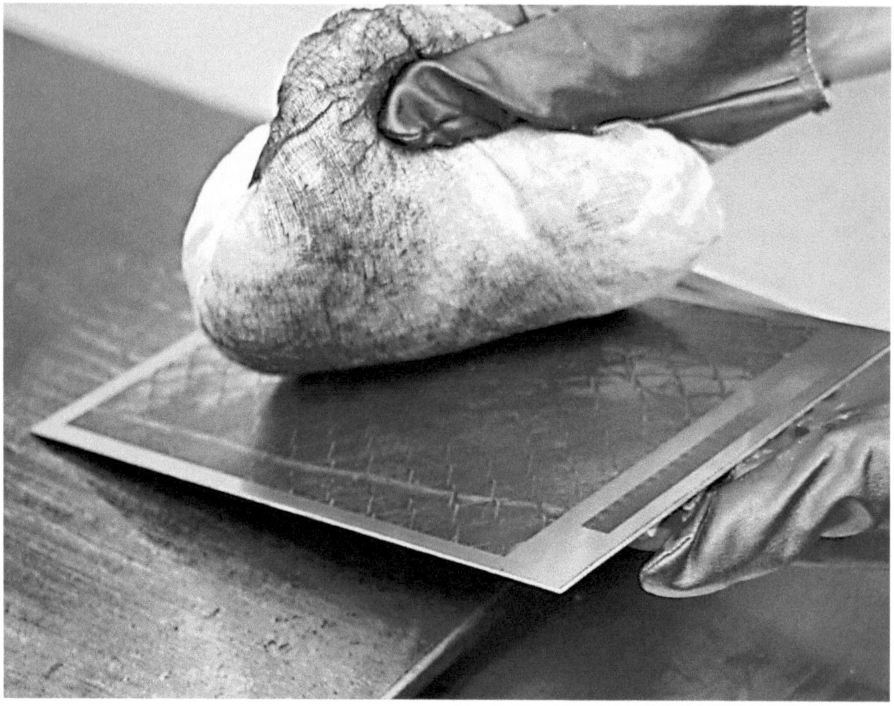

Figure 9-8 A cleaner piece of cheesecloth removes almost all the excess ink remaining on the upper surface of the plate. Use the cheesecloth gently and move it fairly quickly over the surface.

All motions, although quick, should be smooth and gentle with gradual pressure changes. See Color Plate 21.

When printing with colored inks, use a clean piece of cheesecloth or tarlatan if you do not have an inky one with a close color match. Save the used cheesecloth or tarlatan in a plastic bag to keep it soft. Use a separate bag for each color and store in a fireproof place.

At this point you can accept the plate as it is, with a slight plate tone. You can also use a bit of magnesium carbonate (mag), calcium carbonate (whiting), or talc (baby powder) rubbed onto the edge of your hand—with the excess dusted off against your apron—to brighten the highlights even further and totally remove plate tone (Figure 9-10). Be careful not to get any clumps of powder on the plate. If you do get a clump, reapply a fingertip of fresh ink on the spot. Avoid the use of newsprint or paper to wipe the plate because it may lift out too much ink and there is a risk of abrasion.

Once the plate is wiped, clean the edges with a clean rag or one slightly dampened with naphtha or, more traditionally, with a pinch of mag between the fingers (Figure 9-11). Photogravure can be given a very clean wipe to maintain brilliant whites. In fact the plate itself may look overwiped, as though there is very little ink left. Be careful not to actually over-wipe. In order to print all the ink that is left, the plate should be rewarmed slightly before placing it on the press bed.

Printing

Before printing, set up a registration sheet to simplify aligning the plate and the paper. Center a piece of paper with the plate outline and full paper size marked on it onto the press bed. Tape down a larger piece of Mylar over the registration sheet so that cleanup is easy between printings.

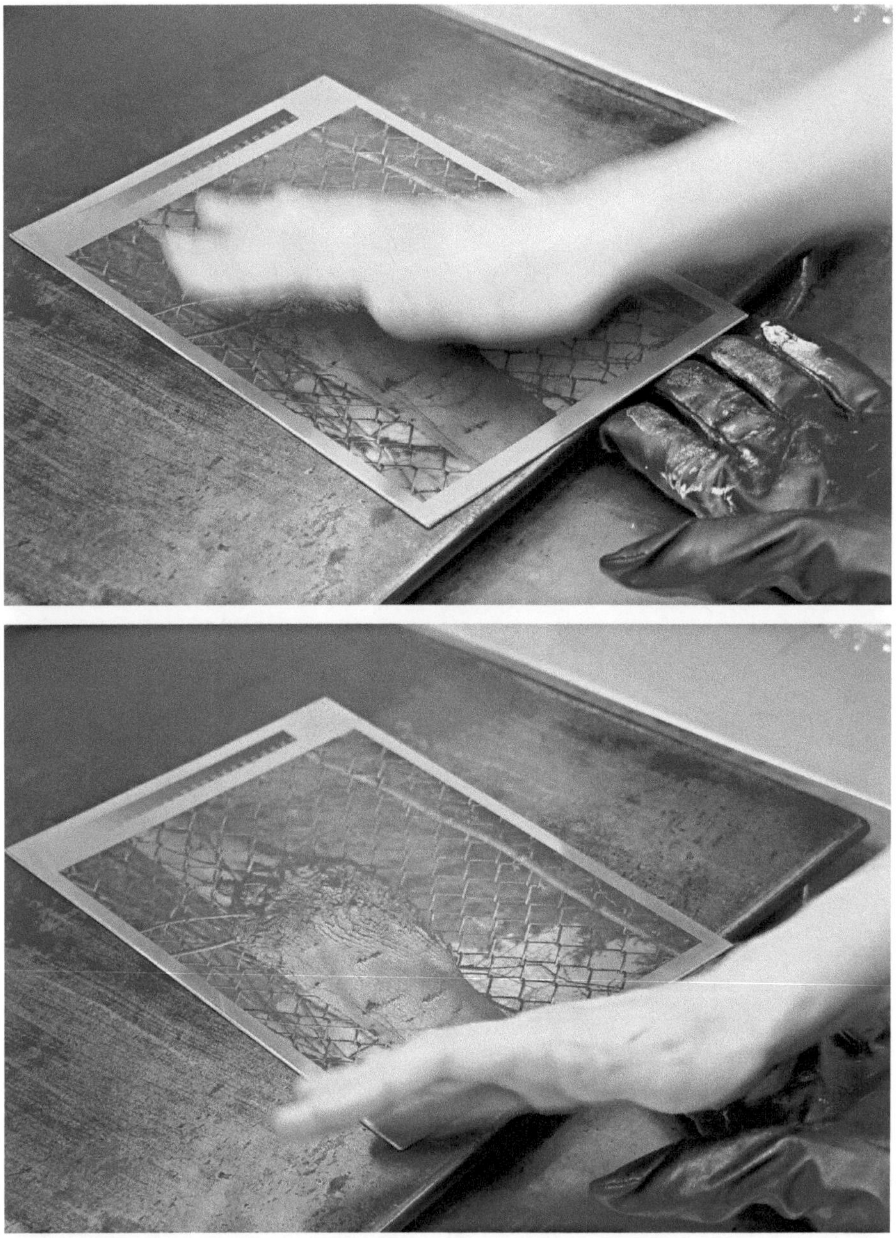

Figure 9-9 This sequence shows how the fleshy part of the palm is used to sweep across the plate and shine the surface. The skin must be in motion at all times when in contact with the plate or a hand print will be left behind.

Set the printing pressure slightly higher than normally used for intaglio printing. Stack the blankets with the thinner blankets next to the press bed and the thickest on the top. Stagger the blankets so that it is easier to roll them under the roller to trap the edge (Figure 9-12). Test the pressure with a piece of soaked and blotted paper and an uninked plate. Run the clean plate and a piece of paper through the press as if you were printing. The paper should show a strong plate mark, even pressure on the left and right edges, and signs of the image showing as an embossment in the paper. Too much pressure is indicated by paper creases, tears at the plate edge, or ripples. This is also a good way to ensure that the pressure is even across the plate. Uneven side-to-side pressure will be

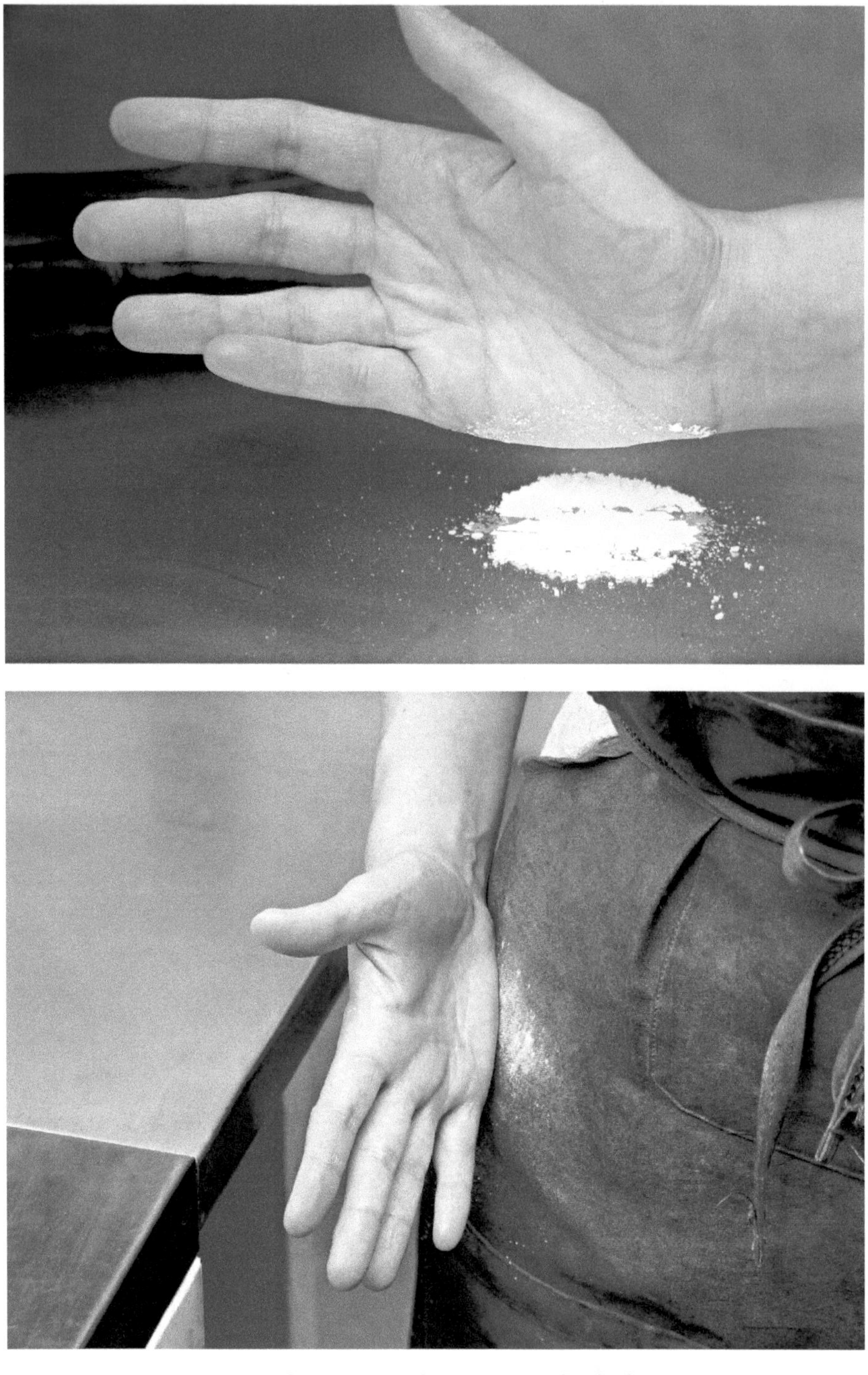

Figure 9-10 Tamp your hand in a pile of mag and rub the excess on your apron. Be sure no clumps of mag are dropped onto the plate. Rerubbing your hand against the white spot on your apron may be enough to recharge your hand with mag.

visible as variations in the degree of edge embossment from one side to the other.

Lay the inked, wiped, and rewarmed plate onto the press bed in its proper position on the registration sheet (Figure 9-13). Make sure your hands are clean before preparing a piece of paper for printing. Drain a

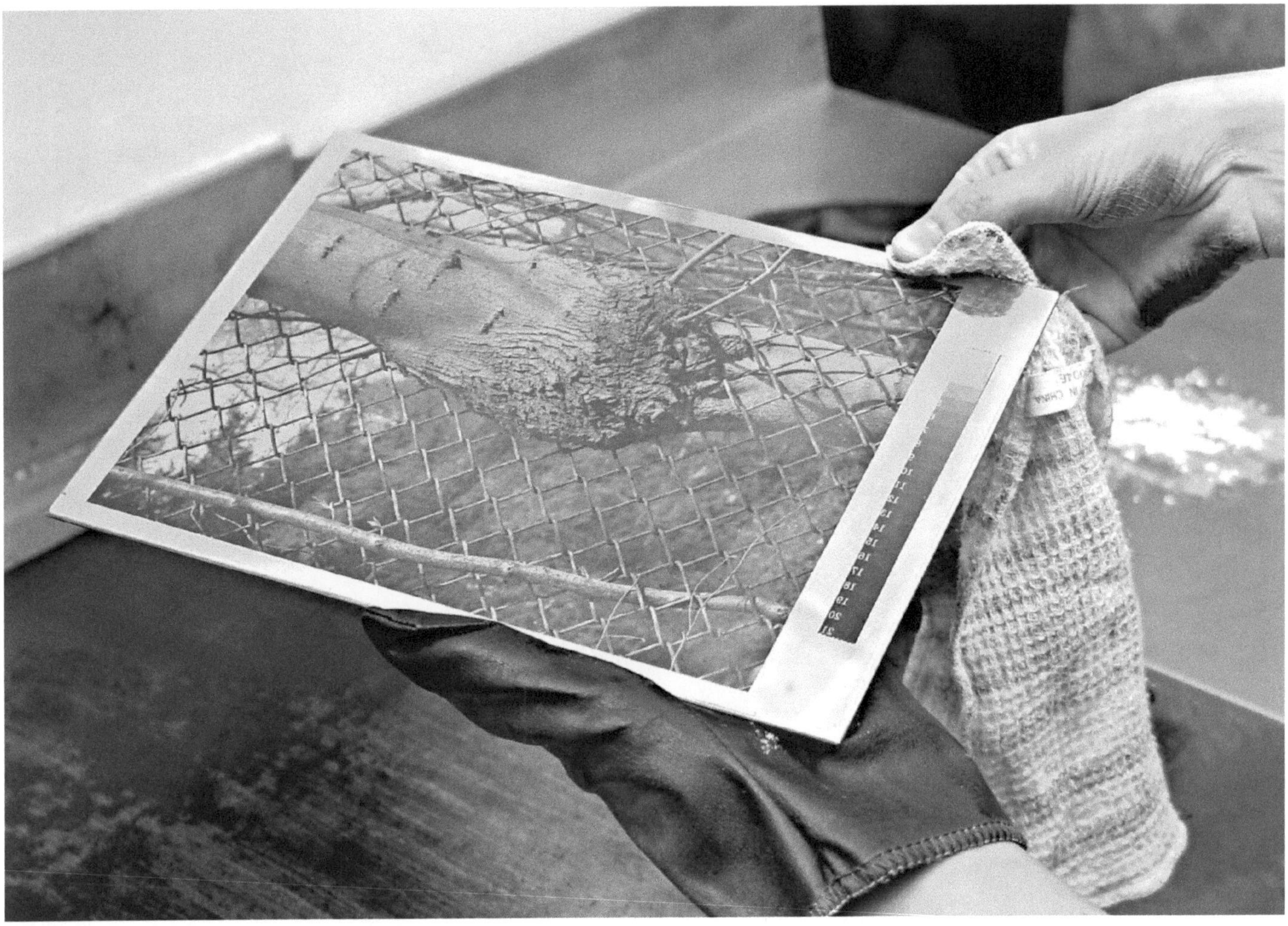

Figure 9-11 During the cheesecloth wipe, some of the excess ink can be removed from the borders using the cheese cloth. After the hand wipe, a dry rag is used. For a pure white border, a pinch of mag can be used for a final edge wipe. Be very careful not to rub the mag, the rag, or the solvents onto the image area. Do a final overall touchup with your hand.

piece of paper and place it onto a clean blotter. Lay a second blotter on top and roll the assembly with a rolling pin or large roller to press the paper into contact with the blotters. Reposition the paper to a dry area and repeat. It is important to be sure that all the water is removed from the paper surface. Uncover the paper and use a clean drafting brush to briskly brush both sides of the paper. This removes specks and debris and fluffs up the surface fibers. Examine carefully. Check the paper for its "good" side or felt side and move to the press bed. You can use two folded playing cards to hold the paper to ensure you don't leave inky finger smudges on its clean borders. The felt or printing side of the paper should face the plate and fall precisely on the lines established by the registration guide (Figure 9-14). Once you have carefully lain the paper down over the plate, cover it gently with the printing blankets. Ideally, an assistant should hold the blankets suspended over the paper by pulling them tight as they are being drawn under the roller (Figure 9-15). This prevents wrinkles and gives a better impression. Use the press's wheel or crank to roll the print slowly and smoothly through the press. Do not pause part way. Pull the blankets back and then gently peel the paper from the plate (Figure 9-16). If there is any tendency for the paper

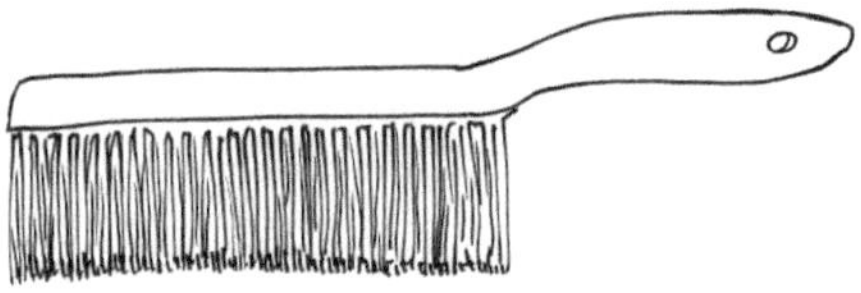

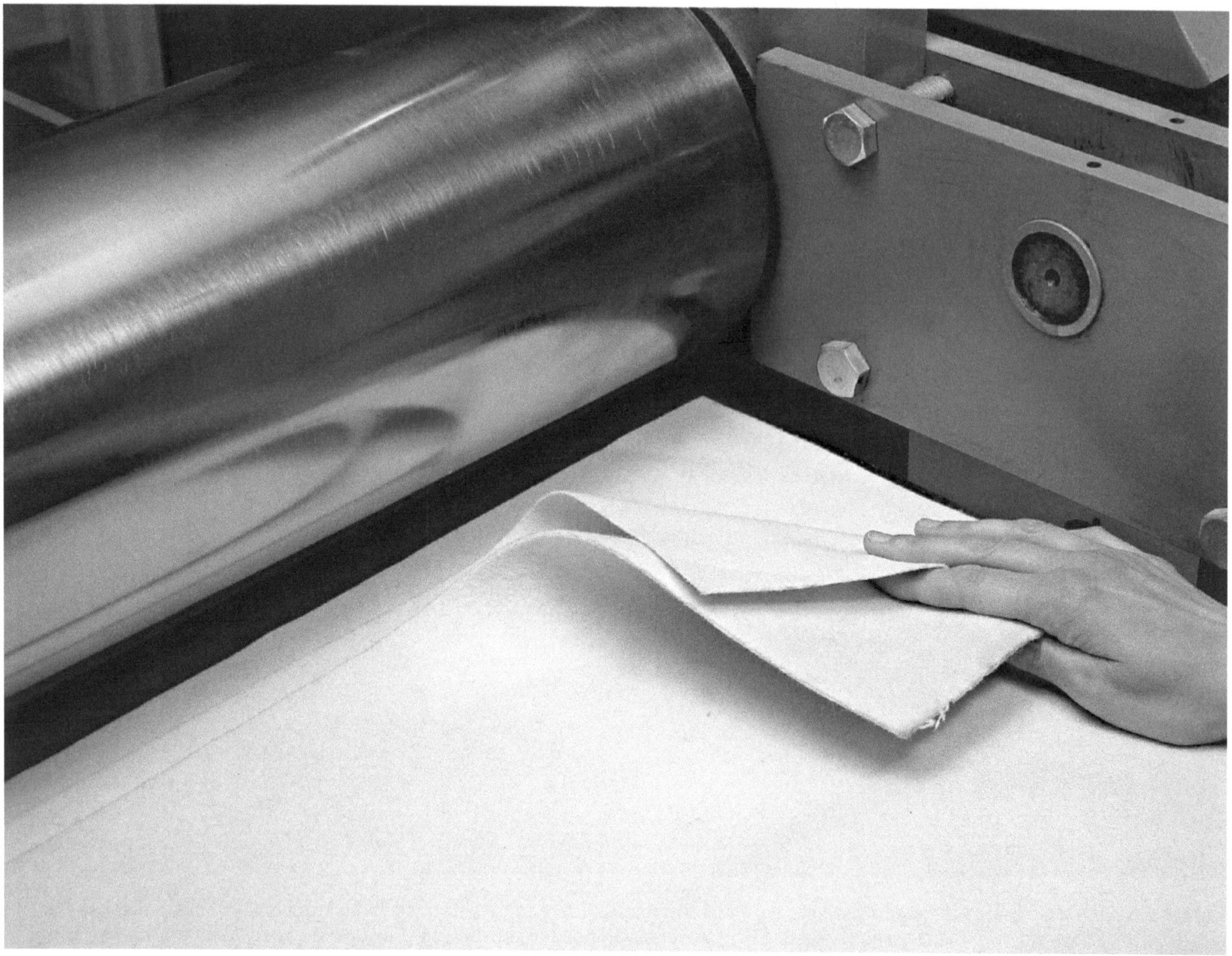

Figure 9-12 The blankets are staggered so that the press roller can climb onto them one at a time. Be sure the thinnest and best blanket is on the bottom.

to stick or pull up fibers, heat the plate on the hot plate before slowly peeling off the print. Cover the hot plate with a clean piece of newsprint first. Be sure to also clean the surface of the plastic over the registration guide before repeating the above procedure for subsequent printings.

Drying the Proof

If there is an extra allowance on all the margins around the image, the print can be pinned, taped, or stapled to a drying board. This method of drying is very quick, but the plate embossment will be reduced. A slower, more traditional method is to place the paper between blotters and fiber boards and then weight it. The plate embossment will be preserved (Figure 9-17). The drying time varies according to paper weight, relative humidity, and movement of air through or around the blotters. We use a system of porous 1.5 cm (1/2″)-thick fiber boards with cotton blotters between them. Change the blotters daily throughout the drying process to speed things up and lessen the rippling effect.

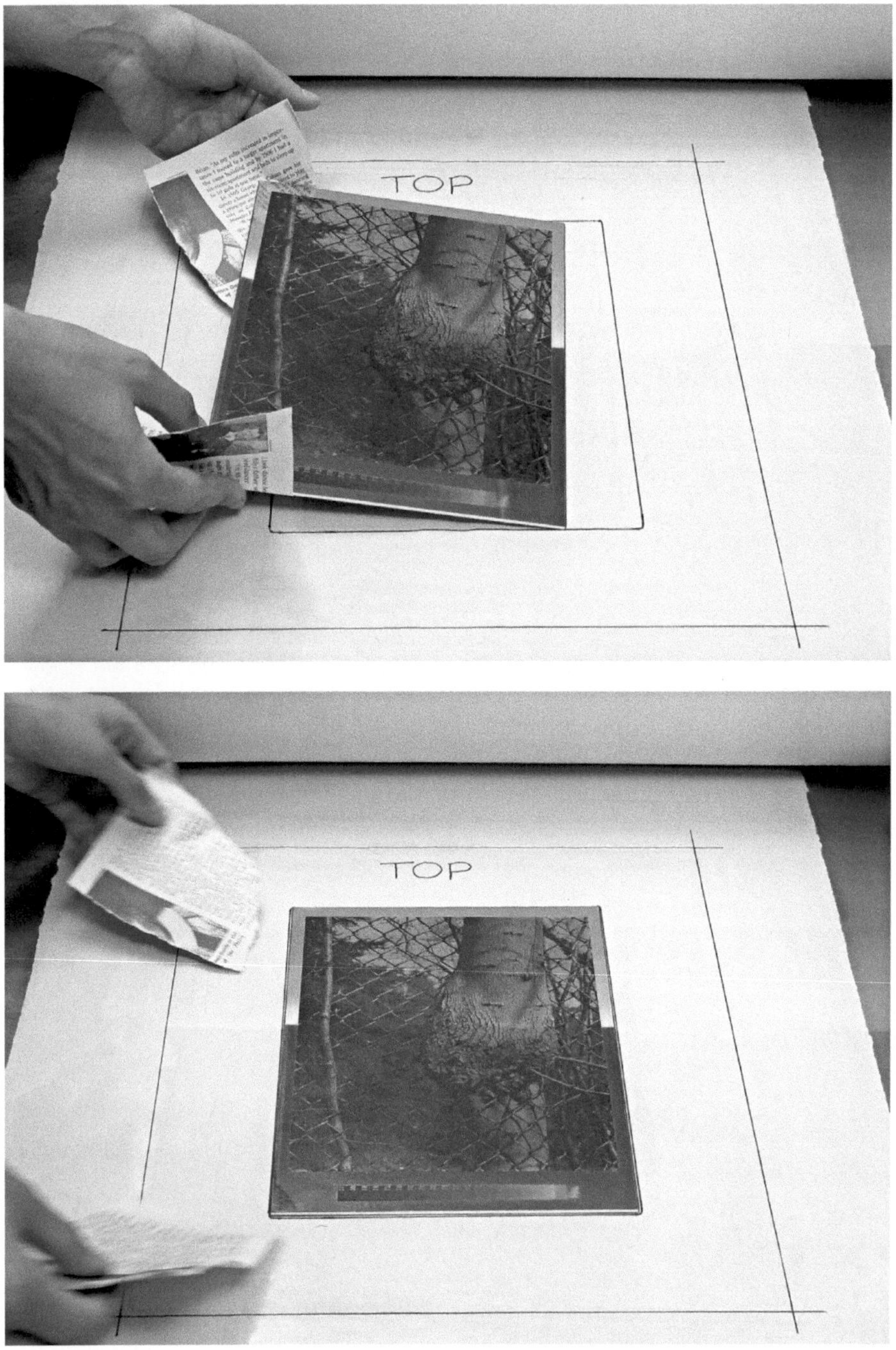

Figure 9-13 Place the plate in its outline on the registration sheet. Be careful not to slide it or it might leave an inky smudge that will transfer to the border of your print.

Cleaning and Storing the Plate

After every printing session, it is important to remove all traces of the ink before it dries. An oily solvent removes most of the ink, but a final cleaning with naphtha is needed to remove oily and inky residues (Figure 9-18). Wear gloves and work under a ventilation hood. After the plate is clean and dry it can be stored briefly before the next printing session.

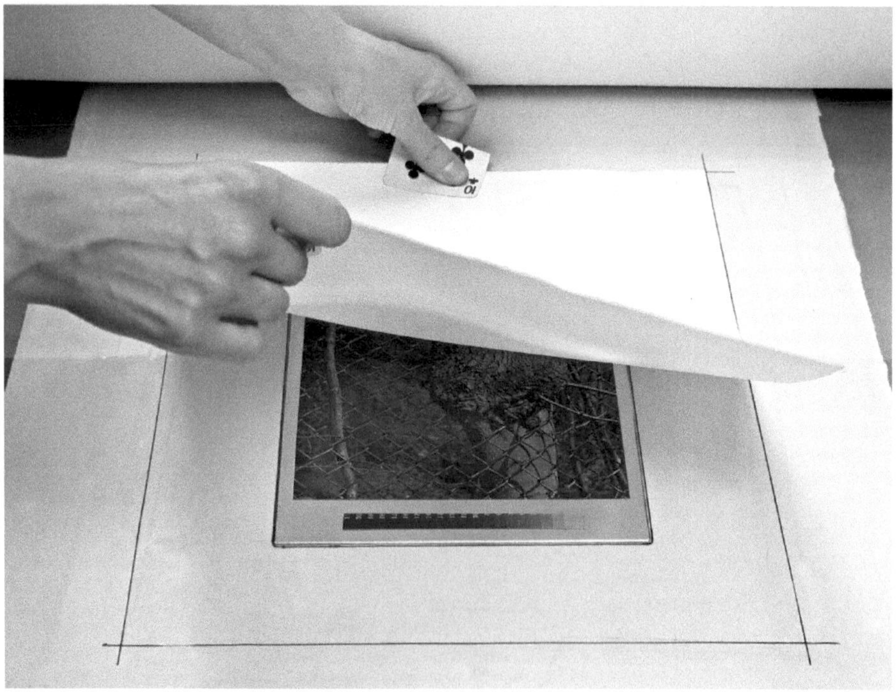

Figure 9-14 Use playing cards to hold the paper as you carefully position it onto the press bed. Do not slide the paper across the surface of the plate. Register it by the leading edge and one side and gently lower it into position.

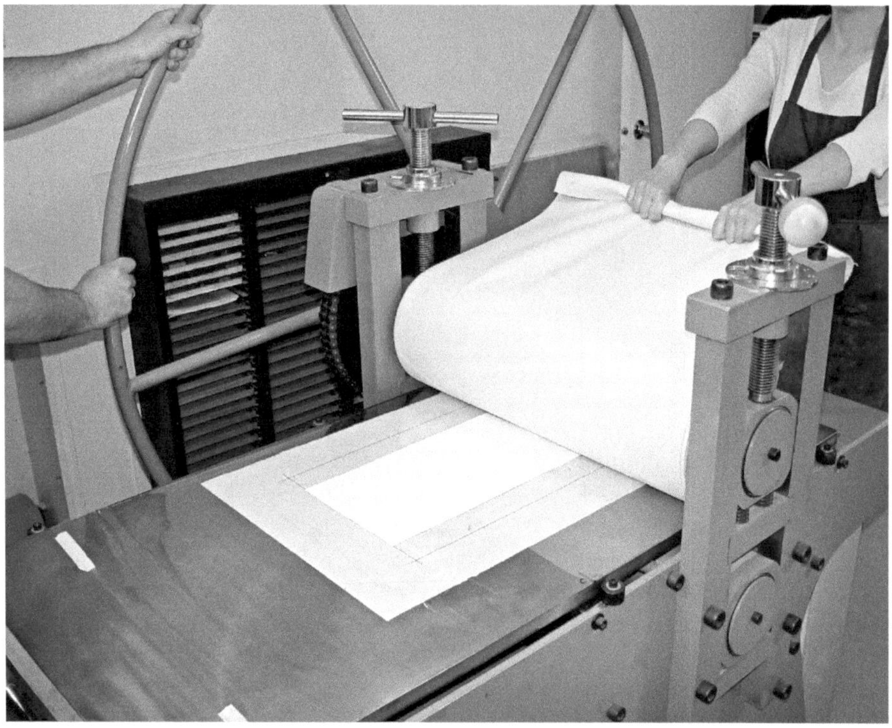

Figure 9-15 By pulling on the blankets, there is less chance of the plate moving or of a crease forming in the paper. The impression is often better.

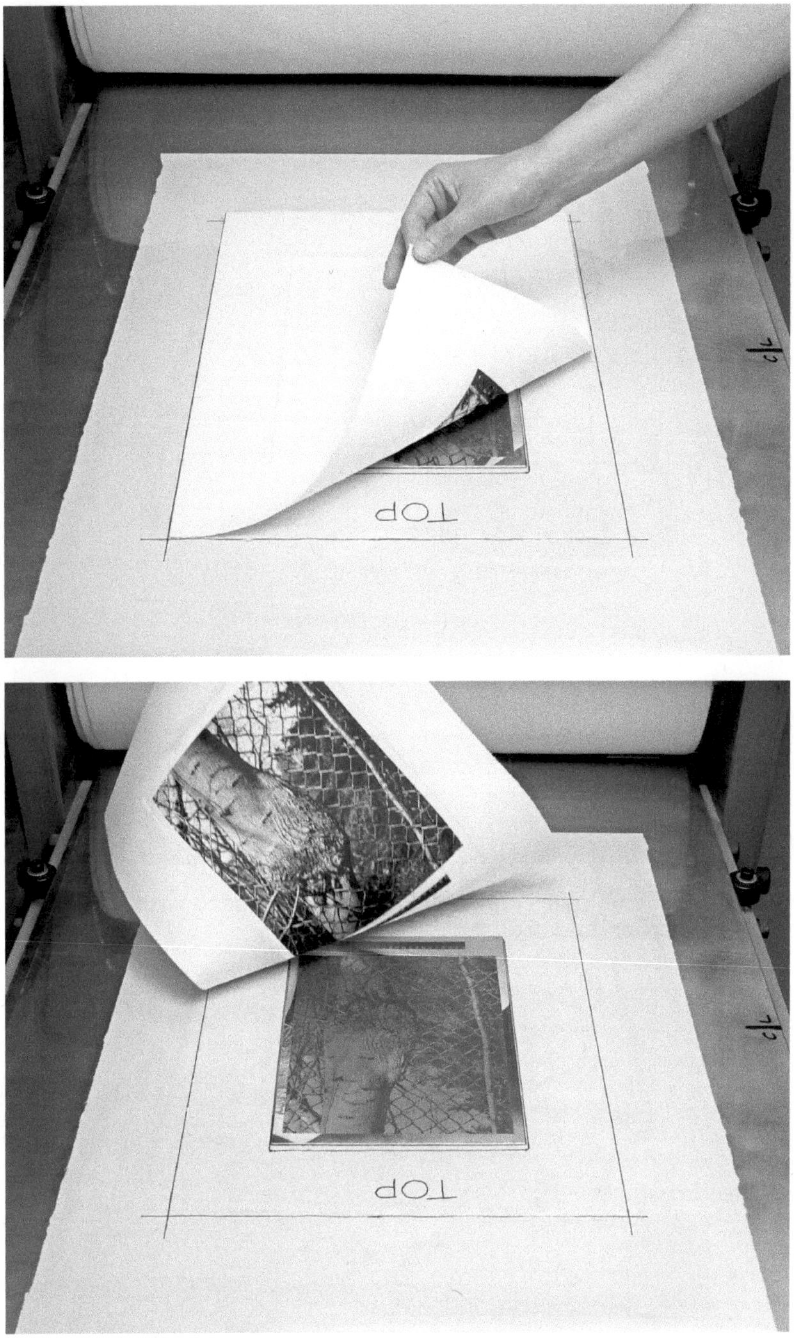

Figure 9-16 Pull the print slowly so as not to lift any paper fibers.

If the plate is to be stored for a week or more, it is best to coat its face with a protective layer to prevent tarnish or oxidization, which can show up in subsequent printings. The best way to protect a plate is to coat it with liquid asphaltum. Asphaltum dries to a thin, impervious layer and remains easily removable with mineral spirits or paint thinner (Figure 9-19). A safer but messy coating alternative is to use Vaseline. A storage folder made of stiff paper or card stock is a good idea to prevent accidental scrapes or marks and protects everything else from the asphaltum or Vaseline. For easy identification label the outside or use an old proof to make the folder.

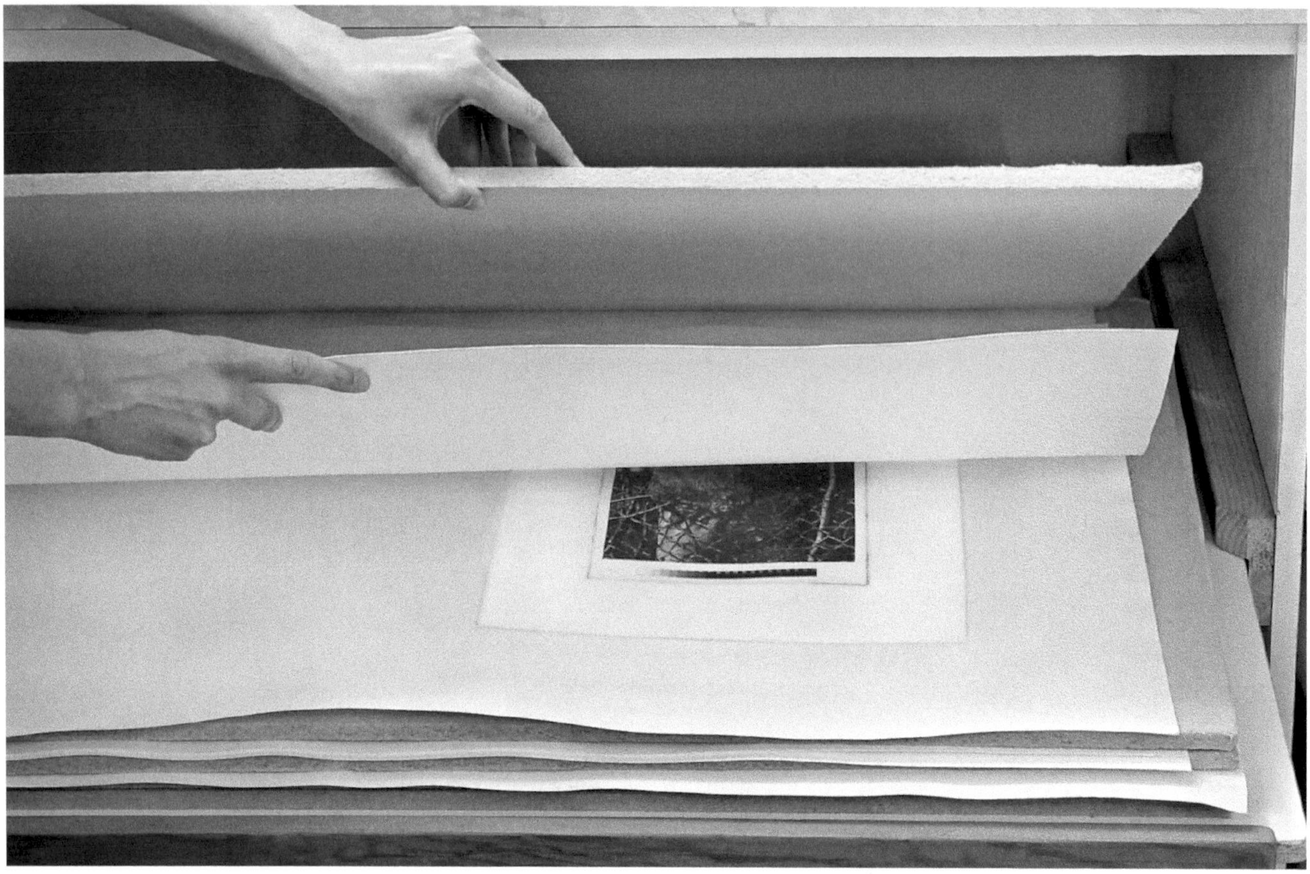

Figure 9-17 Place the print between clean, smooth, acid-free blotters and weight them with fiber boards. Change the blotters daily until the print is no longer cool to the touch, therefore dry.

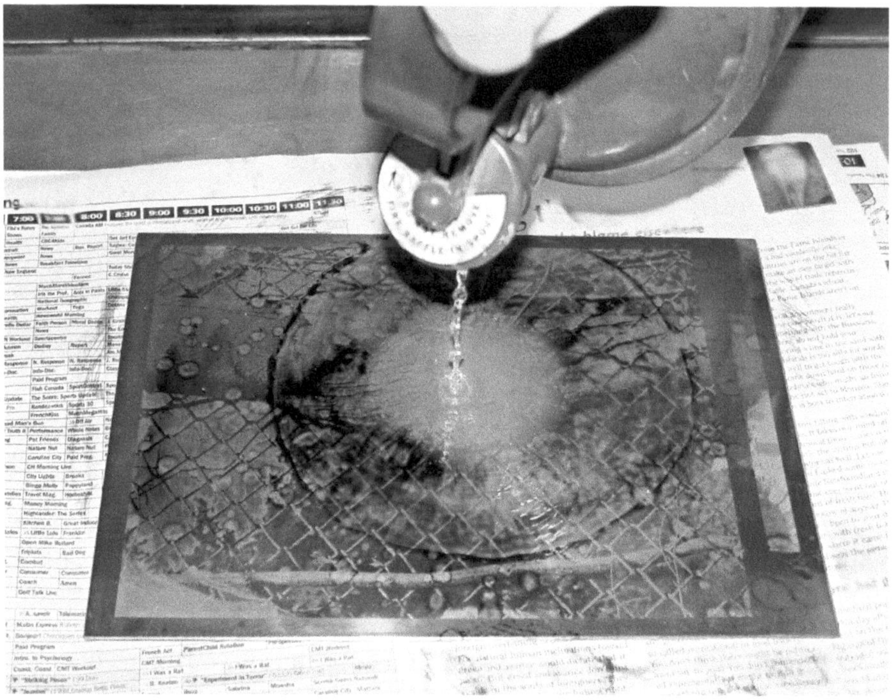

Figure 9-18 Naphtha is used for a final cleanup. Note how it removes the ink left behind by the first solvent.

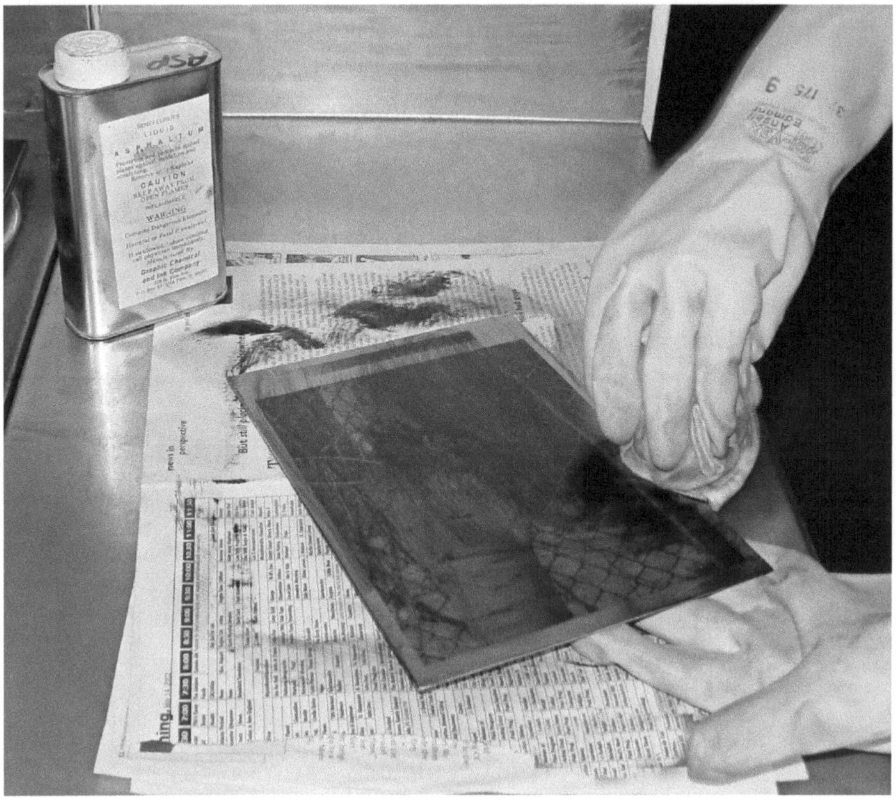

Figure 9-19 Asphaltum is spread in a thin layer to protect the plate. Use gloves and ventilation.

SUMMARY

1. **Make sure the edges of the plate are properly beveled and burnished.**
2. **Have the printing paper prepared—torn to size and presoaked.**
3. **Have the cheesecloth or very well-softened tarlatan prepared—cut to size and balled up.**
4. **Prepare the ink, mixing it to the right consistency.**
5. **Set the pressure of the press slightly higher than for standard intaglio printing.**
6. **Warm the plate on the hot plate for a warm wipe (38°C or 100°F).**
7. **Squeegee or roll the ink onto the plate in all directions. Work the ink into the recesses using a well-inked cheesecloth.**
8. **Continue with the well-inked cheesecloth or soft tarlatan and begin wiping the plate in gentle circular sweeping motions. Be careful not to stop and lift ink from the surface of the plate.**
9. **Once the excess ink is removed, change to a slightly cleaner cheesecloth.**
10. **Use the cleanest cheesecloth for the final stage of wiping. Be sure to use very little pressure—*no* scrubbing—and move the**

cheesecloth in continuous motions over the plate. Rotate the plate as you wipe.

11. **Do a "brisk" but thorough hand wipe for the final cleanup. Magnesium carbonate can be used when hand wiping in order to brighten the plate tone.**
12. **Clean the borders, edges, and plate bevel with a rag or soft paper.**
13. **Rewarm the plate on the hot plate for a "warm" print.**
14. **Use newsprint and an acetate or Mylar registration sheet on the press bed to aid in positioning the plate.**
15. **Blot the wet paper between clean newsprint or white blotters to remove all the water from the surface. Brush in both directions with a clean drafting brush to remove stray lint, dust, and hairs from both sides.**
16. **Lay the blotted printing paper, good side down, over the plate.**
17. **Cover with the printing blankets.**
18. **Roll slowly through the press in an even and steady manner without stopping.**
19. **Carefully lift up the blankets.**
20. **Gently pull the paper away from the plate. If there is resistance, move them together to the hot plate to warm slightly before continuing to pull the paper away from the plate.**
21. **Inspect for printing problems, then set aside between clean blotters to dry. Add weight to keep flat.**

TROUBLESHOOTING

Printing Problems

Degraded Image If the print shows mottled or broken up tones, pale blacks, and a loss of highlight detail, the problem may be dirty blankets, too little pressure, too wet paper, poor paper choice, dry or stiff ink, a cold plate, ink baked into the plate due to high temperatures, and/or poor wiping technique.

Dirty Blankets A common source of trouble is when the blankets are in poor printing condition. After a period of normal use, the blankets may accumulate sizing or starch from the wet paper, which causes them to harden and become less sensitive. They can be dry-cleaned or can be soaked in cool or tepid water in a large sink or bathtub—with or without Woolite or Zero—to remove the sizing. Rinse well. To speed the drying, fold them into long strips and arrange them around the circumference of a washing machine to spin dry for a few minutes. After spinning, they should be spread out flat on a nonmetallic, nonstaining surface or layer of towels to dry.

Plate/Paper Movement If the plate shifts out of place while printing and causes image blur or leaves an inky line, you can use a light spray of photo adhesive on the press bed or plastic sheet just under the center of the plate. If the paper persists in shifting as well, you can tape the leading edge of the paper to the press bed before laying down the blankets. You can also have an assistant hold up the blankets and pull them tight as you roll through because the problem is usually caused by the creeping movement of the blankets. If one side of the print is light and the depth of embossment is different from one side of the print to the other, the press pressure is uneven.

Contrast Problems

Shadow Detail If the shadow detail is dense or too blocked in, use a more translucent ink. If the blacks are pale and the tones of the plate seem light, use a denser, more intense ink. Be careful not to over-wipe the plate. If the shadows are easily over-wiped, try allowing the plate to cool while wiping and warm the cheesecloth on the hot plate instead (Denison 1895/1974, p. 103).

Highlight Detail If the highlight details are too pale or washed out, use burnt plate oil in the ink to leave a little more plate tone. Again, do not over-wipe.

Grainy Print If the print appears too grainy or harsh, the technique called *retroussage* is useful to soften tones and spread a film of ink over the tiny spots of bright bare plate. Dangle an inky cheesecloth or tarlatan loosely over the surface of the plate, being careful to use extremely gentle pressure, if any. Move around in little circular motions to evenly cover all areas that you wish to effect.

Drying Problems

Paper Ripples If the paper ripples after being removed from the drying system then it needs more time to dry. Make sure you change the blotters frequently and feel a print against your cheek to see if it feels cool. If so, it is still damp. When you remove the prints from the blotters, it is a good idea to stack the dry prints, interleave with tissue, and then weight the stack. After a week like this the prints can be safely stored.

REWORKING THE PLATE

Besides the photographic fidelity of the original negative, copper plate photogravure gives you all the flexibility of intaglio plate-making. Mark-making, tonal adjustments, and drastic alterations give unlimited freedom for image manipulation. Intaglio printmaking books can be a wealth of information on specialized tools and their proper use. Reworking a plate can consist of the correction of flaws and subtle modifications to areas of the plate. It can also be drastic and therefore radically alter the image (Color Plates 25 and 26). Some of the printmaking tools you can use for hand working the plate include mezzotint rockers, roulettes, burnishers, scrapers, sandpaper, burins, and etching needles.

Procedure

Before editioning a plate, the step scale and the extra safety borders must be removed. Decide on a final plate size, either leaving an even border between the image and the bevel or cut flush to the image with no border. A borderless image is slightly easier to edge-wipe cleanly. The appearance is quite different, so this is a design choice you have to make. The trimmed plate must be re-beveled and burnished prior to printing. While doing this, it is a good idea to protect the image area by taping down a piece of paper or covering the image area of the plate with self-adhesive plastic shelf lining (Mactac) (Figures 9-20 and 9-21).

Refile and shape the bevel on all sides. This is the final bevel, so take the time to do an elegant job. (See Figure 4-3 in Chapter 4.) After filing with two grades of file, rub the freshly filed edge with fine wet-dry sandpaper on a sanding block to remove all the fine file marks. Carefully clean the plate of all dust and filings. Use a well polished printmakers burnisher to rub the edge of each beveled edge in a side-to-side motion. Use enough pressure to cause the silky finish of the sanded copper to become brilliantly shiny. Slowly rock the burnisher towards and away from the face of the plate as you rub it lengthways along the edge and over the corner. This will smooth the bull nose or bevel into a curved surface with no rough lines or abrupt edges. Pay special attention to the corners and try to make them rounded as well as smooth. Remove the new burr from the back of the plate.

To correct the ubiquitous spotted out pin holes, you need good close-up vision, a loupe, and an extremely sharp etching needle. Locate the spot with the loupe using a light source angled so that these flat-topped spots glare (Figure 9-22). Halogen lights seem to work best. Have a proof print handy to corroborate your find. If the spot is small enough (0.5 mm or less), simply poke a shallow hole in its center to eliminate the white printing tone (Figure 9-23). The depth of this poked hole should be related to the surrounding tonalities to integrate it into the image. If the spot is larger or irregular, stipple with a tight grouping of identical indentations, again produced with the final tonalities in mind (Color Plate 20). Do not try to produce a drypoint mark with its raised burr because this will be lost quickly under gravure's high printing pressures. Make a mark that is below the surface of the plate and resembles the pits and wells already existing throughout the surrounding surface of the plate.

A gravure plate is easiest to repair when the dot pattern is made with rosin or a random-patterned hard-dot screen due to the irregular pattern and granularity of the image and tone. A commercial photogravure screen is regular and its tonalities are so smooth that any handwork done on such a surface will show up, especially in the middle to light areas. Heavy handwork can show up in the print as raised black ink lines or patterns. Do not dig or gouge the plate even in a black area. On the print, these repairs can show up in relief and will often look even denser than the richest tones that they are supposed to blend into.

The plate's bevel or clean border can be foul-bitten by a faulty stop out resist or leaky stripping. To prevent these marks from printing, they have to be burnished out. Be very careful not to scratch or injure the nearby image area. Protect the surface with a piece of Mactac the exact size of the image. You can also use tape and paper. The ideal situation is to prevent these foul-bites in the first place by using a permanent pen line

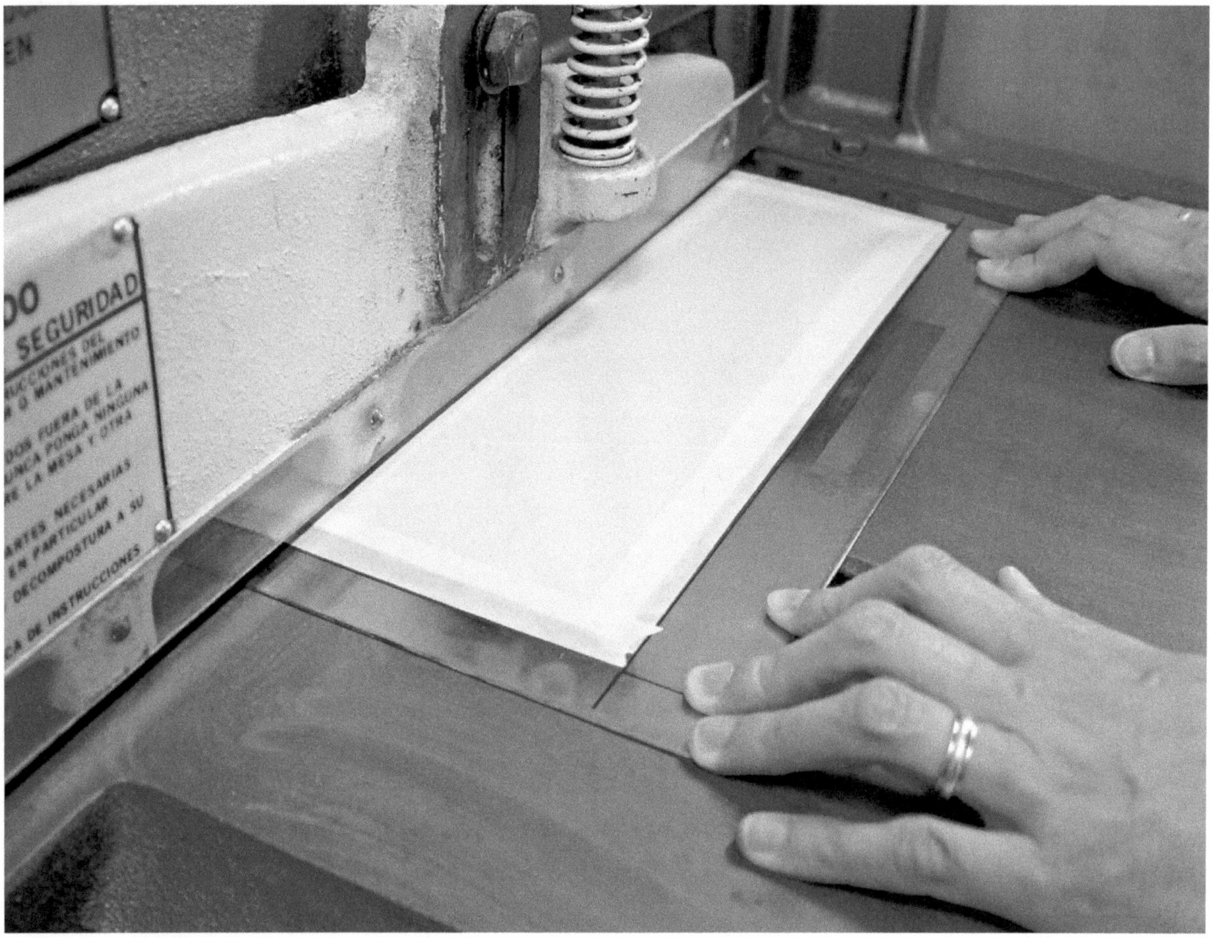

Figure 9-20 Trim a plate with a paper protecting the image from the plate cutter.

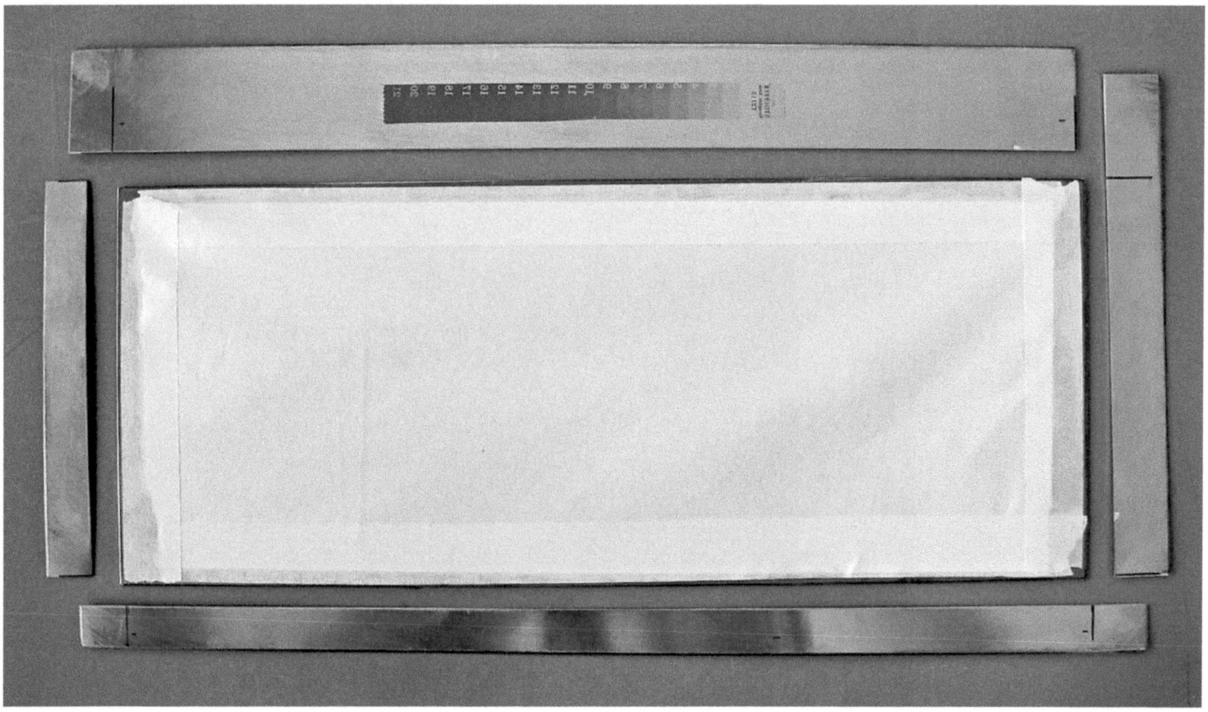

Figure 9-21 A trimmed plate before the edges are re-beveled.

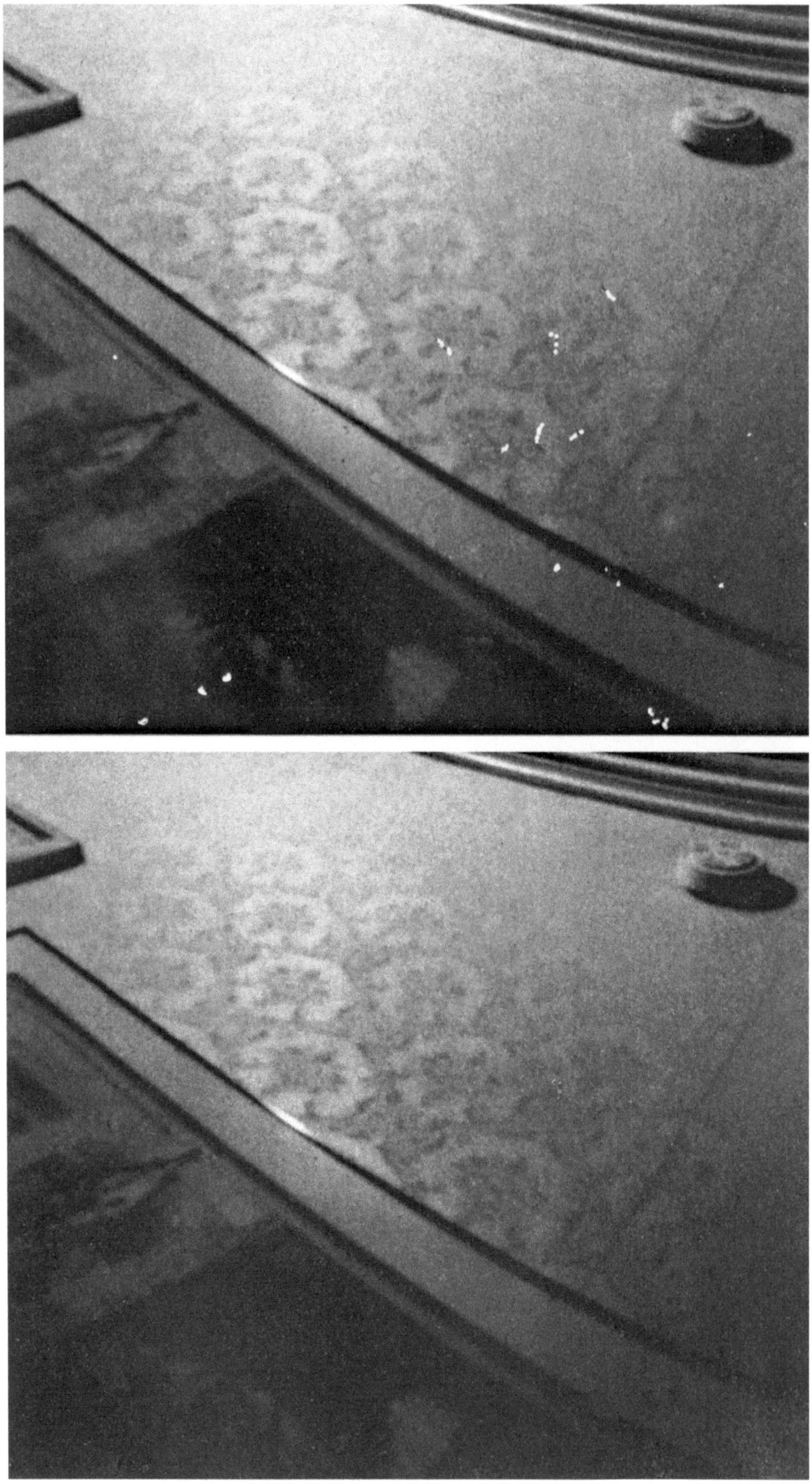

Figure 9-22 Shiny spots left over from spotted out pin holes in the resist. These dots will print white and can be easily darkened to match the background tones. See also Color Plate 20.

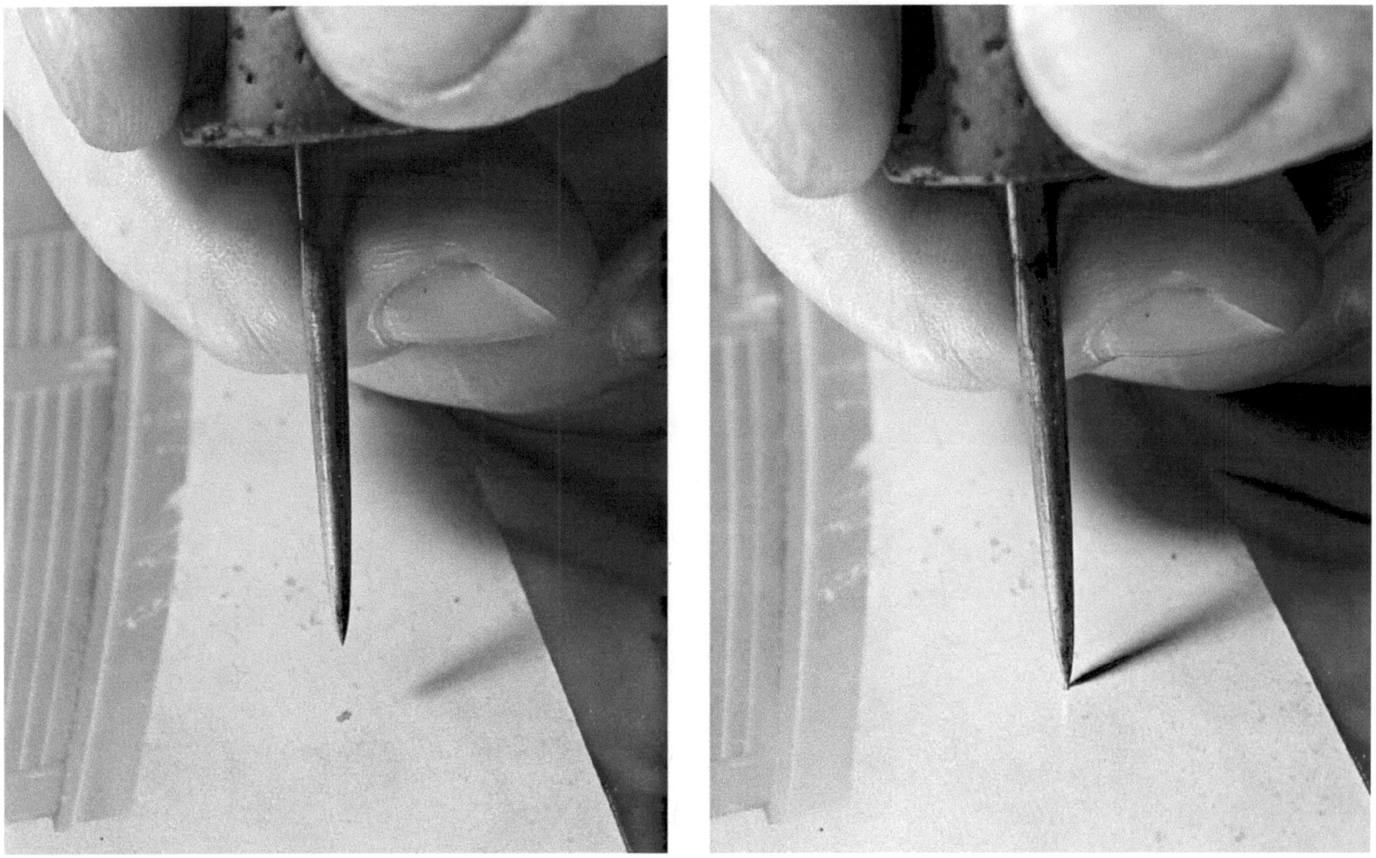

Figure 9-23 Use the needle in a perpendicular up and down motion. Do not scratch.

under the burnished Magic tape. Tape the edges so they overlap onto the contact paper backing. It takes very little extra time to do this and saves a lot of repair time later.

EDITIONING THE PRINT

When you are ready to edition a plate, all the plate repairs should have been completed, the ink testing should have been done, and an appropriate paper should have been chosen, based on proofing and tests. Trace the outline of the final size of the printing paper with the location of the plate on a sheet of newsprint or thin smooth paper. Center this template on the press bed and cover it with a larger sheet of unflawed acetate or Mylar to protect it (Figure 9-24). Do not use tape anywhere the printing paper will fall because the tape will leave an embossment on the print's borders. Have all the paper you need torn to exactly the same size and presoaked, including extra sheets for various proofs and waste. Have the ink mixed according to your tests on trial or state proofs. Mix enough to print the whole edition. It can be covered or wrapped if you don't finish all the printing in one session. A printing assistant is invaluable when editioning, especially as an extra set of clean hands for handling the paper and holding the press blankets.

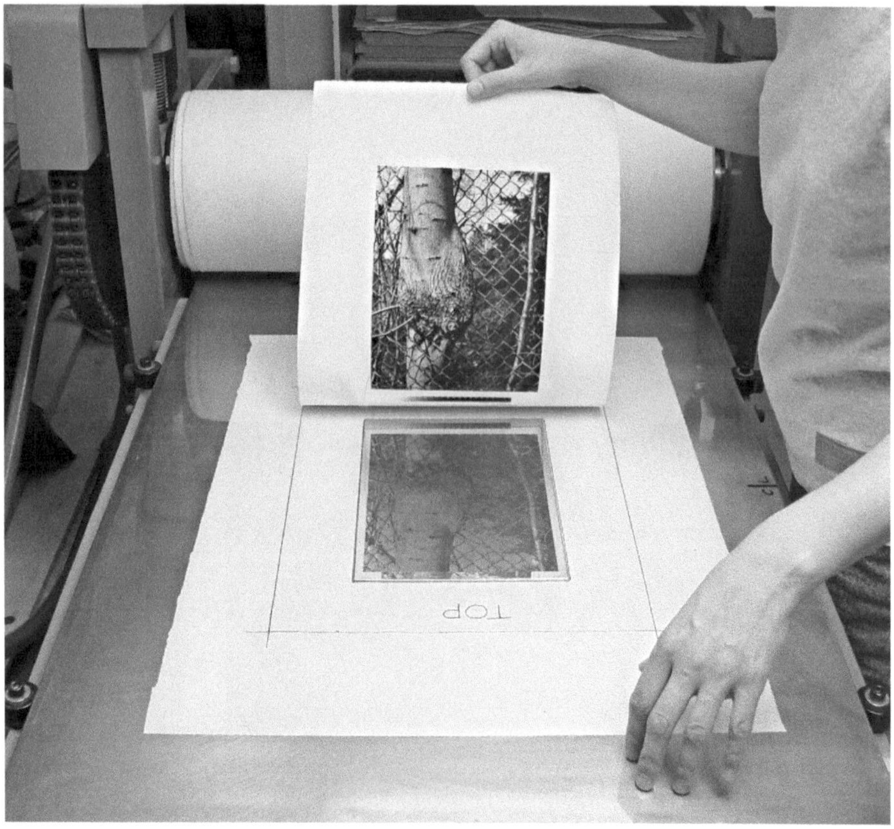

Figure 9-24 The press bed is protected by a Mylar sheet. Under this can be seen a guide for the position of the plate and paper.

If more than one plate is used for an image, the Mylar and template system can be used to register each plate if they are exactly the same size and the printing paper remains trapped under the roller while the plates are exchanged. For extremely accurate registration it may be necessary to calender (prestretch) the paper before printing. See Chapter 10 for more information.

10 Alternative and Historic Methods and Materials

Throughout the history of photogravure many useful techniques, alternative materials, and tricks of the trade were developed and applied successfully and appropriately. Many of these ideas may be the seed for further creative exploration. In this chapter we include some alternatives that may be of interest to curious, creative, and energetic readers. We simply do not have the space to go into greater depth, so if the information here is too brief for serious exploration, we suggest you search out the original or alternative sources. Be prepared to test and adjust all methods to suit your particular situation. Please be aware that issues of health and safety were woefully absent in many early sources, so exercise caution and common sense.

ALTERING POSITIVES BY HAND

Photogravure offers many alternatives to the literal reproduction of a photographic image. You can rework or alter the chosen negative before creating the positive and can then rework the positive before exposing the gelatin resist. The marks you make can potentially add layers of texture and drama to the image. Light scratch marks on the shiny side of the negative can leave little pale lines in the positive image, while scratches through the emulsion result in rough black lines—and vice versa when altering the positive. Drawing materials such as dyes and pens can be used on the surface of the positive, increasing the density and darkening the resulting gravure image. Once again, the densities and contrast range must remain within the limitations required by the photogravure process. You can also chemically bleach areas of the positive to lighten tones or remove information. All of these techniques are extremely difficult to do on film if you hope to keep the hand work invisible, so it is best to use them in a deliberate graphic way. Minor image or flaw corrections (retouching) may be better done by directly reworking the etched plate.

DIGITAL POSITIVES

It is becoming increasingly practical to consider creating your positives digitally. A digitally altered positive would be a seamless method of heavily altering a pre-existing photographic image or creating imagery from digital files. A scanned negative or print can be altered to suit the process by adjusting the levels, curves, and contrast values. The major obstacle is generating a high resolution film positive that will be as smooth as standard continuous tone film. A service bureau can generate the positive film with their high resolution image setters if they are given the right digital file. A lot has been written about digital practice, so it is impractical to go into detail here. A comprehensive source of the kind of information needed to create digital negatives and positives can be found in Dan Burkholder's book, *Making Digital Negatives for Contact Printing* (Bladed Iris Press, 1999). Also see his web site at www.danburkholder.com for more information.

DIRECT GRAVURE

It is entirely possible to create a plate from a completely nonphotographic image (Color Plate 23). Draw or paint on frosted Mylar or any transparent or translucent surface to produce a hand-made positive image. Again, the same contrast range and density limitations would apply as with a photographic positive. Deli Sacilotto calls prints made from drawn Mylar positives *direct gravures* (Sacilotto 1982, pp. 108–109).

Chemically frosted Mylar is smooth and leaves no apparent texture. It has a base density of about 0.16, which is close to the density of Step #2 on a Stouffer Step Scale (0.19), unlike film, which is almost transparent. This must be kept in mind when making a direct gravure. Keep your highlight details above Step #3. Be careful not to etch the step scale down to Step #2 because it would result in an overall gray plate tone. Use Step #3 as background white to be on the safe side.

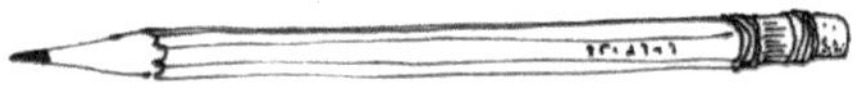

Use any black drawing material that gives you either solid black lines or a tonal or textured gradient. We've tested pencils from 2H to 5B, charcoal, oil pastel, conté crayon, litho crayon, grease pencil, liquid tusche, watercolor wash, gouache wash, felt pen, and markers of various types and transparencies (Figure 10-1). Add a drop of liquid soap to washes so they don't bead up. You can even use rubber stamp pad ink and make marks with your fingers or found rubber stamps. Most dark marks can also be scratched with a needle, scraped with a blade, or smudged with a finger. Draw on a light table in order to get a clear sense of the image development and the detail in the dark textures. Keep these dark areas from becoming overly dense (beyond 1.90), so you can achieve a plate that maintains detail in these areas. For areas denser than that, you'll end up with solid black. To visually evaluate the direct positive, hold it up to room light. Viewing on the light table renders the image tones slightly lighter than they will be in the final print. Viewing by reflected light only renders the tones darker.

When exposing the direct positive, you can flip it so that the image support (backing) is in contact with the gelatin. This maintains the image orientation in the final print, as you drew it. There is, however, a subtle loss of sharpness that is most noticeable in hard-edged lines or scratches.

You can also expose the drawing so that the drawing materials are in direct contact with the gelatin. This method reverses the image orientation but gives the sharpest possible detail. The only risk is that greasy drawing materials may result in a residue on the gelatin. In the wet laydown, the transfer solution may help clean the gelatin surface and prevent flaws.

Expose the screen for 75% to 100% of the determined positive exposure. Use the shorter exposure if the image is linear, longer if it is tonal or especially dark. Then expose the Mylar drawing a little more than normal in order to penetrate the darker textures of some marking materials. This will also ensure that the highlights have enough density to maintain their white. After this point, proceed as you would with any photogravure.

If the drawing is a fine line drawing without subtle tones, gradated washes, or thick dark lines or areas, the screen exposure can be reduced to 50% of the positive exposure. This will prevent the fine lines from being broken up or made rough by the texture of the screen exposure. The etching process is also different for a line drawing whereby you want to preserve the paper white and end up with sharp black lines. In this case, be sure that the positive exposure is long enough to give a good density to the white background without blocking any of the fine lines. Start the etch in a mid Baumé to facilitate a rapid penetration through the lines. Immediately move it to a high Baumé to slow the penetration through the white areas. Keep it in the higher Baumé until the lines are well etched. Move to a lower Baumé only to open up fine lines that did not fully etch. Be very careful not to allow any of the white background to begin to etch. Know the positive density of the white and which step on the scale it relates to so you can anticipate any ferric chloride penetration and stop it before it happens. Unlike the careful progression through the step scale required for photographic positives, the line etch should be abrupt and high contrast. This will produce a clean white background with dark lines. Cartwright (1939, p. 111) states that salt and alcohol can be used in the ferric chloride solution to repress the swelling of the gelatin to minimize background etching. We have never needed to do this.

SAVING A THIN POSITIVE

If the lith film positive is slightly thin though fully detailed, we have found that immersion in Kodak Rapid Selenium toner—diluted normally at 1:25—is one way to increase the overall densities of the image. Recheck the densities and retone rather than overdo it the first time. Keep in mind that color on the positive—either before toning if it is greenish or brown, or after toning if it is brown or purple—has a direct effect on how the sensitized gelatin will react to it. The correlation between density readings and the actual exposure will be changed. Exposure times will be affected, as will contrast. Ideally, all positives should be neutral in color.

STRIPPING ALTERNATIVE

Another method of obtaining a safe edge on the positive is by applying foil Mylar slide masking tape or ruby tape directly to the glossy support side of the film positive. This simultaneously masks, crops, and establishes

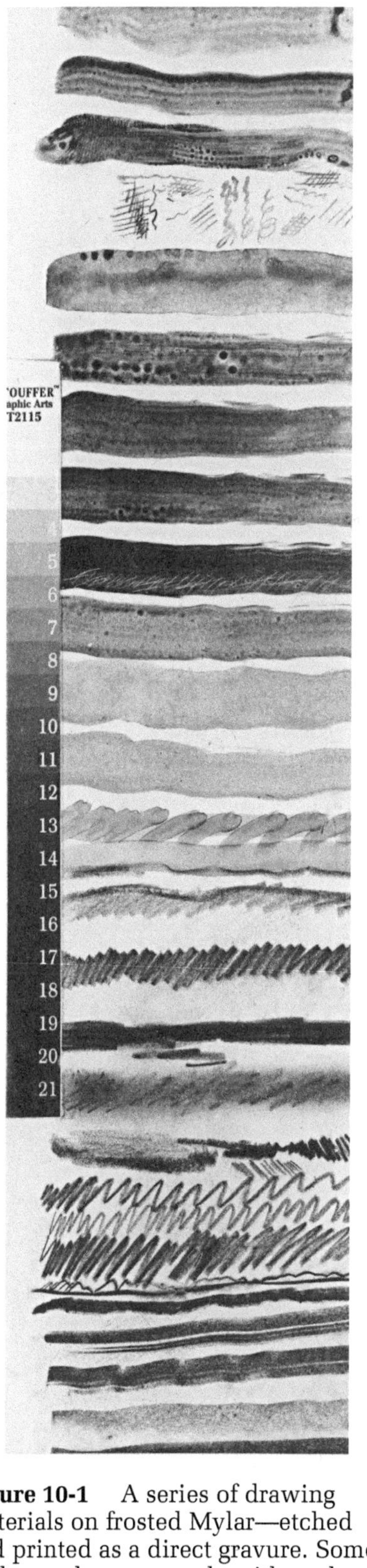

Figure 10-1 A series of drawing materials on frosted Mylar—etched and printed as a direct gravure. Some of the marks were made with such materials as black oil pastel, graphite pencil and stick, charcoal pencil, watercolor and gouache, felt pen, and scrapes with a sharp blade. See also Color Plate 23.

the safe edge and is especially useful for large images. When the positive is placed emulsion side up, one can see the tape through the clear film. This makes it relatively easy to position the sensitized tissue face down onto it in the right area. You can also use a fine permanent marker to outline the precise positioning of the corners of the tissue by marking the film directly.

SENSITIZER ADDITIVES

Autotype's data sheet recommends that the sensitizer's pH range should be 5.3 to 5.5. The pH level affects the contrast. Several sources suggest raising the pH to 6.0 with the addition of 0.3% of 91% liquid ammonia to reduce the acidity of the sensitizer (Mertle & Monsen 1957, p. 335; Cartwright 1939, pp. 70, 94). With a pure grade of potassium dichromate, we have never found this necessary.

When mixing the sensitizer, some sources say that a few drops of Kodak Photo-flo or other wetting agent can be added to aid in the smooth wetting of the resist tissue. The manufacturer recommends against this. To keep the sensitized tissue pliable in excessively dry conditions, the addition of 1% to 4% glycerin to the sensitizer has been suggested (Mertle & Monsen 1957, p. 335). We have had some serious problems with glycerin additives when working in high relative humidity. De Zoete reports that manufacturers advise against any sensitizer additives (1988, p. 34).

ALTERNATIVE DICHROMATES AND CONCENTRATIONS

The standard sensitizer for photogravure is potassium dichromate. Both sodium dichromate and ammonium dichromate can be used but each would require testing because the contrast, sensitivity, and subsequent exposure times would differ greatly. Sodium dichromate is far slower than potassium dichromate, so it would be impractical. Ammonium dichromate is expensive and very flammable, so why use it?

Different concentrations of potassium dichromate will result in different contrasts. On Autotype G35 tissue, the recommended concentration is 3% to give normal contrast in an industrial context. We use 3.5%. Raising the concentration further will increase sensitivity (speed) and will lower contrast. The opposite is also true. Autotype G25 tissue is darker and is described as having more contrast. To attain normal contrast from this tissue, we have used a 5% solution of potassium dichromate. Adjustments can be made with the sensitizer to account for contrast problems if no other means seem to correct them.

ALTERNATIVE WAYS TO ADHERE TISSUE TO PLEXIGLAS AND COPPER

An alternative to using a squeegee to adhere the tissue to the Plexiglas or copper would be the use of a smooth rubber roller, at least as wide as the gelatin tissue being adhered. In this case, one firm, smooth pass of the roller would be required from one extreme edge to the other while the Plexiglas or copper is on a rigid, smooth, flat support. A mangle is also usable for both Plexiglas lay-down when sensitizing and for dry lay-down of the exposed gelatin tissue onto the copper plate.

AQUATINTS: ROSIN VS. ASPHALTUM

If you are not using a gravure screen exposure to create the requisite lands and pits, then you must use an aquatint of either rosin or asphaltum. The advantage to using an aquatint is that it creates an organic surface and, if fine enough, can be less distracting than any mechanical screen. The disadvantage of an aquatint is that the layer of particles prevents the gelatin from having as secure a bond with the copper surface. In the case of a rosin aquatint, alcohol cannot be used at any point in the processing of the plate and tissue because it will dissolve the rosin. For this reason, powdered asphaltum was traditionally used. Today asphaltum dust boxes are rarely found in print shops due to their very messy and toxic nature. Another possible graining material is photocopier toner. This extremely fine powder can be fused to the plate using heat like the other aquatint powders, and can be fused by exposure to solvent fumes in a small enclosed space. It is important to note that there is some question as to the health risks associated with copy toner. Research all materials before using them. It should also be noted that all dusting methods are fire and explosion hazards; a spark can ignite the fine airborne dust within the dust boxes. For a more detailed description of the use, maintenance, and safety issues surrounding dust-grain boxes, refer to the current printmaking literature on aquatint. For more information on dusting and boxes used for photogravure see Cartwright (1939, pp. 182–185) and de Zoete (1988, pp. 49–57, 128–130).

Most print shops use rosin and have only rosin dusting boxes available, so if you decide to use asphaltum or copy toner, you must make one for that purpose alone. A strong, tall cardboard box can be used to construct a usable dusting box light enough for manual shaking. If you wish to build a wooden box, you will have to research other methods of agitating the powder. For small plates, a dusting box should be at least 3 square feet across and at least 3 feet tall. The taller the space above the plate, the greater the volume of airborne dust, making coating easier to control and repeat. A trap door that can be opened and closed without shaking the box and that provides a good airtight seal is needed on the lower portion of the box. An open shelf made of wooden dowels or a metal grate is required inside to hold the plate horizontally a few inches off the bottom. Be sure to tape all seams and joints from both sides with wide durable tape. It is also a good idea to seal the cardboard, both to toughen it and to make it moisture proof. Varnish, paint, urethane, or acrylic medium can all be used. This will keep the asphaltum dry and prevent it from clumping. If the inside gains moisture, it can be removed by leaving a bag of silica gel in the box when it is not in use. This is very important for the production of as fine a coating of evenly spaced particles as possible.

A safer and easier alternative to the asphaltum aquatint is to apply a rosin aquatint. Fox Talbot's method of applying the rosin to the outer surface of the dried gelatin resist just before the etching stage is actually quite workable. The reversal of layers, with the aquatint applied over the resist rather than under it, does not seem to affect the sharpness or resolution of the image's detail. There is another advantage in that the resist has complete, uninterrupted contact with the plate and is less likely to blister or fail. There is more potential for foul-biting the little points of copper, however. Care must be given during the etching procedure to prevent having the mordant migrate under the rosin and subsequently reach the copper below.

The heat used to fuse the rosin is half the temperature needed for asphaltum and doesn't melt the gelatin. The rosin aquatint can be applied in as little as two hours after adhesion if the humidity is lower than 50%. The heat will further dry the tissue. The gelatin resist must then be allowed to reabsorb ambient moisture to gain equilibrium with the room's relative humidity. This may take several hours—preferably overnight. If etched too soon, the absorption of the conditioning bath will be delayed and possibly irregular.

The average rosin aquatint produces a surface of lands and pits roughly equivalent to a 200+ lines/inch screen. The asphaltum aquatint can be equivalent to a 300+ lines/inch screen. Although this finer surface is advantageous for maintaining detail, it increases the risk of foul-biting during the etch and may be somewhat more difficult to apply. Asphaltum is, however, tougher and less likely to foul-bite by falling off during the adhering or etching process. Its finer resolution also allows the asphaltum resist to retain greater detail and smoother tones when applied and etched properly. Combining both resists, one after the other, and then melting them together is possible and is supposed to make a beautiful grain pattern (Blaney 1895, p. 28), though we have yet to try this.

Although we describe fusing the aquatints using a hot plate, it is also common to fuse the grain pattern with a torch from the underside of the plate. This requires more skill but may be worth looking into if you don't have access to a printmaker's hot plate. It would be very risky, however, to try this when there is a resist on the plate. Another technique would be to use an oven. This may be the most controllable method of all.

APPLYING AN ASPHALTUM AQUATINT

After the plate has been polished, edged, degreased, and brightened it is ready for an asphaltum aquatint. First, set the hot plate to 230 to 260°C (450–500°F) and allow it to heat up. Then, shake or tumble the asphaltum box vigorously to cause the asphaltum dust to become airborne inside the box. Set the box down and rap the sides and top to dislodge any large loose clumps of powder. Place the plate, face up, on a piece of mat board an inch or two larger than the plate. In order to get the finest grain possible, wait 4 minutes, and then slide the mat board and plate into the box to the center or toward the back of the rack. Carefully close the door and wait for 4 to 5 minutes. Slowly and carefully remove the plate and mat board from the dusting box (see Figure 10-2). Keep away from drafts. Check the coverage of the plate with a loupe. Be careful not to breath on or touch the delicate coating of dust. If much less than 40% of the plate is covered, repeat the dusting process. Carefully put the plate aside away from drafts. Reshake the dusting box, wait for 4 minutes again, and reinsert the plate for another few minutes. As much as three or four layers of asphaltum dust can be applied to the plate to reach a 50% coverage. Once complete, lift the plate slowly and carefully onto the hot plate. Be very careful not to slide the plate across the surface of the hot plate. This could allow particles from previous hot plate use to contaminate the asphaltum layer. Contamination can also occur from dusty tools, nearby fixtures and cabinets, and dirty lab coats.

The asphaltum will become shiny as it melts. It will probably also smoke as it fuses to the face of the copper; ventilate the area because this

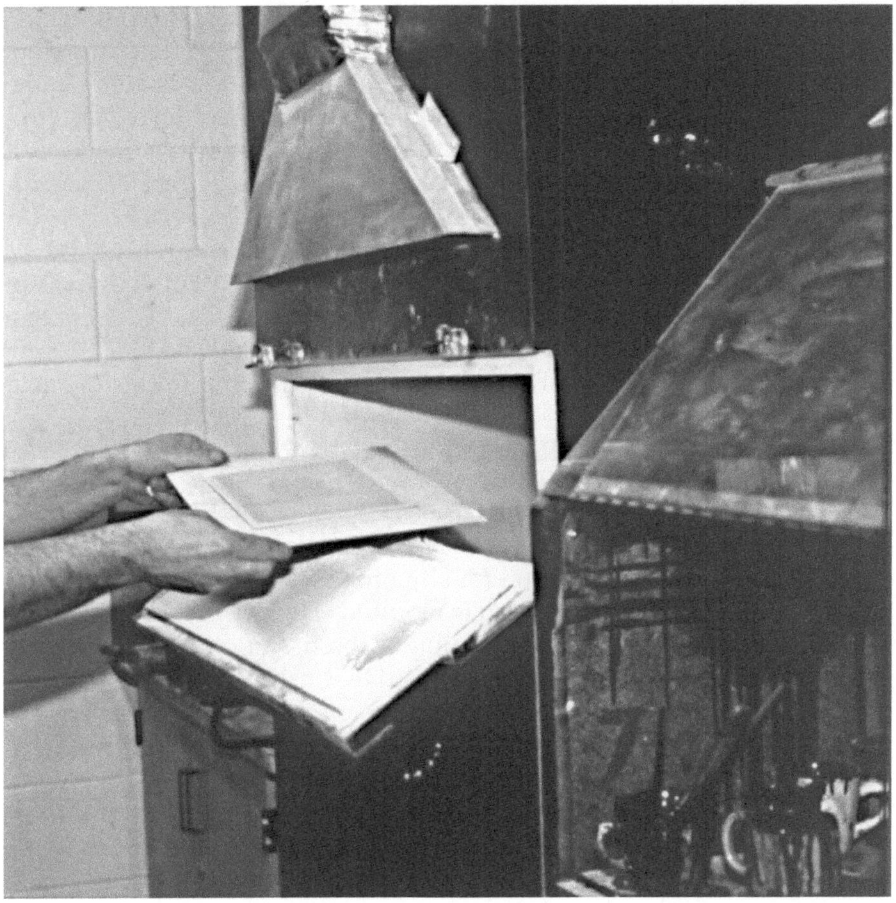

Figure 10-2 The freshly dusted plate is carefully removed from the rosin box prior to fusing on a hot plate.

smoke is dangerous. The copper will tarnish or darken as it is heated. After the asphaltum has fused, remove the plate and place on a cool surface to allow the heat to be drawn out of the copper. After it is cool and the resist has hardened, carefully inspect with a loupe. There should be a fine, even texture with about a 40–50% coverage from corner to corner over the entire surface of the plate. Sacilotto recommends as much as 66%. If everything is fine, brighten the plate using the standard brightener to remove the tarnish and then proceed to the next step: adhering the exposed tissue.

APPLYING A ROSIN AQUATINT

The procedure for the application of a rosin aquatint is very similar to the application of the asphaltum aquatint just described. Note that some rosin is ground very fine in a blender, whereas other print shops may use coarser rosin, which therefore contains fewer fine particles. This will affect the waiting times and coverage (Figure 10-2). The dimensions of each rosin box may differ. This makes a difference to the number of spins, the waiting time, and the final quality of the coverage. All steps are similar to asphaltum aquatint except for the location of the rosin layer and hot plate temperature, which is lower: 115 to 130°C (225 to 250°F). It is possible to apply the rosin over the already applied and dried gelatin resist. In this case, once the rosin is applied do not use brightener, just re-acclimatize, stage the plate, and then proceed to the etching step.

Etching a plate that uses an aquatint is very different from etching one that uses a screen exposure. The start of the etch is quicker, the progression is different, and there is a greater risk of foul-biting the darkest areas. Work cautiously, do not over-etch, and keep accurate records, especially of aquatint coverage.

It should also be noted that an aquatint can be reapplied to specific areas of a previously etched plate in order to correct and darken flat areas of tone that may be too light. A good example of this is the re-etching of a solid black background that was foul-bitten or under-etched. See the section "Correcting Flaws and Reworking the Images" later in this chapter.

THE DRY LAY-DOWN METHOD OF ADHERING GELATIN TISSUE TO THE PLATE

The dry lay-down method works well with rather large and cumbersome sizes of gelatin tissue. It is ideal for situations where multiple plates are being made and exact registration is required. The dry lay-down method allows for a very accurate positioning of the tissue onto the plate and causes no distortion of the original positive. Curling is kept under control because the tissue is hinged into place on the surface of the copper while dry, before moisture completes the adhesion process. As with wet lay-down, development can commence immediately after lay-down. Some believe that the dry lay-down technique is the fastest and most dependable method of adhering the gelatin tissue to the copper plate (Sacilotto 1982, p. 121). It is often made easier by using a mangle—a device from an old wringer/washer.

A relative humidity of less than 45% can cause serious problems with dry lay-down due to excessive curl when the exposed gelatin tissue is too dry. It becomes next to impossible to squeegee the gelatin tissue without causing a fracture, crease, or fold. It is a lot easier if the gelatin tissue has been stabilized to an ideal relative humidity of near 60%.

After the tissue has been trimmed and exposed to the positive (and screen) and the copper plate has been degreased, brightened, and dried in preparation for lay-down, place the dry plate face up on a blotter and center the gelatin tissue, face down, on the plate. Attach the farthest edge of the gelatin tissue to the copper with a wide strip of masking tape. Ensure that the tape does not cover any part of the image area because its presence in the lay-down may affect the resist below it. Use compressed air to blow any debris out from between the tissue and the plate. Hold the free end of the tissue up with one gloved hand and pour a quantity of distilled water at 21°C (70°F) (Cartwright 1939, p. 89) along the taped edge under the tissue and pool it in the center of the plate. The quantity of water must be enough to reach both sides and all corners of the tissue as it is squeegeed. Be very careful not to introduce any air bubbles. Immediately squeegee or roll the tissue down onto the plate from the taped edge while slowly lowering the free edge of the tissue as you roll or pull the squeegee towards yourself. Do this in one smooth and continuous motion with firm pressure, using a squeegee or roller that is as wide as the gelatin tissue. A few extra strokes in each of four directions if required seems to do no harm. Blot the water that is expressed at the edges. There is an immediate bond between the gelatin and the copper as the gelatin absorbs the residual moisture. Note that the temperature of the water used for the dry lay-down process is slightly higher than for the wet lay-down process.

Autotype's G35 Gravure Pigment Paper product data sheet recommends that after the tissue is well adhered and blotted you should submerge the plate and tissue in an 80% alcohol bath (IMS 55 O.P. 80%) for at least 2 minutes. Do not do this when a rosin resist has been applied to the copper. At this point, the tissue is ready for the development wash.

Summary for Dry Lay-Down

1. **Prepare a degreased and brightened copper plate and place it face up on a blotter resting on a firm flat surface. If the plate has had an asphaltum aquatint ground applied to it, it should be brightened and dried just before this stage. (It is not advisable to use the dry lay-down technique on a plate with a rosin aquatint because of the use of the alcohol bath that follows.)**
2. **Place the freshly exposed tissue face down on the clean copper plate and tape down the back of the top edge to the copper surface.**
3. **Lift the tissue by the leading edge and pour a small quantity of 21°C (70°F) distilled water along the taped edge and over the first third of the plate, edge to edge.**
4. **Firmly draw a squeegee or roller from the taped edge over the entire surface as the tissue is lowered into contact with the copper.**
5. **After squeegeeing, pat the edges dry with paper towels.**
6. **Submerge for 2 minutes in an 80% alcohol bath. Omit this step if there is a rosin aquatint.**
7. **Fill a tray with 43°C (110°F) water. Set the timer for 15 to 20 minutes.**
8. **Quickly immerse the plate and proceed as per standard development.**

ALTERNATIVE MATERIALS FOR STAGING THE PLATE

A traditional method of staging a plate is to use an old-fashioned architect's ruling pen and liquid asphaltum to scribe a straight and hard-edged line around the image. Outside this edge, the remainder of the plate can be painted with asphaltum or stop-out varnish to the beveled edge using a small brush. This must be totally dry before etching can begin. Backing and handles are still required. When re-etching a plate's background, asphaltum is used to stop out all areas of the image that must remain untouched by subsequent etching. Use a fine brush and not too much solvent. Overly dilute asphaltum may migrate onto part of the image area you do not want protected.

STEEL FACING THE PLATE

Steel or chrome facing a copper plate is the only sure way of preventing plate wear during the printing process. Steel facing with mild steel is very different from chrome or nickel plating. The latter gives a much harder surface but is a much more toxic process. The former is easier, safer, and can be easily removed and repeated as the need arises. In both cases, the extremely thin and flawless electroplate necessary to print without loss of detail is achieved with only extremely careful technique. This is beyond the scope of this book. Many fine art prints are editioned in low enough numbers to make steel facing unnecessary. If you intend to make large editions, facing would be useful in order to prevent the image from degrading through wear. Steel facing involves specialized equipment and supplies. Suffice it to say that if you plan to make large editions well in excess of 30 or 40 prints and need to rework and proof the plate a large number of times as well, it would be advisable to look into steel facing your plates.

CORRECTING FLAWS AND REWORKING THE IMAGES

If dark undifferentiated shadow areas over-etch and *crevé*, or foul-bite, they may print up lighter than other dark areas nearby. Lateral etching causes the loss of lands or high points in a grained or random screened gravure or the loss of part of the cross-hatched web of a commercial screen gravure. Without these high points the darkest areas of the plate would hold less ink. Normal wiping would remove the ink necessary for a full black.

To darken an area of flat tone that is not dark enough on the gravure plate, you can reapply a traditional rosin aquatint. First, thoroughly clean the plate using naphtha and acetone to be sure that all ink residue has been removed. Degrease the plate with TSP. Rinse well and dry. Coat the whole plate with a fine rosin aquatint. Fuse and cool. Use liquid asphaltum and a fine brush to stop out all areas you do not want to re-etch. Stage and back the plate, and then re-etch in a 42° Bé solution of ferric chloride reserved for printmaking use (not for use with gelatin resists). Be careful not to over-etch. An aquatint laid over an existing tone will have the effect of darkening the tone more quickly than if applied to a bare plate. For more detailed information, refer to standard printmaking books on aquatint.

If a pale or white area on the plate needs to be darkened slightly, you can use sandpaper. It takes some trial and error to determine which grade, but a general guideline is to use 1000 to 1500 when adding subtle tones to highlights, 600 to 1000 to add tone to light to mid-tone areas, and 320 to 400 to add darker tones. The sandpaper is placed grit side against the plate, and then this sandwich is run through the press, using slightly lighter than printing pressure. The sandpaper can be cut or torn to create a specific shape that is carefully laid over the corresponding area of the plate.

An even more controllable system involves using clear self-adhesive plastic shelf liner (Mactac) as a protective stencil layer. First, lay the clear contact film over the entire plate and give yourself registration marks. Using a fine-tip marker, draw the shape or shapes where darker tones are to be applied. Remove the Mactac from the plate. Use an X-acto blade and cut out the shapes, making the openings just very slightly larger than the desired final shape. Return the Mactac to the plate, using your registration marks to align it correctly. Lay a full sheet of sandpaper, grit side down, on top of the contact film and run the whole thing through the press. For a slightly darker tone, run the sandwich through the press more than once, moving the sandpaper slightly each time. This is a very effective way of creating smooth tones, although they do not stand up to a great number of printings. You should be able to get 10 to 15 prints before the newly created tones begin to diminish.

To darken tones dramatically, you can use a mezzotint rocker. Using this method will create a series of dots that will overwhelm the etched image. If you want to create passages of rich, velvety black, it is superb. You can use the mezzotint rocker with the stencil layer just described, but be careful that the rocker does not penetrate the Mactac. Rockers come in a variety of sizes and grades. For extensive information on how to rock the plate and how to work into a mezzotint surface, we recommend Carol Wax's book, *The Mezzotint: History and Technique.*

If you want to lighten tones in a given area without changing adjoining passages, you can use a protective stencil layer as described and then polish through the openings with Brasso. The amount of pressure you use and the amount of time you polish will control how much you lighten the existing tones. As in the case of any new technique, it is always best to be gentle in your initial attempts and pull proofs. Once you go too far, the etched image is lost.

A final use for the protective stencil layer is a variation that permits you to add complex linear drawing that is tonal, rather than a drypoint or etched line. Lay the clear Mactac on top of your plate and make registration marks, but also make sure that the Mactac is securely attached to at least one edge of the plate. Using a fine-tip felt marker, draw the imagery you wish to add. Peel back the Mactac, making sure that one edge remains adhered to the plate. Lay a piece of sandpaper face down over the area where the new drawing is and then carefully lay the Mactac back in place. Use a burnisher or similar blunt hard-tipped tool and retrace your felt marker drawing. This will transfer the sandpaper texture into marks on the plate surface. The amount of pressure, the broadness of the drawing instrument, and the grade of sandpaper will determine the resulting tones (Figures 10-3 and 10-4).

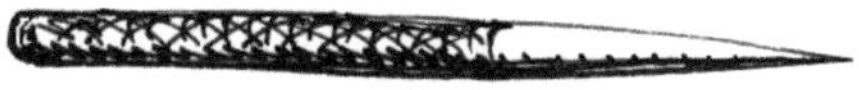

Other options for complex hand work involve the use of traditional intaglio tools: etching needles to create line work, burnishers to lighten tones, roulettes to add tone and texture, and even engraving burins and scrapers. The two last tools are the most difficult to control and integrate into a photogravure plate's surface. If you are already familiar with drypoint or mezzotint, it will be a simple matter to apply this skill to working on a photogravure plate. If you have never worked with these techniques, they require practice to develop skill and sensitivity so we suggest experimenting on test plates to familiarize yourself with the effects. For extensive information on using these tools, refer to standard intaglio texts.

ALTERNATIVE PRINTING PROCEDURES

A 30-second soak can be used with some soft or delicate papers. This seems to work quite well in that the surface is soft enough to pick up the plate's detail. The core does not get too wet so it acts as its own backing sheet to resist a blanket texture. An alternative precaution that helps give a very clean impression is to calender the paper by running the damp paper through the press prior to actual printing. This technique is vital when printing in register with more than one plate. It stretches out the paper so that registration is more accurate. It is important to ensure

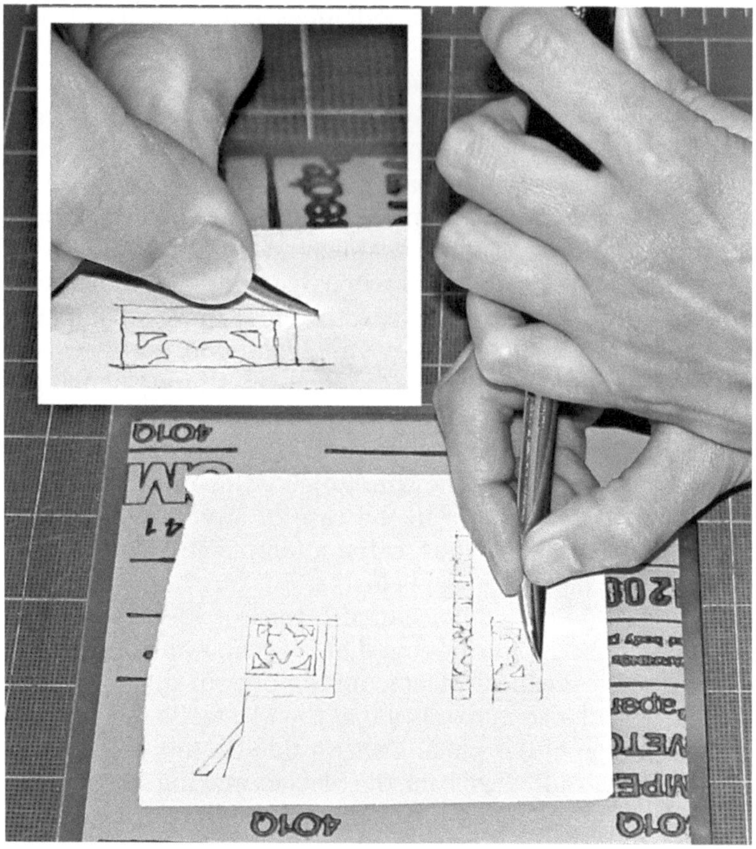

Figure 10-3 Use a blunt stylus or burnisher to draw through a Mylar overlay and the sandpaper beneath in order to create a localized tone on the plate.

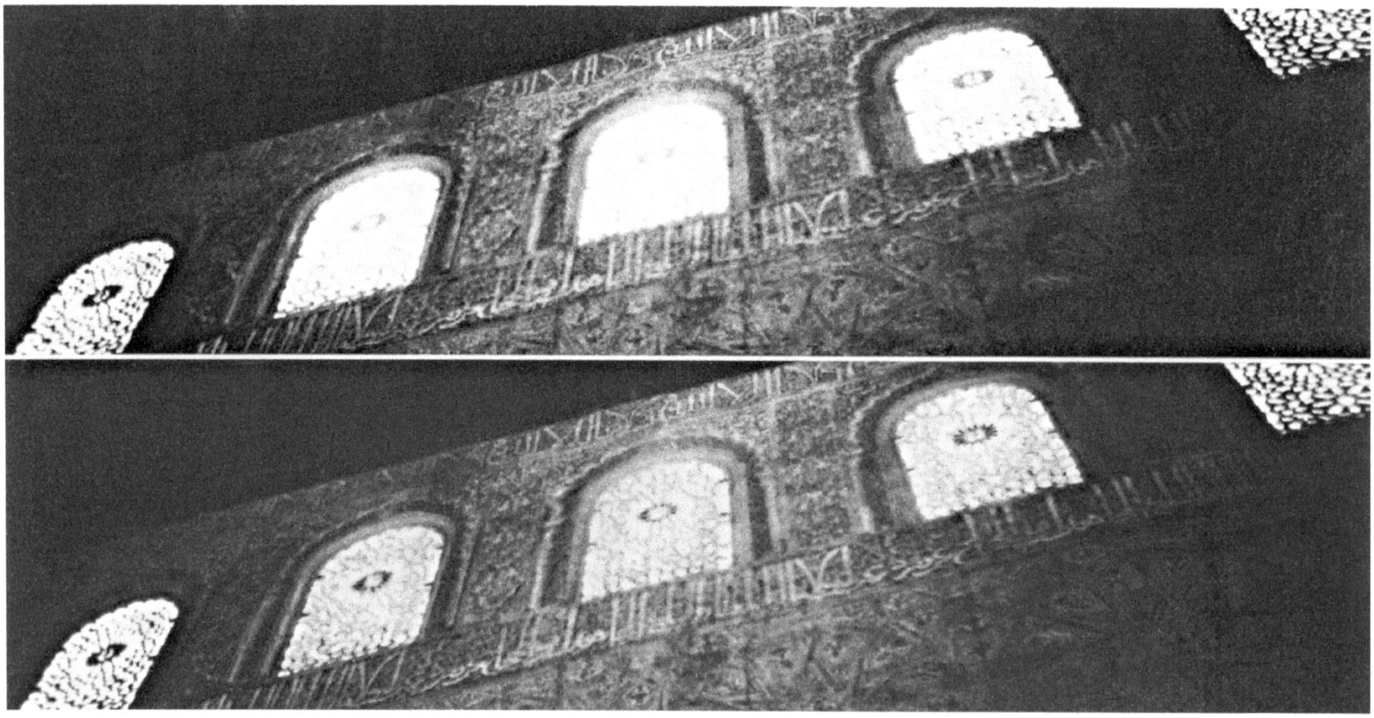

Figure 10-4 A sample of how sandpaper technique was used to fill in burned-out window detail.

that the grain of the paper is running in the same direction as the travel of the bed under the roller. The opposite direction would cause a greater amount of paper stretch. To calender the paper place a large, beveled, clean, smooth, blank metal plate on the press bed. Position the printing paper on it, good side down, and cover with the usual arrangement of blankets. Run through the press at a reduced pressure to remove all vestiges of water from the paper. It will still be soft—the main reason for the presoak in the first place. Set up the inked photogravure plate and reset the pressure. Be sure to brush the surface of the paper with a clean drafting brush before printing to remove lint, hairs, or particles that may have been picked up during the blotting stage.

We have found that double printing may improve the blacks but often causes some blurring or softening of the image. When double printing, run the paper through the rollers until just past the far edge of the plate. As soon as the roller drops off the plate edge and before it leaves the paper, reverse the direction and completely run through again. If printing problems are the result of a press that cannot exert enough pressure or blankets that are not smooth enough, running through a second time can help produce a good print. The print may be darker than one would expect, though. Adjust your ink accordingly.

À LA POUPÉE INKING

You can use more than one ink on the plate as a way to control subtle differences in color or ink intensity. This is *à la poupée* inking, and is frequently used in color intaglio printing. Examples of different uses of *à la poupée* inking include: a) different colors—vivid or subtle—applied to different objects or areas within the plate, like hand-coloring or selective toning; b) different ink qualities needed in different areas, such as transparent ink in one area to show more shadow detail and intense ink in another to maintain density or rich detail; or c) warm and cool blacks used to exaggerate the sense of advancing and receding planes within the image.

Mix the inks that best suit the various areas of the plate, adjusting the color and transparency for each. Apply each with a small rubber squeegee, being careful to keep the inks contained in their respective areas. You can even use cotton swabs to place color in very small areas. Use the cheesecloth to wipe the plate as per standard procedure, but carefully wipe the various areas discretely so that the inks do not simply blend and nullify the effect. Blend edges where appropriate, though. By the time you have removed all the surface ink and have reached the final hand wipe, you can wipe overall, without risk of ink spreading from one area to another. Print normally. Be sure to keep detailed notes and diagrams so that you can repeat the process again.

CHINE COLLÉ

Chine collé is a printing technique that permits the layering of more than one paper. Traditionally, a very thin paper is adhered to a heavier backing paper. The thin paper would be cut to the image size and give the effect of a tone or color shift from the backing paper. The papers normally

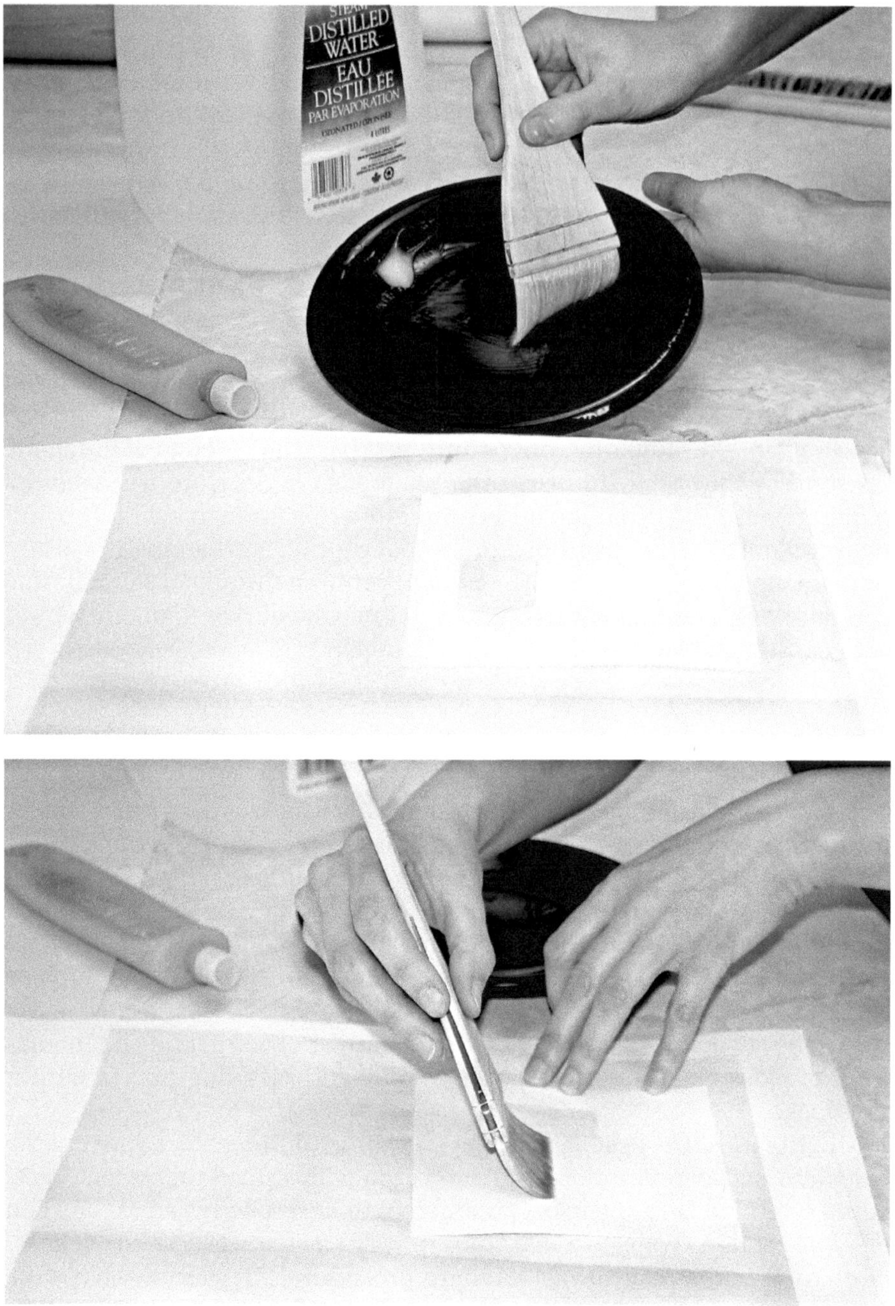

Figure 10-5 Mix the paste with distilled water to make a somewhat thin creamy solution. Apply a thin layer to the back side of the *collé* paper, brushing the paste from the center to the edges.

used are extremely sensitive and give very smooth, subtle prints (Color Plates 22 and 24).

What follows is a very simple set of instructions that you can use to start with. Select the *chine collé* paper and cut it to the desired size or shape. You are not restricted to following the plate shape; it can be used selectively within the image. Keep in mind that the paper will be laid good side down on the plate. Remember this inversion when cutting out shapes. Note the window shape cut out of the *collé* paper in Figure 10-6. See also Color Plate 24 for the final version. Because the paper will be

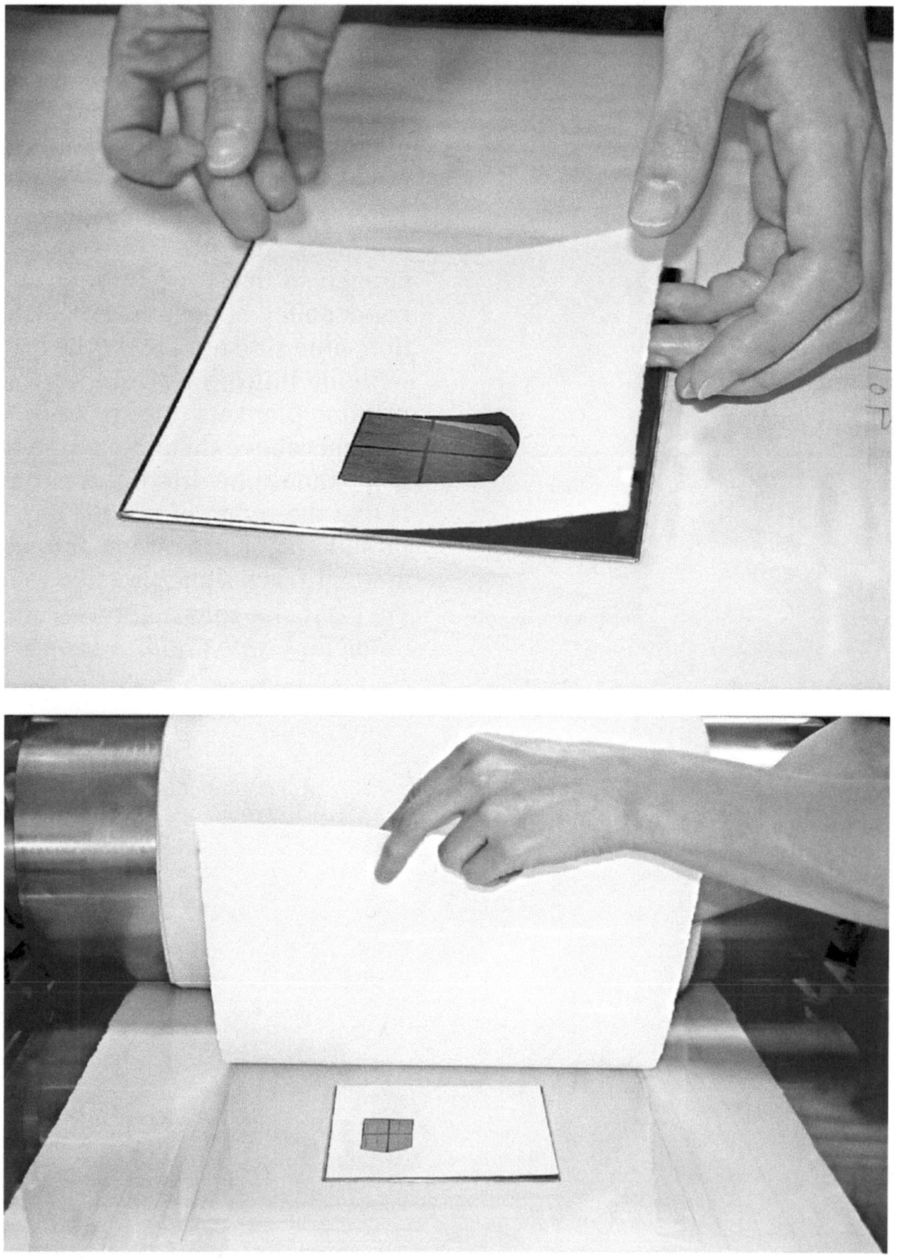

Figure 10-6 Carefully position the pasted *collé* paper—paste side up—onto the plate. Apply no pressure. Lower the backing paper carefully so as not to shift the *collé* paper. Roll through the press with someone stretching the blankets clear of the paper.

misted with water, you might find that you need to cut it very slightly smaller than the desired size to compensate for moisture expansion. (Or in the case of an opening, cut the hole slightly larger.) Prepare your paste ahead of time. Wheat paste is stronger than rice paste, but in most cases the rice paste is perfectly adequate. Either can be bought in powder or premixed form. If you have the powder form, it must be cooked in advance to prepare the stock paste. With either type of paste you will need distilled water, a small plate, and a soft brush for mixing and applying the paste. Mix enough water with the paste to create a creamy solution that is easy to brush (Figure 10-5). Prepare and soak your backing paper following standard procedure. Ink and wipe the plate and lay it on the press. At this time, blot your backing paper.

Lay the *chine collé* paper on clean blotters or newsprint and mist both sides with water. Blot away surface water. Hold the paper in place good side down and brush a thin, even layer of the paste on the back. Always start from the center of the paper and work out toward the edge (Figure 10-5). You must not get any paste on the printing surface, or it will adhere to the plate. You need to coat the whole paper, but keep the layer thin enough so that it will not ooze out while printing. Lay the *chine collé* paper on top of the inked plate, being careful to place it in the exact position, glue side up. Lower the backing paper very carefully on top of this without shifting it. Great care must also be taken when lowering the printing blankets (Figure 10-6). We find it is best if someone holds the blankets above the plate while the other person operates the press. The print should be dried normally. The advantage to using this technique is that the right *chine collé* paper tends to give a smoother print. It can provide color or surface shifts, which add new visual elements. This method takes some practice to do properly but is well worth learning. For extensive information on *chine collé* history and techniques, we recommend *Chine Collé: A Printer's Handbook*, by Brian Sure.

11 Directions for the Home Manufacture of Carbon Tissue for Photogravure Printing

by Sandy King

The home manufacture of gelatin carbon tissue is not extraordinarily difficult, and with a little practice a very satisfactory gelatin tissue can be created. This chapter gives detailed instructions for the preparation of gelatin carbon tissue, which can be used to make photogravures.

First, a note about gelatins. Gelatin is extracted from many sources, including the hide, skin, white connective tissue, and bones of animals. It is useful in carbon photography and in gravure work because it absorbs water and swells with increased temperatures until it reaches a melting point, at which point it forms a colloid; when again cooled, this colloid will set, even at low concentrations, and the cycle can be repeated. Certain chemicals, including chrome alum, potassium alum, formalin, and the various dichromates (ammonium, potassium, and sodium) serve to either harden, reduce, or altogether eliminate the ability of gelatin to absorb water.

Most commercially available gelatins, including some edible gelatins, work reasonably well for making carbon tissue for gravure work. However, for consistent results I recommend the use of a gelatin with a Bloom index of between 175 and 250; the instructions in this chapter are designed for gelatins in this particular range.

When making tissue it is important to use a percent solution that allows time to evenly coat the tissue but that also sets within less than 2 minutes. It is almost impossible to perfectly level the plate glass on which the coating operation takes place. Therefore, if the pigment solution takes too long to set it will tend to settle in a thicker layer toward the low end of the carrier.

Although you can vary the amount of coating solution used to coat a carbon tissue of a specific size in order to adjust for working environments that are warmer or cooler than the norm, the following amounts should be considered standard for a room temperature of about 21°C (70°F) and relative humidity of 50 to 60%: 50 ml for a 20 cm × 25 cm (8″ × 10″) sheet, 100 ml for a 28 cm × 36 cm (11″ × 14″) sheet, and so on.

BASIC TISSUE FORMULA

Distilled water:	1000 ml
Gelatin, 175–250 Bloom:	80–100 grams
Refined sugar:	25 grams
Pigment:	10 grams Cal-Tint II colorant, Venetian Red (or substitute any good quality tube watercolor)

One of the most interesting features of the carbon process is that virtually any color desired for the final image can be obtained by mixing pigments while making the carbon tissue. However, for photogravure work we have attempted to find a color that approximates the rust/brown color of commercial Autotype tissue; to my eye, Venetian Red appears to be such a color. For our tests we used the Venetian Red available in a line of Cal-Tint colorants; however, you should be able to substitute any good quality tube watercolor of the same color, although any substitution will require an adjustment in the amount of pigment needed to achieve the same amount of saturation.

PREPARING THE PIGMENTED GELATIN SOLUTION

1) Stir 80 grams of 175–250 Bloom gelatin into 900 ml of distilled water at about 21°C (70°F), and let the mixture stand for about an hour. During this period the gelatin will absorb water and form a gel.
2) After an hour, place the container of the above mixture in warm water at around 43 to 49°C (110–120°F) and allow the solution to liquefy completely.
3) When the gelatin solution has liquefied and reached a temperature of about 40.5 to 43°C (105–110°F), stir in 25 grams of plain white sugar, and allow to dissolve completely.
4) Now add the pigment. Pour 10 to 15 ml of the warm gelatin solution into a glass mortar and then add the required pigments. Grind with a pestle for a minute or so until the solution appears uniform. The amount of pigment added to the solution affects tissue contrast: A low contrast tissue is made by using enough pigment to produce a tissue coating that is just barely opaque. Adding more pigment to the solution results in higher contrast tissue.
5) Next, add the dispersed pigment in the mortar to the container of gelatin solution, and stir in enough water to make 1000 ml of pigmented gelatin solution. Stir gently for a couple of minutes to achieve maximum dispersion of the pigment.
6) Place the container of gelatin–pigment solution in water at about 38 to 43°C (100–110°F) and allow it to sit for about 2 hours, or until most of the air bubbles have dissipated.
7) The final step in tissue manufacture is coating a suitable paper base with the gelatin–pigment solution. The base should be relatively thin but must have good wet strength. A very good and inexpensive paper carrier is white, unpasted wallpaper, available in rolls 20.5 inches wide by about 14 yards in length. Many smooth drawing papers can also be used.

THE COATING OPERATION

The temperature in the coating room should be at about 21°C (70°F). If much colder than 21°C (70°F), the gelatin will set very rapidly and it will be impossible to achieve a smooth coating, especially in larger sizes; if warmer than 22°C (72°F), the solution will take a very long time to set. The gelatin–pigment solution should be at about 38 to 43°C (100–110°F) when poured onto the carrier. Maintain the temperature of the coating solution throughout this operation. For coating you will need a plastic or wooden frame about an inch smaller in all dimensions than the paper base you plan to coat to serve as a dam to contain the coating solution and keep it from flowing off the paper.

1) Begin preparation for the coating operation by first leveling a piece of plate glass several inches larger on all sides than the largest tissue you intend to coat.
2) Soak the paper base in water for 2 to 3 minutes, then place it on the leveled coating surface, gently squeegee off all surface water, and blot off with a clean towel (Figures 11-1 and 11-2). Next, place the frame over the paper carrier.

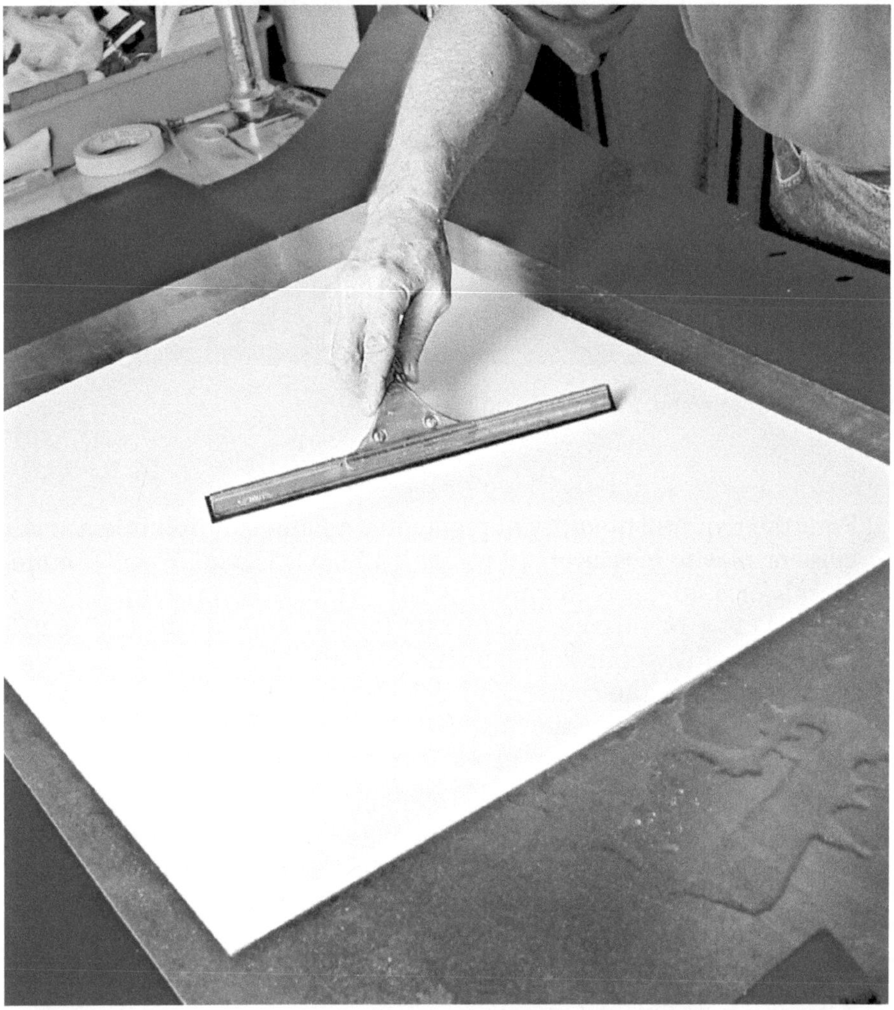

Figure 11-1 Gently squeegee the surface water from the paper. Photo by Sam Wang.

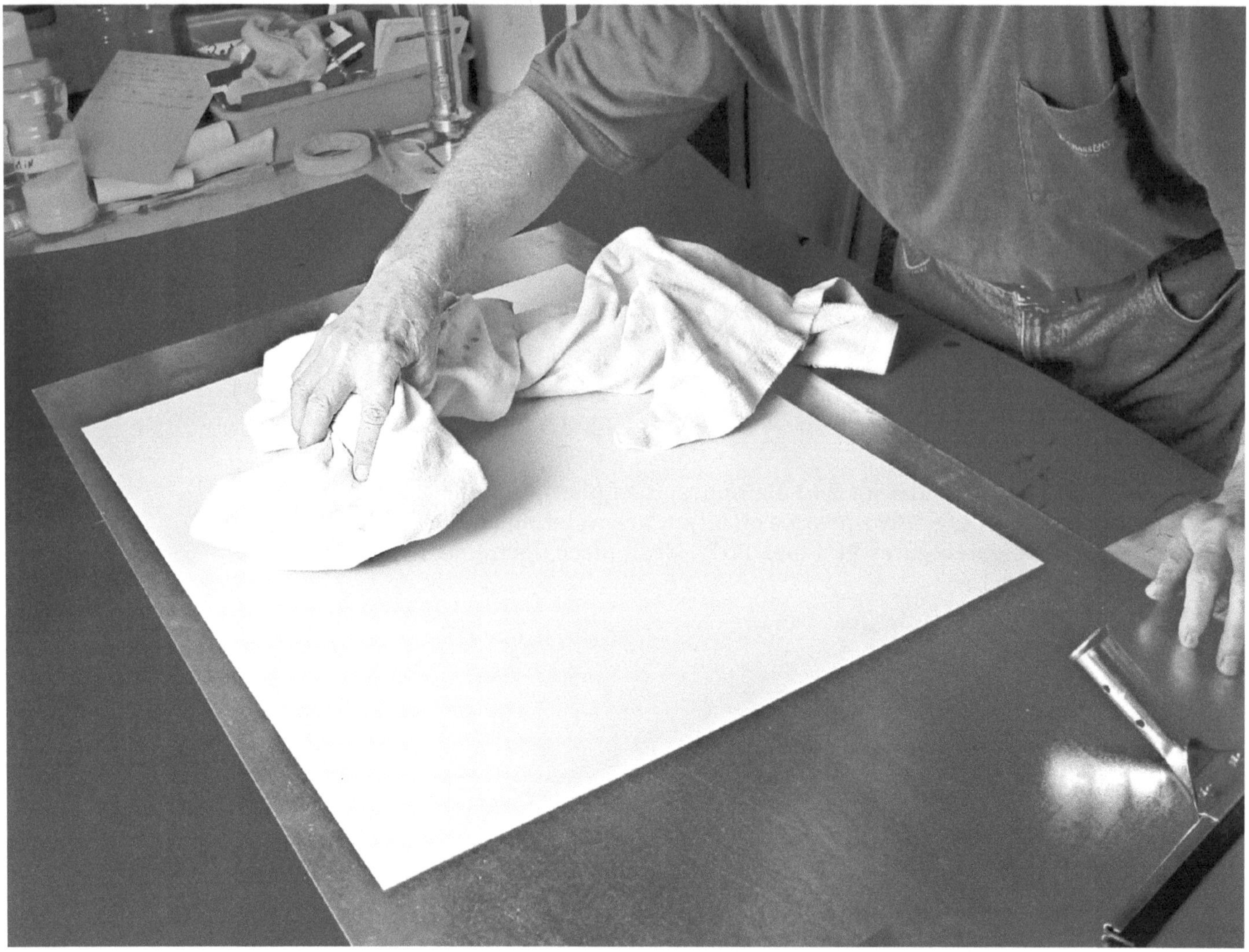

Figure 11-2 Blot the paper with a towel to remove more water. Photo by Sam Wang.

3) Pour the required quantity of pigmented gelatin solution into a small glass or plastic measurer: 20 ml for a 15 cm × 18 cm (5″ × 7″) tissue, 55 ml for a 20 cm × 25 cm (8″ × 10″) tissue, 100 ml for a 28 cm × 36 cm (11″ × 14″) tissue, and 205 ml for a 40 cm × 50 cm (16″ × 20″) tissue. Carefully pour the pigmented gelatin solution onto the paper base, starting in the center, and quickly spread it evenly on the carrier, using a comb or your fingers to even out the gelatin solution on the base (Figures 11-3 and 11-4).
4) If the coating operation creates air bubbles on the surface of the tissue these should be eliminated by immediately misting the surface of the tissue with a spray of alcohol. It is critical that the misting operation step be carried out while the gelatin is still flowing freely because once the gelatin starts to set the drops of alcohol from the misting will form small craters that will not disappear.
5) After you have finished smoothing out the coating, look over the surface carefully and remove air bubbles and any lint, hair, or other foreign

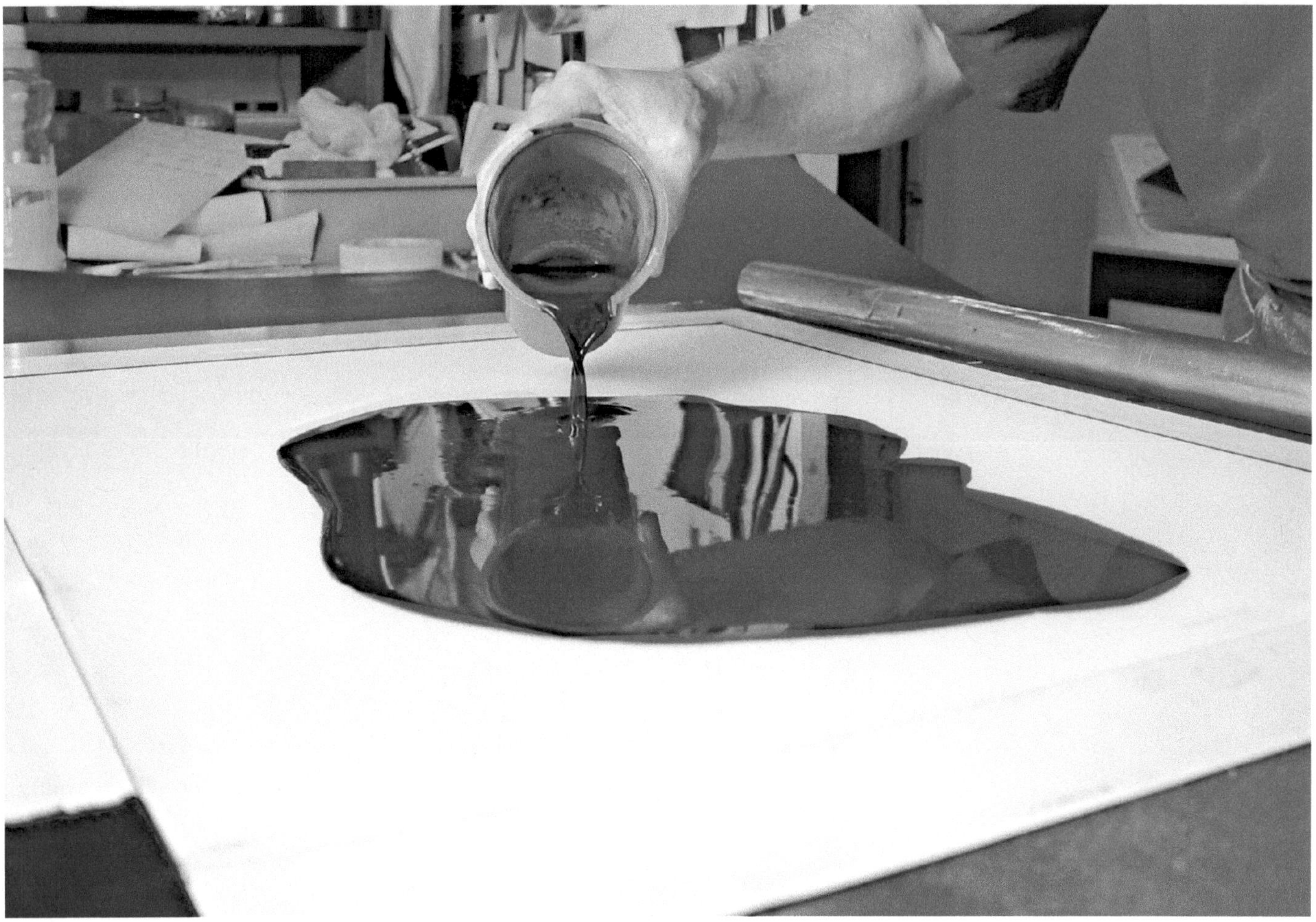

Figure 11-3 Pour the pigmented gelatin solution in the center of the paper. Photo by Sam Wang.

particles that may have settled on the gelatin during the coating operation. You must work quickly because once the solution begins to set it should not be disturbed.

6) The gelatin should set in 10 minutes or less at a temperature of around 21°C (70°F). After it has set carefully remove the tissue from the plate glass and transfer it to a drying screen. Drying will take from 6 to 24 hours, depending on temperature and humidity, the type of carrier used, and the thickness of the tissue coating.

Frame of Magnetic Sign Material

An alternative to a wooden or plastic frame is magnetic sign material. Bob Nugent suggested the use of this material to me and the following information is adapted, with his permission, from a technical article that appears on the Bostick & Sullivan web site.

Magnetic sign material is available in a wide range of sizes up to approximately 0.032 inches in thickness. Use a razor to cut out a frame in the material slightly larger than the size tissue you wish to make.

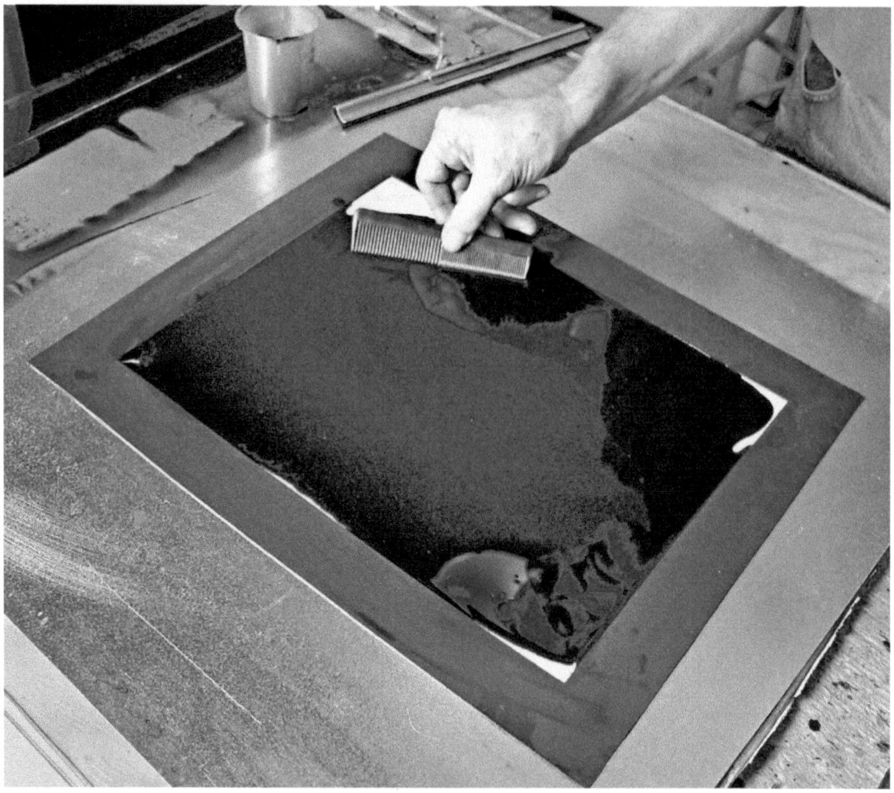

Figure 11-4 Use a comb or your fingers to quickly spread the gelatin evenly all over the paper. Photo by Sam Wang.

You will also need a sheet of steel. It is best to use galvanized steel for this application because plain steel plate will rust and cause all manner of problems. Place the steel sheet on a perfectly level surface. As in the previous directions, first soak your paper for 2 to 3 minutes, and then squeegee it to the steel sheet. Next place the magnetic sign material mask with the cutout over the paper. The magnetic material will stick to the steel through the thickness of the paper with enough force to keep the gelatin from oozing out underneath at the sides (as it tends to do when using other types of frames). For a tissue 25.5 cm by 30.5 cm (10″ × 12″) in size you will need about 65 to 75 ml of pigmented gelatin solution to achieve a wet coating height of 0.032″. This works out to be about the same as the amount of solution recommended earlier for coating with a wooden or plastic frame.

Pour the pigmented gelatin into the corners of the top side first, then let it all flow in, distributing it along the top of the sheet as you go. Using a round iron rod, preheated to about 54.5°C (130°F), squeegee the pigmented gelatin solution down across the paper. The rod needs to be fairly thick—about 3.2 cm (1.25″) in diameter appears to be about ideal—and long enough to completely cover the cutout in the frame. It is important that the rod be thick enough to prevent sagging because this would lead to tissue of uneven thickness. With practice you should be able to distribute the pigmented gelatin solution evenly with just two passages of the rod, one up and one down (Figure 11-5). The warm rod enables you to work slowly, which keeps the gelatin from piling up over the

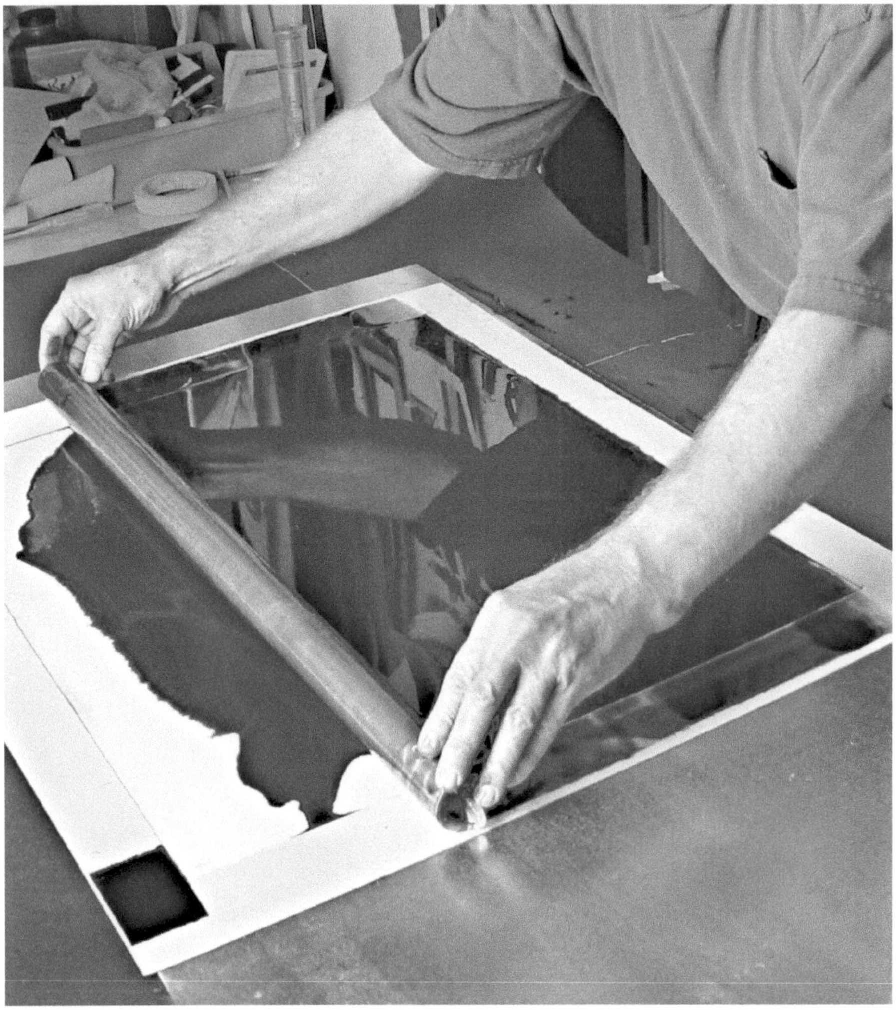

Figure 11-5 One or two passes of the warm rod evenly spreads the gelatin. Photo by Sam Wang.

sides of the rod. The use of the warm rod eliminates another problem—bubbles—because it dissipates any bubbles on contact that may be on the surface of the pigmented gelatin solution. Done correctly, the surface of the tissue will be as smooth as glass after coating. When the gelatin sets, run a knife around the edges of the mask and transfer the tissue to a drying screen.

See Appendix I page 191 for a suppliers list.

Appendices

APPENDIX A—SAFETY CONSIDERATIONS

One of the misconceptions about photogravure is that it is a dangerous process. Like many photographic and printmaking processes, the materials used for photogravure are safe when handled properly. However, a few components of the process require additional caution when working with them. In particular, dichromate powder, asphaltum powder, and ferric chloride powder can cause serious injury or, in the case of dichromates, can be fatal through a one-time acute exposure. Potassium dichromate is the only one of the three that must always be used in the photogravure process. Extreme care must be taken at all times when working with any powder. When a dichromate salt is in solution airborne risk is minimal, but care is still required to prevent all skin contact. Asphaltum powder is a suspected carcinogen and should not be used at all. Ferric chloride powder can cause burns on contact and will react with water to produce toxic and corrosive fumes and exothermic chemical reactions when improperly mixed. It also creates other technical problems in solution, so we recommend that only liquid ferric chloride be purchased.

These and other components of the process can be allergens, can cause sensitivity, or can lead to health problems as a result of chronic exposure. We cannot overstress the importance of eliminating physical contact with any chemical or solvent unless you know it to be absolutely benign. The common routes of entry or contact, and the corresponding protective equipment, are:

Inhalation Wear a respirator or dust mask depending on the material being used. Work in well-ventilated areas with local exhaust whenever possible.

Skin Contact Wear an apron or lab coat, gloves, and a face shield. It is important to wear studio overclothes that you remove so as to avoid accidental contamination after leaving the studio.

Eye Contact Wear a face shield or eye protection when working with any materials that splash or result in fragments that could lodge in your eye.

Ingestion Never eat, drink, or smoke in the studio. Always wash your hands before eating. Remove your lab coat or apron and leave it in the studio before leaving the room.

Notes on Some Personal Protective Equipment

Gloves Be sure they are appropriate for the job. Nitrile gloves are good for most solvents, ferric chloride solutions, and dichromate solutions. Latex gloves are useful for various stages but are a common allergen. Thin vinyl gloves may be a better choice.

Barrier Creams When handling inks and solvents, even with gloves, it is helpful to use a barrier cream as another line of defense against skin absorption.

Dust Mask Wear a fine, well-sealing dust mask when working with airborne particulate matter. Do *not* wear it again while working with solvent fumes.

Respirator Get a respirator that fits properly and be sure that you have the right cartridges for the job. Generally speaking, for solvents and liquid asphaltum fumes, a combination organic vapor/acid fume cartridge will do the job. Do not share your respirator with anyone else. Always store it properly in a hermetically sealed bag. Make sure you check for usage incompatibilities. For instance, if you are working with rosin powders, you should never use the same mask filters for lacquer thinner or alcohol. Remove the dust filters when using the fume cartridges.

Face Shield A full-face visor or shield is the best protection against splashes when working with sensitizer or ferric chloride. Goggles are a second choice because they do not protect the face.

Lab Coat or Apron Ferric chloride will eat away at fabrics and dichromates will soak into them, so be prepared to replace a lab coat regularly. A rubber apron may be more practical but does not protect your arms.

UV Protection Some UV light sources are injurious to the eyes and even the skin. This is especially true of UV-B wavelengths, which are shorter than 320 nm (to 290 nm). UV-C (< 260 nm) is extremely dangerous and is used for sterilizing medical equipment. Use only UV-A because it is safe and it is the optimum wavelength for exposing dichromated colloids (360 to 420 nm). UV-filtered safety goggles, long sleeves, and gloves are recommended if there is a risk of exposure to shorter wave UV radiation. We have been told that the BL florescent bulbs we use are not a problem because they fall into UV-A limits (400 to 320 nm). Call us paranoid, but we still use UV filtering goggles when checking to see if all the bulbs are on.

Material Safety Data Sheets

The most important aspect of studio safety is making sure you are fully and properly informed. This book makes no claim to being definitive. Information on chemical products is constantly being updated.

When you first purchase materials or supplies, request that the supplier send the Material Safety Data Sheets (MSDS). You can also find them online at www.msdssearch.com/msdssearch.htm. You should keep your files up to date, checking them at least every three years. Make sure all your supplies are clearly labeled and stored in appropriate containers and secured from access by children. Use the recommended protective safety equipment.

Reference Materials

There are excellent reference books available on safe studio practices. We highly recommend that you have at least one guide in your studio. For resource books, refer to the Reference Materials bibliography. Check the Internet for MSDS information as well.

Facility Safety

In addition to your health and safety, you must also consider the environment in which you are working. These are a few guidelines to keep in mind.

Never dispose of exhausted dichromate sensitizer solutions or ferric chloride solutions down the drain. They must be brought to a hazardous material disposal facility.

Ferric chloride etches all metal, including stainless steel. It will eventually destroy metal plumbing. Do not pour quantities down your drain. Only the small quantities that result from splashes and clean-up should get into the sink. Flush with lots of water. A sodium carbonate (washing soda) solution will neutralize ferric chloride and is useful as a final bath to end the etching sequence. Do not use any of the plastic etching equipment (trays and any plastic containers in contact with ferric chloride) for photographic purposes again.

When working with solvents, be very careful of the fumes. They can be a fire hazard as well. Do not smoke anywhere near them. Be sure to arrange adequate ventilation for the etching area and especially where you work with solvents, inks, and asphaltum. All solvent-soaked rags should be stored in safety waste cans that are explosion proof. Solvents should be stored in explosion-proof safety cans as well.

Other potential accidents can occur in the studio and print shop. These include physical injury such as cuts from sharp edges on copper and blades, press accidents, and even repetitive stress injury. Common sense is the most effective preventative measure.

APPENDIX B—MAKING A RANDOM-PATTERNED HARD-DOT SCREEN

List of Supplies

- Acid-etched nonglare picture frame glass—high quality, flawless, superfine etch on one side
- Lith film in sheets or off a roll
- High contrast A/B developer

- Other basic photographic chemistry (rapid fix and stop bath)
- A point source or collimated light source
- A vacuum frame with smooth card stock to fit and *perfect* cover glass
- Large sink and trays to process film
- Anti-static cloth or brush, compressed air
- Glass cleaner and lint-free cloth

Procedure

A photogravure screen is a sheet of film with a very fine pattern of opaque randomly shaped squiggles with equal amounts of open and opaque film. Although our first screens were made by contact printing the pebbly surface of Artex Mylar onto lith film, we have found that picture framer's etched nonglare glass works as well if not better and is more easily found. This is especially important because Artex is no longer available, though it may be possible to find an alternative. Do not use the nonglare glass that is smooth on both sides; use etched nonglare glass, which is lightly frosted on one side and shiny on the other. The etched pattern seems finer and more even. The etched glass comes in different qualities or grades of fineness; be sure to find one with as fine and flawless a surface as possible.

The procedure for making screens is straightforward but maddeningly touchy. We cannot overstress the importance of working in an immaculately clean and dust-free environment. Make sure that all glass surfaces are flawless. Take extensive notes and keep everything consistent so that you can accurately repeat or adjust. Be forewarned that making your own screen can take an exhaustive period of testing and seems almost impossible to achieve until you do, finally, get it right. After this, you will understand why the commercial screens are so expensive. In fact, it may not seem like such a bad idea to buy one. For another alternative, see the section later in this appendix on digitally generated screens.

An extremely clean piece of nonglare glass is used to expose a piece of lith film. For a small piece of film, the enlarger can be used at its full height. For a larger piece of film, it is better to work with your vacuum frame set vertically and the enlarger set to project horizontally. The enlarger must provide illumination that is absolutely even. When you are using a vacuum frame make sure that a piece of black mat board or heavy paper covers the pebbly bottom of the vacuum frame and prevents halation. Test that your backing does not create a texture by placing a piece of film on it and running the vacuum.

Under a red 1A safelight, place a piece of absolutely dust-free lith film, emulsion side up, on top of the black backing. Position the glass on top of this with the nonglare side uppermost—away from the film. If you place the frosted side against the film, you will not get a texture. Close the vacuum frame and run the vacuum for a minute or so. Expose this sandwich under an enlarger or other collimated light source so that there is absolutely even illumination across the area of the glass and film. Make test strips at one of the smaller aperture settings on the lens in order to determine the optimum exposure time. This will approximate a point-source light, which is best to preserve the hard dot pattern. Develop the film in a tray of freshly mixed high contrast A/B lith film developer at 19 to 23°C (66.2–73.5°F). Make sure the temperature is stable and repeatable. Temperature variations have a major impact on the density of the texture. Rock the tray slowly in all directions for the full 3-minute development time. At cooler temperatures, you may need to add another 30 seconds. Again, repeatable precision is absolutely important. Record exposure time, f-stop, lens-to-film distance, developer temperature, and development time, so that you can repeat each step exactly for the final sheet.

Mix part A and part B of the developer as you need it with enough volume to cover the sheet of film. Use a 20″ × 24″ tray for a 16″ × 20″ sheet of film with at least 1500 ml of solution. Once mixed, this developer will not keep for more than a half hour or so. In order to maintain consistent results you must replace the developer for each test and for each sheet of film. Some sources suggest still development, but we find slow, steady agitation works best. Getting the film into the developer quickly and evenly is absolutely crucial. Touch one end of the film—face down—to the surface of the developer and roll it down into the tray in one continuous and smooth stroke. Immediately flip it over and agitate to make sure both surfaces are completely wet. Rock the tray slowly but steadily; tilt the tray alternately from side to side and end to end. Make sure the film does not stick to the bottom of the tray. Move to the stop bath quickly at the exact time.

After the sheet is processed, washed, and dried, examine it with a loupe to be sure that the texture is both regular and balanced between light and dark. The overall sheet should look like a light gray tone without any splotches, swirls, flaws, or darker or lighter areas. Place it under a transmission densitometer, and if it is perfect it should read 0.40 ± 0.05. Do not overexpose or overdevelop at this stage because it is imperative to maintain as fine a texture as possible. The next generation will increase the black's density. Use this first generation negative to make second generation positive screens. Contact print this first generation negative onto another sheet of lith film and process the same way. The second generation positive should be a sharp, flawless hard-dot screen with a density of about 0.50 with what appears to be a symmetrical black and white pattern. If your negative has any black dots or flaws, these can be carefully corrected by using a loupe and a blade. Flaws that are clear areas on your negative can be corrected once the positive is made using the same method.

After you have managed to make a perfect negative, store it away carefully so that when you next make screens, you are already halfway there. We have found that the negative first generation is more difficult to make than the positive. Make extensive notes on all aspects so that you can try to repeat the situation. Good luck!

As a guide, here is the setup and procedure that we used to make our most recent screens. We used a Beseler 23CII enlarger with a Schneider Kreuznach Componon-S 2.8/50 mm lens set 165 cm (65 inches) from the vacuum frame. The film was Arista APH lith film and we used Kodak Super RT A/B developer. Our exposure time was 8 seconds with an aperture of f/8. We worked in a 20″ × 24″ tray with 16″ × 20″ film. The development time was 4 minutes in 1500 ml of developer at 19°C. Generally, we found that it took almost 2 minutes before anything would appear during the development, but then it moved along quite quickly and had to be closely monitored. We used a double strength stop bath for 30 seconds and then 4 minutes in rapid fix mixed 1:3 for film. We then washed for 10 minutes and used a wetting agent before letting the film hang to dry. Some precautions to keep in mind are that except for the initial immersion into the developer and stop, the film should always be kept emulsion side up during the processing. Avoid touching the film during development because the heat from your hands will cause streaking and intensified development. Wear thin vinyl or latex gloves and make sure they are dry before you immerse the film into the developer.

To make the second generation positive, we used the same setup of enlarger and vacuum frame. Our exposure was 90 seconds at f/8, with some dodging and burning required to compensate for an uneven first generation negative. The development was done in the same setup for 3.5 minutes at a temperature of 19°C. The remaining process was exactly the same as for the first generation negative.

Generating Digital Screens with an Imagesetter

It is also possible (and far easier) to have a screen made digitally by a service bureau with a high resolution Imagesetter. A rasterized screen at about 50% gray with a resolution of 2000 to 4000 dpi and a random pattern of about 300 lpi should give a fine, flawless screen. The advantages of digital screens, besides being absolutely flawless, are that the clear areas are truly clear, the hard-edged dots have a density of 5.0, and the screens can be made quite large. As long as there are no artifacts, such as banding or mottle, these screens should work well for photogravure. Consult with your local graphics service bureau for more information on this.

APPENDIX C—TESTING FOR CORRECT EXPOSURE WITH YOUR LIGHT SYSTEM

Exposure Tests on Glass

This test eliminates some lengthy and potentially frustrating testing on copper by allowing you to predetermine the correct standard exposure for your exposure system. It also lets you see quite clearly the densities of the gelatin and the amount of effect that a half stop or full stop exposure change can have on densities. This test gives you the chance to familiarize yourself with the procedures of sensitizing tissue, doing an exposure, and adhering and developing the tissue, all without the pressure of working on expensive copper (Denison 1895/1974, p. 43). For the initial exposures, it will be necessary to experiment with a wide range of times. Attempt to achieve both under- and over-exposure. In order to read the results, it is necessary to adhere and develop the tissue onto the glass. Unsuccessful tests can be washed off with a dilute solution of muriatic acid—hydrochloric acid at 20° Bé/31.45% industrial strength—diluted 1:9 in order to reuse the glass. It is worth noting that Crawford (1979, p. 256) recommends that exposure tests be done on copper. There is a superficial hardening of the gelatin from contact with the copper and this should be taken into account. We have used the glass tests to determine a basic exposure range and fine tune the results as we go while using copper.

Method

1) Sensitize a piece of tissue approximately 20 cm × 25.5 cm (8″ × 10″) following the standard procedure. Allow it to dry and stabilize before using. Note: It may take several tests to determine the approximate range before doing the final test, so it is advisable to prepare more than one piece of tissue.
2) Cut the tissue into the appropriate size. To do so, make a mask using goldenrod paper with three to five 1.3 cm × 11 cm (1/2″ × 4 1/2″) openings for Stouffer 21-Step Scales.* Leave approximately 2.5 cm (1″) between each opening. Tape the step scales in, emulsion side up and on top of the mask. Cut your tissue so it covers all the scales and is slightly larger in height and width than the openings for the scales. Lay the sensitized tissue with its emulsion side in contact with the step scales. Lightly tape down the corners to hold the tissue in place while putting the mask, scales, and tissue assembly into your vacuum frame. (*21-Step Scales can be ordered in quantity from Stouffer Industries. See the suppliers list in Appendix I)
3) Prepare a sheet of glass that is at least 2.5 cm (1″) larger than your tissue. Sand the edges with silicon carbide paper so they are dull. Clean both sides well with alcohol.
4) Have the 25% alcohol adhering solution ready at 15°C (60°F). Leave it in the bottle until after the exposure to prevent evaporation.
5) Determine an approximate range for your exposures. The following are starting suggestions for various light systems:
   a) Bank of eight tightly spaced 350 Blacklight (BL) bulbs, 20 watts each, 17 inches from the vacuum frame: 10.5 minutes
   b) 4000 watt metal halide, 5 feet from the vacuum frame: 350 exposure units
   c) NuArc Fliptop Platemaker: 25 exposure units
6) If using the BL bulb system as described in step 5a, do an exposure of 5 minutes, cover one of the openings with goldenrod or black opaque paper, and then do additional 5-minute exposures, covering each step scale in turn. The result is four exposures of 5, 10, 15, and 20 minutes. After you have determined an approximate range, you can narrow it down to a precise exposure.
7) While the tissue is being exposed, degrease the glass with TSP and leave it in a tray of cool water in preparation for adhering.
8) Immerse the exposed tissue in the cool 25% alcohol solution until it lies flat.
9) Transfer the tissue gelatin side down into the tray holding the glass and adhere as per standard procedure.
10) Develop the tissue as per standard procedure.
11) Transfer to an 80% alcohol solution for 5 minutes. Remove, drain, and blow dry.

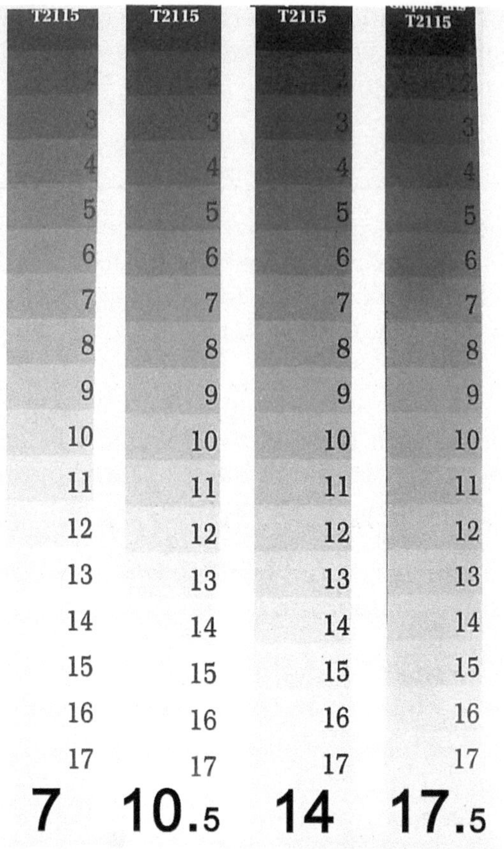

Figure C-1 Four exposures of gelatin resist on glass showing differences in densities for exposures of 7, 10.5, 14, and 17.5 minutes.

Inspect the results. What you are looking for in a correct exposure are clear and distinct steps all the way from #1 to #15. Step #15 will have a very thin layer of gelatin but will still be clearly distinct from the clear glass at Step #16; nothing higher than that is visible. This gives you good density and clear separation in the ranges used for shadow detail (1.71 to 1.85, Steps #12 and #13) and still allows for separation in the very darkest areas and blacks, up to 2.13 or Step #15. It is also important to keep the highlight densities manageable—not too dense—so that they can be etched in a reasonable time frame. More exposure than needed is not better (Figure C-1).

To assess your exposures and determine the best possible time, remember that the difference between each step on the 21-Step Scale equals one half of a stop. Thus, in an exposure of 5 minutes, the density reading at Step #12 would move to Step #14 when exposed for 10 minutes. Dichromated colloids have a straight line response to exposure from actinic light (Gassan 1977, p. 197). Changes in exposure can compensate for overall density differences in a positive if the contrast range is constant. When inspecting the results from the four exposures, we determined that with our equipment an exposure of 10.5 minutes is our correct standard, and this is what we base our exposures on for standard positives.

Making an Exposure Unit

Many sources give plans and instructions for making an exposure unit using 24-inch F20T10-350BL fluorescent bulbs. This unit is common in alternative photographic practice. Check out plans in Carl Weese and

Richard Sullivan's *The New Platinum Print* (1998), Luis Nadeau's *History and Practice of Platinum Printing* (3rd ed., 1994), *Post-Factory Photography*, Issue #5. (Aug. 2000), and more. The most important points are to use the right bulbs and to have a system that turns all of them on quickly and simultaneously. The unit is simply an array of standard 24″ double flo-rescent fixtures wired in parallel to a single switch or grounded plug and positioned tightly together. If heat is generated by the ballasts, a fan and venting are necessary.

APPENDIX D—THE CHEMISTRY OF ETCHING WITH IRON(III) CHLORIDE

Iron(III) Chloride Formulae and the Production of Free Acid

Iron(III) chloride (formerly called ferric chloride) is a salt, that is, it consists of Fe^{3+} ions and Cl^- ions. In solution, each ion is surrounded by water molecules, thus the chloride ions are written as $Cl^-(aq)$. Six water molecules are particularly strongly attracted to the Fe^{3+} ion, thus it is written as $[Fe(OH_2)_6]^{3+}(aq)$, with the "H_2O" reversed to indicate that it is the oxygen atom of the water molecule that is chemically linked to the iron(III) ion. In solution, the following reaction takes place and produces the free acid, correctly represented as the H_3O^+ (aq) ion or hydronium ion. Most early literature refers to the free acid while in solution as HCl.

$$[Fe(OH_2)_6]^{3+} + H_2O(l) \Leftrightarrow [Fe(OH_2)_5(OH)]^{2+} + H_3O^+(aq)$$

Thus, a solution of iron(III) chloride is always acidic and you should not be confused by the fact that the chemical is a salt. This reaction is reversible, that is, it goes both ways: addition of acid will "force" the iron(III) back to $[Fe(OH_2)_6]^{3+}(aq)$, whereas addition of base will result in the production of $[Fe(OH_2)_5(OH)]^{2+}(aq)$.

The reaction described is not the only one that takes place. The iron(III) ion can lose a second and third hydrogen ion. The third step, predominant when a large proportion of base (such as sodium hydroxide) is added, results in the formation of the iron(III) hydroxide sludge, $Fe(OH)_3(s)$.

$$[Fe(OH_2)_4(OH)_2]^+(aq) + OH^-(aq) \Leftrightarrow Fe(OH)_3(s) + 4H_2O(l)$$

However, these are not the only species being formed in the solution. Because we work with a very concentrated iron(III) chloride solution with a high concentration of chloride ions, the chloride ions can replace the water molecules.

$$[Fe(OH_2)_6]^{3+}(aq) + Cl^-(aq) \Leftrightarrow [Fe(OH_2)_5Cl]^{2-} + H_2O(l)$$

As the chloride ion concentration increases, more and more chloride ions replace the water molecules. Under the etching conditions, the species that is probably predominant is $[Fe(OH_2)_4Cl_2]^+(aq)$. The chloro-species are key to the reaction, which is the etching process.

Etching Reactions

Etching copper with iron ions is actually contrary to the direction of normal chemistry because copper is a very unreactive chemical element. However, the iron(III) species acts as both an oxidizing agent and a chloride ion carrier. It is believed that the $[Fe(OH_2)_4Cl_2]^+(aq)$ ion diffuses through the gelatin layer. Upon reaching the copper surface, the iron(III) is reduced to iron(II) while simultaneously the copper is oxidized from copper metal to—initially—copper(I). The chloride ion then reacts with the copper(I) to give first the insoluble white copper(I) chloride and then the soluble $CuCl_2^-$ (aq) ion. This ion is seen as a dark precipitate as it diffuses back through the gelatin along with the iron(II) species. The overall reaction can be depicted as follows:

$$[Fe(OH_2)_4Cl_2]^+(aq) + Cu(s) + H_2O(l) \Rightarrow [Fe(OH_2)_6]^{2+}(aq) + [CuCl_2]^-(aq)$$

The copper(I) chloride is what produces the phenomenon of white etching. When the $CuCl_2^-(aq)$ ion encounters oxygen (in the air or dissolved in the solution) it is further oxidized to green-blue copper(II) chloride, $CuCl_2$. It is the increasing concentration of the copper(II) chloride that eventually turns the etching solution a greenish color.

The literature frequently recommends that the addition of small quantities of hydrochloric acid will rejuvenate exhausted iron(III) chloride solutions. This is because the addition of HCl reintroduces the necessary chloride and H_3O^+ ions, both of which become depleted over the course of etching.

The Role of Free Acid or H_3O^+

The $H_3O^+(aq)$ ion can also diffuse through the gelatin and etch the copper. This reaction produces a gas. This can cause devils and blistering. The formula for this reaction is:

$$Cu(s) + 2H_3O^+(aq) + 2Cl^-(aq) \Rightarrow CuCl_2^-(aq) + H_2(g) + 2H_2O(l)$$

$$\uparrow (2\ \text{“HCl”}) \uparrow$$

However, if $H_3O^+(aq)$ is completely eliminated from the solution, sludge is produced, which results in too small a proportion of the active species $[Fe(OH_2)_4Cl_2]^+(aq)$. Thus, the solution must contain enough free acid to produce the active chloro-species but not so much that there is production of hydrogen gas.

Blue Label 48° Bé Composition

$FeCl_3$:	39.5–40.3%
Free acid:	0.02–0.03%
Sulfate, as $Fe_2(SO_4)_3$:	4.0–4.2%
Copper:	.005 oz. cu/gal.
(Smeil 1975, p. 87)	

Our thanks to Dr. Geoff Rayner-Canham for his contribution to the production of this appendix.

Iron(III) Hydroxide (Sludge) Preparation

The recommended practice is to make enough iron(III) hydroxide sludge to treat one gallon of working mordant. Making the sludge is difficult and messy. When we tried a recipe from one of the older texts, we were unsuccessful at producing any sludge. We are fortunate, however, to have the help of a chemistry department's facilities and technicians.

Start with iron(III) chloride solution at 42° Bé. Using a pH meter, check the pH. It will be very acidic. Pour 100 ml of iron(III) chloride into a 1000 ml glass beaker. Measure out 300–400 ml of 6 mol-L^{-1} ammonia (one-half strength concentrated ammonia). Add the ammonia to the iron(III) chloride while stirring. Keep checking the pH and stirring while adding more ammonia until the pH is up to 8 or 10. Clean the electrode of the pH meter in a solution of EDTA.

Caution: This produces an exothermic reaction (the solution gets hot). Use a fume hood to vent the strong ammonia fumes that are produced.

A brown sludge (the iron(III) hydroxide) will be produced. Let this mixture settle overnight, covered. Pour off the ammonia slowly and carefully. Add 700 ml of water to rinse. Stir and let settle again. Decant and repeat once more to remove as much ammonia as possible. Before adding the iron(III) hydroxide to the etching baths, it is important to remove as much water as possible. The sludge tends to form a clay-like layer over the filter, which prevents any further moisture from going through. However, by allowing the solution to stand for a period of time, you can pour the clear liquid off the top. Do not allow it to stand open to evaporate because this causes *aging* and chemically alters the sludge. The best method is to centrifuge it. Store the pasty sludge in a sealed jar.

Note: The above procedures are best carried out in a fully equipped chemistry lab by trained technicians who know how to handle the dangerous materials involved.

APPENDIX E—EXPOSURE AND ETCH FORM

Exposure and Etch Form

Description of Image: Tree through fence for Book Cover Negative Info: 120 TMY 400	Tissue Exposure Date: June 1, 2002
Title: Penetration Series: —	Etch Date: June 2, 2002

POSITIVE DENSITIES:	DESCRIPTION:	DENSITY READING:	21-STEP SCALE STEP:
Max Black:	Black holes in fence	< 1.90	> #13
Dark shadow detail:	top of tree	< 1.77 1.67	± #12
Important mid tones:	dark bark mid bark	.94 .74	#7 #6-5
Dark highlight detail:			
Bright highlight detail:	bark	.37	+ # 3
Thinnest areas:	brightest sky	.24	#2-3
Other notable areas:			

BL Exposure of new Screen: * 11 ½ min.	Age of Sensitized Tissue: frozen for 2 weeks
BL Exposure of Positive: * 11 ½ min.	Age of Sensitizer used: fresh
Temperature of the various Bé's at time of etch: 21°c	

47° = 47.3° 45° = 45.2° 43° = 43.5° 41° = 41° 39° = 39°

Step #								begins etching *
Time:	60:00	58:00	55:00	53:00	50:00	47:45	46:00	40:30
Baumé:	47°	45°	47°	45°	43°	45°	43°	

Step #	#21	#20	#19-18		#17	#16	#15	#14	#13
Time:	40:15	39:00	38:30	37:00	36:40	34:40	33:00	30:45	28:30
Baumé:	45°			43°					

Step #		#12		#11		#10	#9			#8
Time:	27:00	26:15	26:00	24:15	24:00	22:15	21:15	21:00	19:15	18:35
Baumé:	41°		43°		41°			43°	41°	

Step #		#7		#6		#5	#4		#3	#2
Time:	16:00	15:45	14:45	14:15	12:50	12:40	11:45	10:20	9:25	7:30
Baumé:	39°	41°	39°	41°	39°		41°	39°		OUT

* slow start for etching indicates over exposure **

Figure E-1 A completed form showing how the etch is recorded. The time is noted down for each change of Baumé and each step appearance. In this way, the time between steps can be calculated at a glance. Photocopy the blank form on the following page for use.

Exposure and Etch Form

Description of Image: Negative Info:	Tissue Exposure Date: , 200__
Title: Series:	Etch Date: , 200__

POSITIVE DENSITIES:	DESCRIPTION:	DENSITY READING:	21-STEP SCALE STEP:
Max Black:			
Dark shadow detail:			
Important mid tones:			
Dark highlight detail:			
Bright highlight detail:			
Thinnest areas:			
Other notable areas:			

BL Exposure of_______ Screen:	Age of Sensitized Tissue:
BL Exposure of Positive:	Age of Sensitizer used:
Temperature of the various Bé's at time of etch:	

Step #	
Time:	60:00 58:00
Baumé:	

Step #	
Time:	
Baumé:	

Step #	
Time:	
Baumé:	

Step #	
Time:	
Baumé:	

APPENDIX F—PRINTING INK TESTS

Ink	*Consistency*	*Wiping Characteristics*	*Color*	*Density (on a Normally Etched 21-Step Scale)*	*Permanence (Manufacturer's Claim)*
Charbonnel 55985 Etching Ink	good body, applies well	wipes easily but leaves a plate tone that requires light, vigorous hand wipe	warm black	very dense, last distinguishable visual separation between Steps #9 & #10 on the Stouffer 21-Step Scale	absolutely permanent
Charbonnel Luxe C Etching Ink	good body	moderate to wipe; leaves a plate tone that takes extensive hand wiping to remove	neutral black	dense, last distinguishable visual separation between Steps #10 & #11 on the Stouffer 21-Step Scale	absolutely permanent
Charbonnel F66 Etching Ink	slightly waxy	wipes easily but leaves a plate tone that is difficult to fully remove	warm black	moderate to dense, last distinguishable visual separation between Steps #10 & #11 on the Stouffer 21-Step Scale	absolutely permanent
Daniel Smith #827 Preferred Black Etching Ink	stiff	wipes easily and very cleanly	warm black	moderate to dense, gives a higher contrast print, last distinguishable visual separation between Steps #10 & #11 on the Stouffer 21-Step Scale	lightfast
Gamblin Carbon Black Etching Ink	good body	difficult to wipe; requires vigorous wipe with cheesecloth and extensive hand wiping to remove plate tone	warm black	very dense, last distinguishable visual separation between Steps #9 & #10 on the Stouffer 21-Step Scale	absolutely permanent
Gamblin Portland Black Etching Ink	good body, slightly stiff	wipes easily; leaves light plate tone that wipes off easily with hand wipe	neutral black	dense, last distinguishable visual separation between Steps #10 & #11 on the Stouffer 21-Step Scale	absolutely permanent
Gamblin Portland Cool Black Etching Ink	good consistency	wipes easily and cleanly	cool, blue black	moderate, last distinguishable visual separation between Steps #11 & #12 on the Stouffer 21-Step Scale	absolutely permanent
Gamblin Bone Black Etching Ink	good consistency	wipes very easily and leaves little plate tone	warm black	translucent, last distinguishable visual separation between Steps #12 & #13 on the Stouffer 21-Step Scale	absolutely permanent
Graphic Chemical Intense Black Etching Ink	very loose, oily	moderate to wipe, difficult to remove the plate tone (leaves a film)	warm black	moderate to translucent, last distinguishable visual separation between Steps #11 & #12 on the Stouffer 21-Step Scale	information not available

(*Continued*)

Ink	*Consistency*	*Wiping Characteristics*	*Color*	*Density (on a Normally Etched 21-Step Scale)*	*Permanence (Manufacturer's Claim)*
Graphic Chemical Bone Black Etching Ink	waxy consistency, applies in a very thin layer	very easy to wipe	warm black	very translucent, gives a thin print with very open shadows, last distinguishable visual separation between Steps #14 & #15 on the Stouffer 21-Step Scale	information not available

Some general notes or observations:

1. Daniel Smith makes seven different etching blacks of varying color and consistency.
2. All the Charbonnel inks we tested tend to give a dense print with veiled highlights and blocked shadow detail.
3. Charbonnel provides extensive information on their inks including listing the pigments used. They also make a transparent base.
4. All the Gamblin inks give a very smooth print with smooth, even shifts in the tonal scales.

APPENDIX G—PAPER CHART FOR PHOTOGRAVURE PRINTING

Highly recommended, great for editioning: *Name of paper*	*Weight g/m²*	*Sized (S) or Unsized (W)*	*Soaking Time*	*Cost Rating*
Alcantara	160	S (light)	Short	High
Arches 88	300	W	Mist	Ave.
Copperplate	300	W	Short	Ave.
Coventry Rag	234	S	Long	Low
Gampi tissue (only for Chine Collé)	35	W	Mist	V. high
Gampi Torinoko	95	W	Mist	V. high
Hahnemühle German Etching (cream)	300	S	Long	High
Hosho (thin text weight)	85	W	Mist	V. high
Kitakata (good for Chine Collé)	40	W	Mist	Ave.
Lana Laid (Verge) (text weight)	125	S (light)	Short/mist	Ave.
Lana Gravure	250/300	S	Short/long	Ave.
Magnani Biblos (text weight)	130	S	Mist	n/a
Moulin de Larroque	200	S	Long	Ave.
Ruscombe Margaux Estampe	250	S	Long	V. high
Ruscombe Medoc Velin	250	S	Long	V. high
Somerset Book Wove	175	S (light)	Short	Ave.
Somerset Satin	250/300	S	Long	Ave.
Twinrocker, Yale (text weight)	130	S (light)	Short/mist	V. high

Recommended, appropriate for editioning: *Name of paper*	*Weight g/m²*	*Sized (S) or Unsized (W)*	*Soaking Time*	*Cost Rating*
BFK Rives	250	S	Long	Ave.
Fabriano Artistico	200	S	Long	Ave
Fabriano Tiepolo	250	S	Long	Ave.
German Etching 100 (bright white)	300	S	Long	High
Johannot	250	S	Long	Ave.
Kozo (good for Chine Collé)	40	W	Mist	Ave.
Lanaquarelle (hot pressed)	185	S	Short	Ave.
Magnani Pescia	300	S	Long	Ave.
Tosawashi Gampi Sukiawase	100	W	Mist	V. high
Twinrocker, Printmaking	300	S	Long	V. High
Twinrocker, White Cotton	300	S	Long	V. High
Yumei Juan Silk (imparts a fabric texture)	n/a	S	Mist	V. high

Good for proofing: *Name of paper*	*Weight g/m²*	*Sized (S) or Unsized (W)*	*Soaking Time*	*Cost Rating*
Arches Cover	300	S	Short	Ave.
Arches Platine	310	S	Long	V. high
Arches Text Wove	120	S (light)	Short	Ave.
Domestic Etching	175	S (light)	Short	Low
Fabriano Rosaspina	220	S	Long	Ave.
Lana Royale	250	S	Long	Ave.
Mulberry (good for Chine Collé)	40	W	Mist	High
Opus Domestic Archival	250	S	Long	Low
Rice (generic) (good for Chine Collé)	50	W	Mist	Low

These papers were tested on a gravure plate of fine detail and many tonal subtleties. The judgments were based on comparison of the printed results. Substandard papers were excluded. Test papers using your own printing conditions.

APPENDIX H—THE CONVENTIONS FOR EDITIONING PRINTS

Different Types of Proofs and How to Number the Edition

What follows is a brief description of the types of proofs and prints that would most frequently occur when printing a photogravure print. These designations are traditionally written in the title line under the image (always in pencil) along with the title, date, and signature. How much of this information is included on this line is optional.

In-Process Proofs

State Proofs (S/P) These are a record of the evolution of a plate. They include the first proofs and each printing after the plate has been reworked and the progress is checked.

Trial Proofs (T/P) Once the plate has been finalized, proofs are taken while working out the details of printing such as ink and paper combinations.

Color Proofs (C/P) When making a color print, these are the proofs done to verify the correct color ink.

Edition Proofs

Consecutive or Progressive Proofs When using more than one plate, these proofs are the record of each of the separate plates, and also a record of the successive combinations of the plates.

Artist's Proof (A/P) These prints are the same as the edition but are reserved for the artist's use. They normally do not exceed 10% of the total edition size. This convention arose from working with a publisher, where the artist would receive a portion of the edition in the form of the artist's proofs. It is still used, even by artists printing their own editions. In this case, however, it has evolved to often indicate a print that varies minutely from the final edition by factors such as slight alterations in ink or paper. Artist's proofs should always be of the same high quality as the edition.

Printer's Proof (P/P) When working with a printer, it is a professional convention to provide them with a proof for each edition they print. This proof is identical to the edition.

Cancellation Proof (CP) This is a convention whereby after the final print is done, the plate is defaced and then the cancellation proof is pulled, thus recording that the plate can no longer be printed. It is a record to ensure that the edition is limited to the stated amount.

Edition Prints

Bon à Tirer (B.A.T.) This is the print that indicates the required image quality and characteristics that the edition must follow. It is used by the printer as the standard in gauging whether a print meets the requirements of the edition.

Studio Print or Shop Print (SP) When working in a print shop or studio, it is standard practice to leave at least one print for the archives. Some shops require more prints, depending on the size of the edition. This print is identical to the edition.

Hors du Commerce or Not for Trade (HC or NFT) These are prints that are not to be sold.

The Edition

The prints that comprise the edition are numbered. This is written in pencil on the print itself as a fraction, with the lower number representing the total number of prints in the edition and the top number separately enumerating each print. For example, an edition of three prints would be indicated by the three copies being designated 1/3, 2/3, and 3/3. These are also called the impression numbers.

It is possible to print an edition that consists of varied manifestations of the plate matrix. This varied edition makes use of the printing permutations available with any plate. Each print in a varied edition is deemed to be equally successful as an image, but the prints are not identical one to the other. The numerical designation system is the same as for an edition except that Roman numerals are used: I/IV, II/IV, III/IV, IV/IV.

There is literature available that describes all the various types of prints that can make up an edition. For further information, we recommend *Code of Ethics for Original Printmaking* by Nicole Malenfant and Richard Ste-Marie, published by Conseil québécois de l'estampe, 2000.

APPENDIX I—SUPPLIERS

Specialty Materials and Sources for Photogravure

Prices provided for key specialty items as a rough guide. Prices will vary over time. (approximate 2002 prices in US$ unless otherwise noted)

Item	*Description*	*Price: USD*	*Source*	*Phone, Fax, and Email*
Autotype gravure tissue	Autotype G.35 Gravure Pigment Paper (Carbon Tissue) Item #2039	$310.00/roll 95cm × 2000cm (37.5″ × 65′)	Autotype Americas, Inc. 2050 Hammond Drive Schaumburg, IL 60173-3810 www.autotype.com	PH 847-303-5900 PH 800-323-0632 FX 847-303-5225
Hydrometer Glacial acetic acid Potassium dichromate	Hydrometer, 39 to 51 Degree Baumé range Catalogue no. 11-571E	CDN $26.50	Good Health and Safety 3393 Colonial Drive Mississauga, ON L5L 5B9	PH 877-828-1611
Hydrometer Glacial acetic acid Potassium dichromate	12″ Baumé Light Hydrometer, 39–51 Degrees Baumé Catalogue no. LKE8512LT	$14.81 each	Wilkem Scientific Ltd. P.O. Box 301 Pawtucket, RI 02862 www.wilkem.com	PH 800-766-5676 PH 401-723-1840 FX 401-724-8760
Hydrometer for alcohol	12″ Hydrometer, Specific Gravity and Baumé Light, 600-1.000 SG, 100-10 Bé Catalogue no. LKE8799.6	$16.03 each	Wilkem Scientific Ltd. P.O. Box 301 Pawtucket, RI 02862 www.wilkem.com	PH 800-766-5676 PH 401-723-1840 FX 401-724-8760
Copper	Mirror finish, 18 gauge .040 thickness, 36″ × 96″	$305.00/single sheet $295.00/two to five sheets	C.G. Metals Inc. P.O. Box 672 Nyack, NY 10960 www.cgmetals.com	PH 845-358-8364 FX 845-358-0274
Lith Film	Arista APH—halftone lith film item #5202200	$249.00 for 20″ × 200′ roll	Freestyle Photographic Supplies 5124 Sunset Blvd. Los Angeles, CA 90027 www.aristagraphics.com www.freestylesalesco.com www.freestylephoto.biz	PH 800-292-6137 PH 323-660-3460 FX 800-616-3686 email: info@ reestylephoto.biz
Bergger continuous tone film	BPFB18 film for enlarging negatives to positives	Contact supplier for current prices	Bergger Products, Inc. 5955 Palo Verde Dr Rockford, IL 61114 www.bergger.com	PH 815-282-9876 FX 815-282-2982 email: sales@bergger.com

(Continued)

Item	*Description*	*Price: USD*	*Source*	*Phone, Fax, and Email*
Developer for making continuous tone lith film positives	Clayton P20 Paper Developer Item #36204	$21.99 per gallon	Freestyle Sales Co. 5124 Sunset Blvd. Los Angeles, CA 90027 www.freestylephoto.biz www.freestylesalesco.com	PH 800-292-6137 PH 323-660-3460 FX 800-616-3686 email: info@ freestylephoto.biz
Press blankets	Woven wool: 1839E-16D is 1/16″ thick 1982-38C is 1/8″ thick Sold by the yard in various widths	1/16″ thick: 24″ wide is $34.70/yd. 36″ wide is $52.72/yd. 1/8″ thick: 24″ wide is $49.40/yd. 36″ wide is $82.12/yd.	Charles C. House & Sons P.O. Box 158 19 Perry St. Unionville, CT 06085	PH 860-673-2518 FX 860-675-8956
Ferric chloride	Rotogravure Iron 48 Degree Baumé	$40.00 for 4 liter container plus dangerous goods shipping surcharge	Wilkem Scientific Ltd. P.O. Box 301 Pawtucket, RI 02862 www.wilkem.com	PH 800-766-5676 PH 401-723-1840 FX 401-724-8760
Ferric chloride	48 degree Baumé ferric chloride solution 40% weight by volume	CDN $40—49/liter (approximate) plus dangerous goods shipping surcharge	Good Health and Safety 3393 Colonial Drive Mississauga, ON L5L 5B9	PH 877-828-1611
Stouffer 21-Step Scale	The famous 21-step transmission guide, Item # T2115	$5.40 + s&h	Stouffer Industries 922 Cleveland St. Mishawaka, IN 46544 www.stouffer.net	PH 574-234-5023 FX 574-232-7989 email: info@stouffer.net
Printmaking papers	various	varies	New York Central Art Supply 62 Third Avenue New York, NY 10003 www.nycentralart.com	PH 800-950-6111 PH 212-477-0400 FX 212-475-2513 email: sales@nycentralart.com
Printmaking papers	various Sole supplier in Canada for Ruscombe papers.	varies	Opus Framing & Art Supply 1677 West 2nd Avenue Vancouver, BC V6J 1H3 www.opusframing.com	PH 800-663-6953 PH 604-736-7535 FX 604-731-3519
Printmaking inks and papers	various	varies	Daniel Smith P.O. Box 84268 Seattle, WA 98124-5568 www.danielsmith.com	PH 800-426-6740 PH 800-426-7923 FX 800-238-4065 FX 206-224-0404 email: sales@ danielsmith.com

Gelatin Tissue Supplies (Chapter 11)

Item	*Description*	*Source*	*Phone, Fax, and Email*
Knox unflavored gelatin	Available in a package of 4 to 32 envelopes, each containing about 7 grams.	any grocery store	
Gelatin		Artcraft Chemicals, Inc. P.O. Box 583 Schenectady, NY 12301 www.artcraftchemicals.com	PH 800-682-1730 PH 518-355-8700 FX 518-355-9121 email: artcraft@peoplepc.com
Gelatin and pigments, potassium dichromate	Gelatin, pigments in dispersion, colors for textiles and airbrush, general photographic supplies with an emphasis on alternative photo process supplies. Workshops on alt photo processes, including making gelatin tissue offered.	Photographers Formulary P.O. Box 950 Condon, MT 59826-0950 www.photoformulary.com	PH 800-922-5255 FX 406-754-2896 email for info: lynnw@blackfoot.net
Pigments		Daniel Smith P.O. Box 84268 Seattle, WA 98124-5568 www.danielsmith.com	PH 800.426.6740 PH 800.426.7923 FX 800.238.4065 FX 206-224-0404 email: sales@danielsmith.com
Pigments	Cal Tint II and Mixol pigments	Douglas and Sturgess, Inc. 730 Bryant Street San Francisco, CA 94107 www.artstuf.com	PH 888.278.7883 FX 510.235.4211 email: sales@artstuf.com
General art supplies and magnetic sign material	Magnetic sign material	Dick Blick Art Materials P.O. Box 1267 Galesburg, IL 61402-1267 www.dickblick.com	To place an order: PH 800.828.4548 FX 800.621.8293 Customer service: PH 800.723.2787 Product information: PH 800.933.2542 email: info@dickblick.com

Reference Materials (Bibliography)

TECHNICAL AND PROCESS-RELATED SOURCES

Anchell, Steve and Bill Troop. *The Film Developing Cookbook.* Burlington, Mass. Focal Press, 1998.

Bennett, Colin N. *Elements of Photogravure: Photo Printing from Copper Plates.* 2nd Edition. Boston: American Photographic Publishing Co., 1927. Also found reproduced in *Nonsilver Printing Processes.* Edited by Peter C. Bunnell. New York: Arno Press, 1973.

Bennett, Colin N. *Elements of Photogravure: Photo Printing from Copper Plates.* 3rd Edition. London: The Technical Press Ltd., 1935.

Blaney, Henry R. *Photogravure.* Scovill's Photographic Series. New York: Scovill, 1895.

Burkholder, Dan. *Making Digital Negatives for Contact Printing.* Carrollton, Tex. Bladed Iris Press, 1999.

Cartwright, H. Mills. *Photogravure: A Text Book on the Machine and Hand-Printed Processes.* 2nd Edition. Boston: American Photographic Publishing Co., 1939.

Cartwright, H. Mills. "The Causes of Mottle in Photogravure," *Penrose's Annual* 34, (1933). pp. 75–77.

Crawford, William. *The Keepers of Light: A History and Working Guide to Early Photographic Processes.* Dobbs Ferry, N.Y.: Morgan & Morgan, 1979.

Denison, Herbert. *A Treatise on Photogravure.* London: Iliffe and Sons, 1895. Reprinted Rochester, N.Y.: Visual Studies Workshop, 1974.

De Zoete, Johan. *A Manual of Photogravure.* Haarlem, The Netherlands: Joh. Enschedé en Zonen, 1988.

Foster, Kenneth C. "Photogravure and Carbon Printing from the Same Type of Resist Tissue." M.S. thesis, East Texas State University, 1982.

Gassan, Arnold. *Handbook for Contemporary Photography.* 4th edition. Rochester, N.Y.: Light Impressions, 1977.

Kolb, Gary P. *Photogravure: A Process Handbook.* Carbondale and Edwardsville: Southern Illinois University Press, 1986.

Kraft, James N. "An Historical and Practical Investigation of Photogravure." M.F.A. thesis, University of New Mexico, 1969.

Malenfant, Nicole and Richard Ste-Marie. *Code of Ethics for Original Printmaking.* Montréal: Conseil québécois de l'estampe, 2000.

Mertle, J. S. and Gordon L. Monsen. *Photomechanics and Printing: Practical Information on Platemaking and Presswork by Recognized Procedures.* Chicago, Ill.: Mertle Publishing Company, 1957.

Nadeau, Luis. *Modern Carbon Printing.* 2nd Edition. Fredericton, N.B.: Atelier Luis Nadeau, 1996.

Rothberg, Samuel W. (Sandy). *Photogravure Handbook.* Chicago: Privately published technical manual, 1976.

Sacilotto, Deli. *Photographic Printmaking Techniques.* New York: Watson-Guptill Publications, 1982.

Schaffert, Roland M. et al. *The Ferric Chloride Etching of Copper for Photoengraving.* Columbus: Photo-engravers Research Inc., 1949.

Smeil, Oscar. *Technical Guide for the Gravure Industry.* edited by Oscar Leiding. New York: Gravure Technical Association Inc., 1975.

Soemarko, Dave. "Lith Film in Continuous Tone," *Post-Factory Photography* 2 (1998). pp. 37–40.

Sullivan, Richard and Carl Weese. *The New Platinum Print*: Working Pictures Press, Ltd., Co., 1998.

Sure, Brian. *Chine Collé: A Printer's Handbook.* San Francisco: Crown Point Press, 2000.

Wax, Carol. *The Mezzotint: History and Technique.* New York: Harry N. Abrams, Inc., 1990.

Wilkinson, W. T. *Photogravure.* London: Iliffe and Sons, 1890.

Wood, Franklin. *#19: Photogravure.* Series: *Printing Theory and Practice.* Series ed. John C. Tarr. London: Sir Isaac Pitman & Sons, Ltd., 1949.

A FEW HISTORICAL SOURCES

Annan, Thomas. *Photographs of the Old Closes and Streets of Glasgow, 1868–1877, with a Supplement of 15 Related Views.* (Introduction by Anita Ventura Mozley.) New York: Dover Publications, Inc., 1977.

Baldwin, Gordon. *Looking at Photographs: A Guide to Technical Terms.* London: British Museum Press, 1991.

Buckland, Gail. *Fox Talbot and the Invention of Photography.* Boston: David R. Godine, 1980.

Coe, Brian, and Mark Haworth-Booth. *A Guide to Early Photographic Processes.* London: Hurtwood Press, 1983.

Crawford, William. *The Keepers of Light: A History and Working Guide to Early Photographic Processes.* Dobbs Ferry, N.Y.: Morgan & Morgan, 1979.

Curtis, Edward S. *The North American Indian: The Complete Portfolio.* New York: Taschen, nd.

Goodman, Jon, Pietro Sarto, Malcom Daniel and Florian Rodari. *Graver La Lumière: L'héliogravure d'Alfred Stieglitz à nos jours.* Switzerland: Fondation William Cuendet & Atelier de Saint-Prex, 2002.

Newhall, Nancy. *P. H. Emerson: The Fight for Photography as a Fine Art.* New York: Aperture Inc., 1975.

Roth, Andrew. *The Book of 101 Books: Seminal Photographic Books of the Twentieth Century.* New York: PPP Editions with Roth Horowitz LIC, 2001.

Steiglitz, Alfred. *Camera Work: The Complete Illustrations 1903–1917.* New York: Taschen, 1997.

Weaver, Mike. *Alvin Langdon Coburn: Symbolist Photographer, 1882–1966. Aperture #104.* New York: Aperture Foundation, Inc., 1986.

HEALTH AND SAFETY RESOURCES

Material Safety Data Sheets: online at www.msdssearch.com/msdssearch. htm.

McCann, Michael. *Artist Beware.* 2nd Edition New York: Lyons & Burford, 1992.

Rossol, Monona. *The Artist's Complete Health and Safety Guide.* 2nd Edition New York: Allworth Press, 1994.

Shaw, Susan and Monona Rossol. *Overexposure: Health Hazards in Photography.* 2nd Edition New York: Allworth Press, 1991.

Contributors

Sandy King
Sandy King is a professor of Spanish at Clemson University in Clemson, South Carolina. King received a Ph.D. from Louisiana State University in 1971. King is a photographer, photohistorian, and alternative process printer. King is the author of *El impresionismo fotográfico en España: Una historia de la técnica y de la estética de la fotografía pictorialista*, trans. José Luis Gil Aristu, in *Archivos de la Fotografía*, Vol. IV, No. 1. Zarautz (Spain): Photomuseum (Argazki Euskal Museo), 2000; *Schmidt de las Heras: Fotografías 1944–1960*. La Coruña (Spain): Xunta de Galica, 1999; *El Libro del Carbón: Introducción y Guía de Trabajo para la Impresión en Carbón Monochromo*, trans. Luis Segura Sellés. Novelda (Spain): Cuadernos de Fotografía Alternativa, 1998; and *The Photographic Impressionists of Spain: A History of the Aesthetics and Technique of Pictorial Photography*. New York; The Edwin Mellen Press, 1989. King edited with Manuel Estebanez Consuegra *El Bromóleo Monocromo: Guía y Manual de Trabajo. Cuadernos de Fotografia Alternativa*. Novelda (Spain), 2001; *La Goma Bicromatada: Procedimiento Básico, Cuadernos de Fotografia Alternativa*. Novelda (Spain), 1999; and *The Book of Carbon and Carbro: Contemporary Procedures for Monochrome Pigment Printmaking*. Self-Published, Greenville: Permanent Light Systems, 2000. King has written numerous journal publications on photographic aesthetics and printmaking, including articles on carbon printing in the Sept/Oct and Nov/Dec 2001 issues of *Photovision*.

Jon Goodman
Jon Goodman has been practicing photogravure full time since 1976. Initial funding was provided by a fellowship from the Thomas J. Watson Foundation. Subsequently he worked with *Aperture* and the Paul Strand Foundation to produce photogravure portfolios of the early work of Paul Strand, Edward Steichen, and early British photography. Since 1984 he has operated a studio, Jon Goodman~Photogravure, devoted to producing editions in photogravure for publishers, artists, photographers, and museums. His work can be found in many public collections including

the Museum of Modern Art, The Metropolitan Museum of Art, and the Bibliothèque National in Paris. His studio is located in Florence, Massachusetts.

Steven Dixon

Steven Dixon was born in Woodstock, New Brunswick, in 1960 and educated at Mount Allison University (B.F.A.), Queen's University at Kingston (B.Ed), and Arizona State University (M.F.A.). Since 1986, Steven has worked in the Printmaking Division at the University of Alberta, Edmonton, Canada. He has an extensive international exhibition record and has been an invited guest artist/lecturer at Palacky University in the Czech Republic, Northern Illinois University in the United States, and the University of British Columbia and Capilano College in Vancouver, Canada.

Lothar Osterburg

Lothar Osterburg is an active artist, teacher and master printer in photogravure. He completed art school in Germany before moving to San Francisco in 1987. His work has been shown internationally from Germany to Japan, and is exhibited regularly in the United States. His awards include a prize at the 6th International Graphic Triennial in Frechen, Germany and several residencies at the MacDowell Colony, Bogliasco Foundation in Italy and the Virginia Center of the Arts.

As a master printer he has worked in several printshops throughout the United States, including Crown Point Press, where he started working in photogravure. Since 1993, he owns and operates a photogravure and etching workshop in New York City, where he worked with artists such as Lorna Simpson, William Wegman, Judy Pfaff, Ruth Thorne Thompson, McDermott and McGough, Zoe Leonard and Adam Fuss.

Besides teaching numerous workshops in photogravure throughout the United States and Canada (Cooper Union, Anderson Ranch Arts Center, Cleveland Institute of the Arts, Utah State University, Pyramid Atlantic, Graff...etc) he currently is visiting professor at Bard College, Cooper Union and Columbia University.

Glossary

acid resist—Any substance that prevents the action of an acid or mordant on the surface of a plate.

actinic light—Light that possesses the radiant energy necessary to produce chemical changes in light-sensitive photographic emulsions. In orthochromatic film and dichromate emulsions, the UV and blue-green end of the spectrum has the greatest effect.

agitation—The regular or intermittent movement of a plate, film, or solution to allow fresh chemicals to periodically or constantly reach the material being processed.

à la poupée—(French, "with a doll") The use of different daubers of different colors to ink small areas of a plate so as to distribute several colors over different areas of the plate.

alcohol—*See* ethyl and isopropyl alcohols.

ammonium dichromate—A toxic chemical salt used to sensitize organic emulsions such as gelatin or gum arabic to the hardening effect of actinic light. Similar to potassium dichromate and sodium dichromate, but more sensitive to light. Can be absorbed through skin contact to cause dichromate poisoning. Flammable.

anti-halation backing—A dye or colloidal silver layer on photographic films that prevents reflex halation during exposure to light.

aquatint—An intaglio printmaking process in which rosin or asphaltum powders are dusted and then fused onto a metal plate to produce a fine granular resist against etching in order to give a printed tone or texture.

asphaltum—A complex hydrocarbon mixture used in intaglio printmaking as an acid resist. Its powdered form can be used to make fine aquatints. Its liquid form is used as a resist to stage and retouch a plate before etching. Inhalation of the dust can cause respiratory irritation. A suspected carcinogen.

bath—A tray of acid solution or other mordant for etching the plate.

Baumé—A system of measurement that describes the density or specific gravity of a solution in degrees Baumé (° Bé); after Antoine Baumé.

Baumé hydrometer—The floating, graduated instrument invented by Antoine Baumé that was designed to measure the density or specific gravity of a solution.

Bé—Abbreviation of Baumé.

bite—The corrosive action of an acid or mordant on an intaglio plate.

black light—The trade name for a fluorescent light that produces light rich in the ultraviolet region of the spectrum. In gravure it can be used to expose the positive onto gelatin tissue in order to obtain good shadow detail. (Use industry code BL, not BBL.)

blankets—The set of woven wool blankets used over the plate/print on an intaglio press that absorb and distribute the pressure from the rollers and help push the paper's fibers into the recesses of the plate in order to pick up all the ink. They include the pusher, the cushion, and the sizing catcher, in that order, from roller to paper/plate. Usually superior to felts. *See also* felts, press blankets.

bleed—Images that are trimmed or printed to the edge of the paper.

blood warm—The temperature to which the copper plate is brought for wiping and printing—approximately body temperature. *See also* warm wipe.

body—The viscosity or flow characteristics of an ink used in printing.

Brasso—The trade name of a metal polish used to remove tarnish and polish a copper plate.

brayer—A single-handled rubber, leather, or gelatin roller used in printmaking in various ways, including to apply ink to a plate, stone, or woodblock.

brightening—A chemical step after degreasing the copper plate where a brightener (see page 50) is applied to remove the last vestiges of tarnish or scum from the surface of the plate.

burin (graver)—A tool for engraving wood or metal. It has a square or lozenge-shaped shaft and a wooden handle that is held against the palm of one hand.

burning in—During exposure, an area is given more light than the base exposure in order to darken it in proportion to the rest of the image area.

burnisher—An intaglio tool with a smooth curved and rounded shaft set in a wooden handle. It is used to polish or rub a metal plate in order to remove scratches or lighten tones.

burnishing—Using a burnisher to rub and smooth an area on a plate.

burr—On a metal plate, the curled ridge of metal created when scratching a line or mark with a drypoint tool or with a mezzotint rocker. The ridge holds ink and produces a line or mark that is characteristically soft and fuzzy. Also, the sharp protruding knife-edge created when filing or burnishing the bevel on a plate.

calcium carbonate—($CaCO_3$) Used with ammonia to degrease a plate. *See also* whiting.

calender—In industry, a series of rollers over which paper is passed to smooth its surface. In printmaking, to pass the paper through the press over a blank plate to remove all water and smooth the printing surface.

carbon tissue—An orange-colored, gelatin-coated paper sensitized and used in photogravure to transfer a positive image to a copper plate as an acid resist. Its color comes from iron oxide or burnt sienna pigment (originally carbon black pigment, hence its name) suspended in the gelatin. *See also* gelatin tissue.

cells—Individual etched wells on a gravure plate formed when a gravure screen is used to establish a pattern of walls and openings. Each cell is etched to a different depth and holds a different amount of ink, resulting in a complete tonal scale when printed.

chine collé—A printmaking process by which a layer of thinner paper is bonded to another heavier paper using an adhesive paste and the high pressure of the intaglio or lithographic press. Usually done simultaneous to the printing of an image on the thin paper.

cobalt drier—A toxic additive to printing ink that shortens drying time.

cold wipe—When the copper plate is wiped at ambient temperature, without the use of the hot plate. Results in a print with more contrast.

collimated light—Light rays emanating from a light source that are parallel rather than scattered or diffused and therefore cast a hard-edged shadow that is the same size as whatever interrupts the light rays.

color sensitivity—The sensitivity of different films to different segments of the color spectrum. *See also* orthochromatic, panchromatic.

commercial film—Photographic film, usually insensitive to red light (orthochromatic) and capable of reproducing continuous tone images.

contact print—General photographic term describing any print or film reversal made by direct contact printing in a vacuum or contact printing frame (i.e., not enlarged or reduced).

contact printing frame—A glass-fronted and vacuum- or pressure-backed frame in which contact prints are exposed through film originals.

continuing action—The continuing aftereffect of exposure to actinic light, which causes the sensitized gelatin to harden or tan further even when no longer being exposed to light. Most pronounced during the first hour after exposure with a dramatic decrease after that. Also known as continuing reaction.

continuous tone—A negative or positive film or print exhibiting a gradual and complete gradation of tonalities from clear to dense or white to black without a screen or dot pattern.

continuous tone positive—In photogravure, the film image used to expose the gelatin tissue in order to translate the original tonal information onto the gelatin resist. Often used in conjunction with a gravure screen positive.

contrast—The tonal difference between the light and dark areas of a negative or positive.

contrast range—The amount of variance in optical density between highlights and shadows, measured using a densitometer and determined by subtracting the density reading of the positive's highlight detail from the density of the positive's shadow detail. Also known as density range.

contrasty—Weak, washed out highlights and heavy, blocked in shadows where detail is restricted to the few grays in between.

conventional gravure—A photogravure made using a continuous tone positive and a conventional gravure crossline screen of uniform square cells, and etched to various depths to achieve a full tonal range.

conventional screen—A crossline gravure screen with rulings from 60 to 300 lines per inch. Opaque dots are square with clean transparent lines at ratios of 1:1 to 5:1. Used to create a pattern of hardened gelatin in a light-sensitive resist. *See also* gravure positive screen.

copy camera—A camera used in the graphics arts for making halftones, separations and line shots of originals or paste ups. *See also* process camera.

crevé—*See* foul-biting.

crop—To designate an area of an image that will not be part of the final composition and will be cropped off or eliminated.

cushion—The middle of the three felts or press blankets used on an intaglio press.

damp box—An airtight box used to dampen paper in lieu of soaking or to make editioning paper consistent throughout.

damping—Soaking the fibers of the printing paper prior to printing in order to soften them.

dark effect or reaction—In light-sensitive coatings, the hardening or fogging action that takes place over time and without light, especially when the humidity or temperature is higher than ideal.

deckle (edge)—The raw, ragged, untrimmed edge of a piece of handmade or mold-made paper.

definition—The fine, sharp detail rendered in a pictorial image.

degrease—To remove all traces of oil or grease from the surface of the copper plate or Plexiglas drying support.

densitometer—An instrument used to measure the optical density or light stopping characteristics in specific areas of photographic film. Also known as a transmission densitometer.

density—The optical density or the light stopping characteristics of film positives or negatives. Also an indication of the covering power of ink on paper.

density range—*See* contrast range.

dichromate solution—A solution of dissolved dichromate salts used to sensitize gelatin tissue or gum arabic. *See also* ammonium dichromate, potassium dichromate.

diffusion—The scattering of light rays as they pass through a translucent or optically imperfect material.

direct gravure—A photogravure made from a positive created by drawing or painting on frosted Mylar rather than using a photographic positive image on film.

distilled water—Chemically pure water used in all gelatin tissue processes to avoid contamination caused by the impurities usually found in tap water.

drier—A substance (siccative) that speeds up the drying process of inks.

dry lay-down—A method for mounting the exposed gelatin resist onto the surface of the copper plate using very little water and no pre-soak. Water is applied to the copper plate under the dry gelatin tissue and rolled down with a rubber roller or stiff squeegee.

drying support—A sheet of Plexiglas used to dry the sensitized gelatin resist tissue to a mirror-smooth finish.

drypoint—A nonacid intaglio technique in which the plate is marked directly using a needle or other pointed tool in order to create a scratch and its characteristic burr.

dusting—The application of an aquatint ground onto a metal plate with either powdered asphaltum or rosin. Also known as grounding.

dusting bag—A small bag made of multiple layers of muslin, silk, or nylon stocking filled with powdered asphaltum or rosin and shaken to dust a plate in the making of an aquatint ground.

dusting box—A sealed box in which metal plates are coated with asphaltum or rosin dust by placing them within the box after the dust has been suspended in the air within the box. Also known as a tumble box.

Easy Wipe Compound—The trade name of a greasy compound added to ink to facilitate the wiping of the plate when inking.
electroplating—The deposition of one metal onto the surface of another using electrodes. The purpose is to give the base metal the surface properties of the metal used for electroplating.
emulsion—The light-sensitive layer on photographic films or plates.
etch or etching—(*vt.*) The application of an acid or mordant to specific, unprotected areas of a metal plate in order to create ink-holding depressions, patterns, or lines.
etch—(*n.*) Any acid or corrosive salt used to etch metal. In photogravure, the ferric chloride solutions. *See also* mordant.
etching—(*n.*) An intaglio print produced from a metal plate on which an image is made through the use of a mordant or etch.
etching press—An intaglio printing press in which the printing paper and the inked intaglio plate are placed on a metal press bed and passed between upper and lower rollers under high pressure. Also known as an intaglio press.
ethyl alcohol or ethanol—(C_2H_5OH) A grain alcohol (usually denatured) used at various stages of the gravure process to help prevent pinholing in the resist and to speed drying through the displacement of water.
exposure—The quantity (time and intensity) of light that effects a light-sensitive material; also, the product of this light.
extender—An additive to printing ink to reduce its opacity. Usually a clear varnish or transparent white. Also known as transparent base.
false biting—*See* foul-biting.
felts—The set of pressed wool blankets used over the plate/print on an intaglio press that absorb and distribute the pressure from the rollers and help push the paper's fibers into the recesses of the plate in order to pick up all the ink. They are usually of lesser quality and sensitivity than woven blankets. *See also* blankets.
felt side—The top or face of a sheet of paper.
ferric chloride—($FeCl_3$) A salt of iron that acts as a mordant for copper and is used as the mordant in photogravure. Can cause burns and respiratory irritations. Wear gloves and vent fumes. (Also known more correctly as iron(III) chloride and, historically, iron perchloride.)
ferric hydroxide—The sludge produced when a strong ammonia solution is added to ferric chloride stock solution. This sludge is then used to neutralize the free acid in the working solutions of ferric chloride. (Also known more correctly as iron(III) hydroxide.)
film base—The transparent base or support upon which film emulsions are coated.
film base plus fog—The transmission density of the processed but unexposed film, which includes the density of the base material plus the chemical fog associated with normal processing.
flashing—Exposing a light-sensitive emulsion to even overall light to raise its base exposure to threshold. Subsequent exposure (of an image) will then have an immediate effect on densities but at a slightly reduced contrast. *See also* threshold.
flat-plate gravure—The hand-pulled, one-at-a-time process of printing photogravure images using an intaglio press. This differs from rotogravure, which is a high-speed continuous printing of gravures, usually for commercial applications.

fog—Additional overall or localized density or tone on light-sensitive emulsions that has no connection to the intended image. Can be caused by light leaks, unintentional overall exposure to light, incorrect safe-light, exposure to X-ray or IR radiation (heat), chemical faults, or dark reaction.

foul-biting—The eating away of lands or cell walls by lateral etching. The loss of these high points results in loss of detail and tonal reversal in the darkest, most affected areas. Also known as crevé or undercutting.

free acid—Also known as hydrochloric acid (HCl), properly described as H_3O^+ (aq), and produced within a ferric chloride solution as it reacts with water.

gelatin—An animal byproduct that, in its pure form, is used as an emulsion to suspend light-sensitive salts on photographic films and papers, and for gelatin (carbon) tissue in photogravure.

gelatin resist—A layer of gelatin on a copper plate which has been tanned to interpret in contour the pattern of highlights and shadows of a photographic image and that results in an acid resist of varying thicknesses due to those differences.

gelatin tissue—An orange-colored, gelatin-coated paper sensitized and used in photogravure to transfer a positive image to a copper plate as an acid resist. Its color comes from iron oxide or burnt sienna pigment suspended in the gelatin. *See also* carbon tissue.

generation—A distinct stage in the production and translation of an image. The first generation is usually the original negative, the second is a positive, the third (in gravure) is the gelatin tissue resist, the fourth is the copper plate, and the fifth is the print.

glacial acetic acid—Concentrated acetic acid (CH_3COOH) that is used in a diluted form for brightening copper plates and in various photographic processes. Can cause skin burns and respiratory irritation.

goldenrod paper—A yellow-orange paper used to protect light-sensitive materials from exposure to actinic light. Used to frame transparencies for exposure onto film or gelatin tissue.

grain—The distribution and size of silver particles in a photographic emulsion or image. Can also describe the fineness of the aquatint ground on a plate or the directional quality of papers, especially machine-made papers.

graver—*See* burin.

gravure—Shortened form of the word *photogravure*.

gravure positive screen—A cross-line gravure screen with rulings from 60 to 300 lines per inch. Opaque dots are square (diagonally) with clean transparent lines at ratios of 1:1 to 5:1. Used to create a pattern of hardened gelatin in a light-sensitive resist. Also known as conventional screen and line screen.

gray scale—A positive or negative scale divided into steps of increasing density from brightest highlight value to darkest shadow density. *See also* step scale.

ground—An acid-resistant covering applied to metal plates. A fine acid-resistant powder distributed over the surface of a plate and melted to form an aquatint or a liquid painted onto a plate before etching to control the areas being etched.

halation—The loss of sharpness and the spread of densities evident in films and light-sensitive materials when focused light bounces back through the support layers to re-expose the emulsion layer.

halftone—A photomechanical printing process by which the continuous tone image is translated into a fine grid of evenly spaced dots of identical tonality but varying sizes and shapes according to the intensity of the tone they represent.

handmade paper—Fine paper that has been pulped and molded by hand.

hand pulled—(hand proof) A print made on an etching or lithographic press in which dampening, inking, and taking the impression are done manually; usually describes prints made by artists.

hand wiping—When an intaglio plate is inked, the finishing technique of wiping the plate with the heel of the bare palm. *See also* rag wiping.

handwork—Any marks or work done to an intaglio plate manually using drypoint tools, roulettes, or engraving burins.

hard (-dot)—Hard denotes excessive contrast in photographic terms. A hard-dot is a halftone dot, negative or positive, with sharp, high contrast edges. *See also* soft (-dot).

high contrast—A wide ratio existing between the density or value of a highlight and the density or value of a shadow area on a positive or negative image.

highlight—The dense areas of a negative or the clearest areas in a positive, which give the impression of brightness in an image.

highlight detail—The areas on film that correspond to the brighter or lighter parts of the original scene. These would read as visible detail, texture, or pale tone.

hot plate—A heated flat metal surface used to warm the copper plate to aid in the inking, wiping, or printing of a plate by making the ink more fluid. Also used to fuse a dust-grain aquatint onto a plate.

hydrometer—An instrument for measuring the specific gravity, strength, or density of a liquid.

hygrometer—An instrument for measuring the relative humidity of the air at a given temperature.

hygroscopic—The ability of a substance to take up or absorb moisture from the surrounding air.

image exposure—When the gelatin tissue is exposed through the positive in order to create a three-dimensional record of the details and tonalities within the hardened gelatin resist. Also known as positive exposure.

impression—An imprint on paper. A proof or a print.

inking—Applying ink to an intaglio plate by forcing it down into the recesses of the plate with a dabber, roller, or squeegee.

inking knife—A short, stiff putty knife used to mix and soften printing ink on the inking slab.

inking slab—A piece of plate glass or marble used as a mixing surface for working, altering, or mixing inks, oils, additives, colors, and the like in preparation for inking the plate for printing.

insolubilization—The tanning effect on an organic colloid sensitized with a dichromate salt upon exposure to UV light.

intaglio—One of the four major divisions of printmaking (intaglio, relief, planographic, serigraphic) whereby an image is made with engraved or etched lines or textures on a metal plate. After wiping, these marks hold the ink below the surface of the metal.

intaglio press—*See* etching press.

isopropyl alcohol or isopropanol—($CH_3CHOHCH_3$) Also sold as rubbing alcohol. It should *not* be used for any stages of the gravure process because its properties and additives introduce severe problems during development of the resist.

jeweler's rouge—A fine rubbing compound used to smooth and polish metal plates.

lay-down—A method for mounting the exposed gelatin resist onto the surface of the copper plate.

lands—The high points on the surface of an intaglio plate that have not been etched and that help hold ink within the texture they create on the plate.

line screen—*See* screen, conventional screen, and gravure positive screen. Also lpi.

lines to the inch—A descriptive designation of how fine the screen pattern is on crossline, halftone, conventional, or gravure positive line screens.

Mactac—The trade name of a self-adhesive plastic sheeting usually used as a decorative covering or shelf paper.

magnesium carbonate—($MgCO_3$) A fine white powder used to increase the body of printing inks. It is also used on the hand to lighten or remove all plate tone during the final hand wiping stage when inking a plate. Also known as mag.

mask—In photogravure, an opaque border around the positive to block the transmission of light outside the image area.

methyl alcohol or methanol—(CH_3OH) A toxic form of alcohol. Do not use in its pure form.

mezzotint—(*la manière noir*) An intaglio technique in which a plate is worked from dark tones to light. The black plate tone is initially produced by roughening the plate with a mezzotint rocker. Middle tones and highlights are created by burnishing and smoothing the plate's tooth.

mezzotint rocker—A finely toothed and chisel-shaped drypoint tool that is rocked back and forth on the surface of a metal plate to create a texture that will print as a rich dark tone.

mezzotint screen—A halftone screen made up of a reticulated random dot pattern rather than a regular grid-like pattern.

micron—One millionth of a meter. A standard measure of the gravure cell depth. (25.4 microns equal one thousandth of an inch.)

middle tones—Tones that fall between the highlights and shadows of an image.

modified single-bath etching—The etching method which starts with a single tray of high Baumé ferric chloride that is progressively diluted with water during the etching process, thereby producing the range of Baumés needed to etch all the tones.

moiré—The pattern of circles and arcs that occur when two regular patterns or screens are superimposed.

moldmade paper—Paper that has been molded into single sheets.

mordant—An acid or salt solution having a corrosive effect on metal. *See also* etch.

multiple-bath etching—An etching method that employs several trays of ferric chloride at different Baumés. Each bath is used progressively to bring out the different steps on the gray scale. *See also* modified single-bath etching.

negative—Processed photographic film on which the tones of the original are reversed so that the highlight areas are recorded as dark opaque areas and the deepest shadow areas are clear.

Newton rings—When two polished surfaces are pressed against each other, such as film base against glass, the irregular pattern of rings and swirls that appears in prismatic color.

opaque—Not permitting the passage of light. Also an iron oxide paste used to spot out the pinholes on graphic arts films.

open biting—Biting that occurs on large unprotected areas of the plate where no lands or textures are retained or created.

orthochromatic—Film emulsion that is sensitive to most of the visible spectrum except red.

panchromatic—Film emulsion which is reactive (sensitive) to all colors of the visible spectrum.

penetration—In photogravure, the initial moment when the mordant reaches and acts on the surface of the copper plate after migrating through the gelatin resist.

perchloride of iron—Archaic term used to describe ferric chloride, now properly called iron(III) chloride.

pH—A scale of 0 to 14 that expresses the acidity or alkalinity of a solution, 7 being neutral. Readings lower than 7 are considered acid; higher are considered basic or alkaline.

photoengraving grade—A commercial grade of copper plate on which the surface is highly polished and backed with an acid-resistant coating.

photoetching—A halftone intaglio process, usually on zinc plates, that breaks up the continuous tone gray scale into halftone dots of varying size but of uniform ink intensity.

photogravure—An intaglio process using copper plates that are etched through a gelatin resist to varying depths in order to print the continuous tone gray scale into a virtually continuous tone translation of varying intensities of grain-sized ink deposits.

photosensitive—A substance that has been treated or coated with a chemical compound that will change chemically when exposed to light.

pinholing—Small holes in the solid areas of a resist or film that are the result of the failure of that film or coating to form a complete coating.

pits—In photogravure, the small holes or depressions in the plate created by etching and which hold the ink in order to define detail and tone in the image when printing. Also known as wells.

planographic—One of the four major divisions of printmaking whereby an image is made utilizing the property of greasy ink and water repelling each other and printed from a planar or smooth surface. Also known as lithography.

plate—A general term used in photography to describe sheet film, and in photogravure to describe the intaglio copper plate.

plate cutter—A guillotine cutter used for cutting and trimming metal plates.

plate mark—The imprint or embossment of the beveled edge of a plate on the paper of an intaglio print.

plate tone—The visible trace of tone or color obtained by the thin film of ink left on the polished nonimage areas of a plate after wiping.

Plexiglas—The trade name of a sheet of methacrylate plastic. Also known as Lucite and Perspex.

point source light—A light source emanating from the smallest possible point, thereby casting the sharpest shadow edge.

polishing—The method of obtaining a fine mirror finish on a copper plate.

positive—The second generation image in the photogravure process, which contains the full tonal scale in the same relationship as the original scene; opposite from the negative.

positive mottle—A regular pattern created during the exposure of the positive onto the sensitized gelatin tissue. It is exacerbated by low vacuum pressure and sometimes coarse film grain.

potassium dichromate—(KCr_2O_7) A toxic chemical salt used to sensitize organic emulsions such as gelatin or gum arabic to the hardening effect of actinic light. Similar to ammonium dichromate, but less sensitive to light and less costly. It can be absorbed through skin contact and cause dichromate poisoning.

precipitation—The formation of insoluble material when two clear solutions are mixed or when the saturation point of one solution is reached.

press blankets—Also known as printing blankets. *See* felts, blankets.

process camera—A graphic arts copy camera used for process work (with panchromatic materials). *See also* copy camera.

proof—A trial print.

proofing—Used instead of "proving" to denote the operation of pulling proofs at various stages of the plate.

pulling a proof/print—To hand print a proof or print with a manual etching or lithographic press.

pumice—Porous volcanic rock used as an abrasive and polish when finely ground and mixed with a lubricant. *See also* rottenstone.

pusher—The upper felt or press blanket that absorbs wear from the top pressure roller.

Putz Pomade—The trade name of a metal polish used to clean the copper plate after etching.

rag wiping—The plate wiping technique that uses the cheesecloth or tarlatan only. *See also* hand wiping.

register—To align the paper and plate so that successive prints are in exactly the same position on the paper. This is especially critical when multiple plates are printed in succession on the same support.

register marks—Cross hairs, T-bars, or other marks that aid in the registration between plates, films, and printing paper.

relative humidity—The ratio or percentage of water vapor in the air compared to the amount required to saturate it at the same temperature.

relief—A printmaking technique in which the printing is from a raised inked surface. In photogravure, relief describes the contours and three-dimensional appearance of the resist on the copper plate.

repoussage—Leveling the front of a scraped or bitten plate by the application of tape to the back of the plate.

resist—In photogravure, the gelatin layer hardened onto the copper plate so as to control the action of the acid on the copper in relation to the exposure information within the resist.

resist failure—The breakdown of the gelatin resist during the etching process, thereby allowing uncontrolled etching or foul-biting.

resolution—The degree of sharpness and detail of the image on a negative, positive, resist, plate, or the printed result.

retroussage—Passing a soft cloth over the intaglio lines of an inked plate to pull a little ink out of the lines to spread it over the surface of the plate, the result being a softer print.

rocker—*See* mezzotint rocker.

rolling up—Charging or covering with ink from a roller. In gravure, an acid-resistant ink can be used to act as a resist on the high points before re-etching to raise the contrast.
rosin—Powdered rosin is dusted and fused to a plate to act as an acid resist for an aquatint or a photogravure. A respiratory irritant when inhaled.
rotogravure—The commercial process of high-speed continuous printing of gravures from etched copper cylinders on a rotary press.
rottenstone—A very fine abrasive powder used for polishing when mixed with a lubricant. *See also* pumice.
roulette—A burred or textured roller set in a handle. Used to create a grain tone by rolling on the surface of the copper plate.
safe edge—The border on a gravure resist that surrounds the image area. This area is masked during the image exposure. Any border left around an image that allows for handling and fingerprints so there is no damage to the image.
safelight—Darkroom illumination specifically filtered to prevent fogging by the action of light on specific photosensitive materials.
scraper—An etching hand tool with three sharp edges coming to a point. Used to scrape the surface of a plate or to cut off burrs that project above the surface.
screen—A fine network or pattern of lines or dots that break up tonalities in a continuous tone negative or positive in order to make printing with ink possible. *See also* gravure positive screen and conventional screen.
sensitize—To make chemically sensitive to light.
sensitized gelatin tissue—Gelatin (carbon) tissue that has been immersed in a dichromate solution to make it photosensitive.
sensitizer—In photogravure, a 2% to 5% dichromate solution that is soaked into the gelatin tissue to make it photosensitive.
set-off—The transfer of ink from a freshly pulled print to the sheet placed over it.
shadow—The clearest area of a negative or the dense areas in a positive, which gives the impression of darkness in an image.
shadow detail—The areas on film that correspond to the darker parts of the original scene. These read as visible dark detail, texture, or differentiated dark tones.
single-bath etching—A method of entirely etching photogravure plates in a single ferric chloride bath.
sizing—Gelatin or starch impregnated or coated on paper to make it stronger and to lessen absorption.
sizing catcher—The press blanket or felt nearest the printing paper. Its purpose is to soak up moisture and sizing from the damp paper and prevent it from hardening the other press blankets.
slip sheet—A sheet of protective paper or tissue placed between prints to protect them from set-off.
sodium carbonate—(Na_2CO_3) In photogravure, used to neutralize the activity of an etchant such as ferric chloride. Also known as washing soda or soda ash.
soft (-dot)—Soft denotes a low contrast or flat appearance in photographic terms. A soft-dot is a halftone dot, negative or positive, with graduated or vignetted edges. *See also* hard-dot.

specific gravity—The comparison of the weight of a given volume of a material with an equal volume of water.

spotting—Touching up the white dust spots or pinholes on a dark print, or covering transparent pinholes in an opaque area of a film image. Also covering spots and pinholes in a resist with liquid asphaltum or stop-out varnish.

squeegee—In screen printing, a stiff, sharp, square-edged rubber strip set into a horizontal wooden handle of various widths.

staging—The application of etch-proof barriers such as tape, Mactac, liquid asphaltum, or stop-out varnish to a plate prior to etching in order to mask and protect nonprinting areas of the plate from the mordant.

step scale—A piece of film exposed in graduated increments to form equally spaced steps of increasing density from clear to maximum. Also known as step tablet, gray scale, tone scale, step wedge, or tone wedge.

stop-out varnish—A liquid acid resist painted or spotted on a plate to mask areas and prevent etching.

stopping-out—The application of a stop-out (varnish) to a plate. Also known as staging.

stripping in—Adding or correcting film, piece by piece, with tape and goldenrod paper in the final positive assembly.

surface roll—Ink applied to the relief surface of an intaglio plate with a hard roller.

tack—The stickiness of ink.

tarlatan—A loose-weave, starched, cotton cheesecloth used for rag wiping intaglio plates.

threshold—The point at which a light-sensitive emulsion begins to form a latent exposure that will read as density after development. Light exposure below threshold will not result in a visible effect until its duration or intensity reaches threshold. Post-threshold exposure will have an immediate effect on ultimate density. *See also* flashing.

tipped (in)—A print or any separate paper glued by one edge or the corners onto a bound page of a book.

tool marks—Tooling that shows up as visible marks on the print.

tooling—Hand work to correct blemishes, add detail, or adjust tones on an etched copper plate.

translucent—Partially transparent, therefore able to allow the passage of light while diffusing it at the same time.

transparent—The ability to transmit light without diffusion.

transparent base—*See* extender.

trisodium phosphate (TSP)—A strong base used to degrease copper or Plexiglas plates.

ultraviolet (UV)—Short wavelengths just beyond visible light at the violet end of the spectrum. Effective as an actinic light source.

undercutting—Also known as *crevé* or underbiting. See foul-biting.

undersheet—Any sheet placed on the press bed under the plate.

vacuum printing frame—A contact printing frame that uses negative pressure to press the film/tissue/paper sandwich tight against the glass during exposure. *See also* contact printing frame.

viscosity—Resistance to flow; the opposite of fluidity.

wall—The side or division between cells in an etched copper plate.

warm wipe—Keeping the inked plate warm while it is wiped.

washing soda—*See* sodium carbonate.

water bath—A surrounding tray of temperature-controlled water that maintains an inner tray of liquid balanced to a specific temperature.

water mark—A design, text, or logo formed in moldmade paper to identify its maker.

waterleaf paper—Unsized paper.

well—A single gravure cell.

wet lay-down—A method for mounting the exposed gelatin resist onto the surface of the copper plate by sliding the pre-soaked gelatin tissue onto the surface of the distilled water that covers the plate.

whiting—Calcium carbonate ($CaCO_3$). A multi-purpose fine white powder used in printmaking. Can be used to dry the hand when doing a hand wipe in order to obtain minimum plate tone. Also used with ammonia to degrease a plate. *See also* magnesium carbonate.

wiping—*See* rag wiping and hand wiping.

Index

About the Authors

David Morrish

David Morrish is an artist and teacher living and working in Corner Brook, in the eastern Canadian province of Newfoundland and Labrador. He is an Associate Professor teaching photography in the visual arts program at Sir Wilfred Grenfell College, Memorial University of Newfoundland and served as Head of Visual Arts then chair of the visual arts program between 1995–1997. He has taught university level photography in Calgary, AB and Sackville, NB before moving to Corner Brook in 1989.

David received his B.F.A. degree in Visual Arts in 1981 from the University of Manitoba in Winnipeg after studying architecture and carpentry for many years. In 1985 he received his M.F.A. degree in Visual Arts from the University of Calgary, Alberta.

David has been exhibiting photographic work in Canada since 1978 but has exhibited photogravure prints exclusively since 1996. In 2001–2002, solo exhibitions included "Locomotive Torpor" (SNAP Gallery, Edmonton, AB) and "Photogravures, 1996–2001" (CCFM Gallery, Winnipeg, MB) and a two-person show, "Animalia" (James Baird Gallery, St. John's, NL). Participation in juried or curated group exhibitions in many countries have been ongoing and include: "No Such Animal" at the SAW Gallery in Ottawa, ON, 2003; the 2000 Marion McCain Atlantic Art Exhibition, Dalhousie Art Gallery, Halifax, NS; the 2001 International Print Triennial in Kanagawa, Japan; the First and Second Biennales internationale d'estampe contemporaine de Trois-Rivières, QC; the 1st Biennial International Miniature Print Exhibition, New Leaf Editions/Dundarave Print Workshop, Granville Island, BC (winning the Fourth Place Honorable Mention); "Relativities," the 4th British International Miniature Print Exhibition, a touring exhibition organized by Off-Centre Gallery, Bristol and Loughborough University School of Art and Design. He also participated in the "5th Sapporo International Print Biennale," Sapporo, Japan, and the Fourth Kochi International Triennial Exhibition of Prints, Kochi, Japan, all since 1999. Much of this work can be seen at David's website: www.deadcat.ca.

In 2001 David was the recipient of a Project Grant from the Newfoundland and Labrador Arts Council, in 2000 and 1998 he received

Artistic/Creative Grants from the Office of Research, MUN, and in 1999 was awarded a Canada Council Mid-career Creation/Production Grant, this being his third Canada Council Grant. His works are held in several collections across Canada.

Marlene MacCallum

Marlene MacCallum is an artist and teacher living and working in Corner Brook, NL since 1990. She has been teaching visual arts at the university level since 1985 and is currently a Professor teaching printmaking in the visual arts program at Sir Wilfred Grenfell College, Memorial University of Newfoundland. From 1997–2000, she was chair of the visual arts program.

Marlene received her B.F.A. degree in Studio Arts in 1981 from Concordia University, Montréal, Québec and her M.V.A. degree in Printmaking in 1984 from the University of Alberta in Edmonton.

Marlene has been making photogravures since 1994 and exhibits her photogravure prints and book works nationally and internationally. Between 1985 and 2003, her work was exhibited in more than 75 solo, invited, and juried group exhibitions in 13 countries. She was an invited panelist for a symposium on book arts at Connecticut College in New London and at "Sightlines," a symposium on printmaking and image culture, at the University of Alberta. She has been a visiting artist and lecturer in Canada, the United States, Northern Ireland and Brazil. In 1996 she was artist in residence at the University of the Arts in Philadelphia.

Selected awards include Juror's Commendation for a photogravure print exhibited in the Boston Printmakers 2003 North American Print Biennale, Honorable Mention at the Atlanta Book Prize, Nexus Contemporary Art Center, Atlanta, Georgia in 2000 and Grand Prize Winner for the "Biennale Internationale d'Estampe Contemporaine de Trois-Rivières" in 1999, for a group of photogravure prints. In 2001, Marlene was the recipient of a Project Grant from the Newfoundland and Labrador Arts Council, in 2000 she received an Artistic/Creative Grant from the Office of Research, MUN and in 1999 was awarded a Canada Council Creation/Production Grant. Her works are held in over 25 public collections in the United States and Canada.

Marlene MacCallum, along with her husband David Morrish, live in the beautiful Bay of Islands area of Western Newfoundland with their two cats, Phoebe and Norton, and a few badly stuffed creatures used for photographs. In addition to their practice as artists, Marlene and David have been researching the photogravure process since 1993. This book, *Copper Plate Photogravure: Demystifying the Process*, is the result of that research.